T0138950

Genetics *of* Apoptosis

Genetics *of* Apoptosis

Edited by

Stefan Grimm

Max-Planck Institute for Biochemistry,
Martinsried,
Germany

A CIP catalogue record for this book is available from the British Library.

ISBN 1 85996 064 2

BIOS Scientific Publishers Ltd
9 Newtec Place, Magdalen Road, Oxford OX4 1RE, UK
Tel. +44 (0)1865 726286. Fax +44 (0)1865 246823
World Wide Web home page: http://www.bios.co.uk/

Production Editor: Andrew Watts
Typeset by Phoenix Photosetting, Chatham, UK
Printed by Cromwell Press, Trowbridge, UK

Contents

Section II: Cell Biology of Apoptosis

Section III: Cell Death in Model Systems

Chapter 10. *Apoptosis in* Drosophila   **177**
Kristin White

Chapter 11. *Cell-culture systems in apoptosis*   **189**
Stefan Grimm

Chapter 12. *Caspase-independent cell death*   **203**
Marcel Leist and Marja Jäättelä

Abbreviations

AD	Alzheimer's disease
AICD	activation-induced cell death
AIF	apoptosis-inducing factor
ALPS	autoimmune lymphoproliferative syndrome
ANT	adenine nucleotide translocator
APC	antigen-presenting cell
APP	amyloid precursor protein
ASK1	apoptosis signal regulating kinase
ATRA	all-*trans*-retinoic acid
Bax	Bcl-2 associated protein X
BH	Bcl-2 homology
bZIP	basic leucine zipper
CAD	caspase-activated DNase
CARD	caspase recruitment domain
Caspases	cysteinyl aspartate proteinases
CD95L	CD95 ligand
CF	carboxyfluorescein
CKI and CKII	casein kinases I and II
CNS	central nervous system
CRDs	cysteine-rich extracellular domains
CsA	cyclosporine A
CTLs	cytotoxic T lymphocytes
Cyt.c.	cytochrome *c*
DAP	death-associated proteins
DD	death domain
DED	death-effector domain
DIC	differential interference contrast
DIF	differentiation-inducing factor

DISC	death-inducing signaling complex
DKO	double knockout
DPI	diphenyleneiodonium chloride
DTCs	distal tip cells
EDA	ectodysplasin A
ER	endoplasmic reticulum
FADD	Fas-associated DD protein
FLIPs	FLICE-inhibitory proteins
GEFs	guanidine-nucleotide-exchange factors
GFP	green fluorescent protein
HCD	hydrophobic C-terminal domain
HIV	human immunodeficiency virus
HMM	hidden Markov model
IAP	inhibitor-of-apoptosis protein
IMM	inner mitochondrial membrane
IMS	intermembrane space
JNK	c-Jun amino terminal kinase
LEI	leukocyte elastase inhibitor
MAC	mitochondrial apoptosis-induced channel
MCV	molluscum contagiosum virus
MHC	major histocompatibility complex
MIM	inner membrane
MMR	1) mismatch repair
	2) mitochondrial membrane permeabilization
MOM	outer membrane
MORT1	mediator of receptor-induced toxicity
MS	multiple sclerosis
NFAT	nuclear factor of activated T cells
NGF	nerve growth factor
NIK	NF-κB-inducing kinase
NK	natural killer
OGD	oxygen and glucose deprivation
OMM	outer membrane
OPG	osteoprotegerin
PARP	poly(ADP) ribose polymerase
PCD	programmed cell death
PDTC	pyrrolidine dithiocarbamate
PKC	protein kinase C
PTP	permeability transition pore
PYD	pyrin domains
ROS	reactive oxygen species
SAGE	serial analysis of gene expression
SH2	Src-homology-2
SODD	silencer of death domains
SR	sacroplasmic reticulum
TCR	T-cell antigen receptor

TG	thapsigargin
THD	TNF homology domain
TNF	tumor necrosis factor
TNFR	tumor necrosis factor receptor
TNFR-CRD	TNF receptor cysteine-rich domain
TNH	TNF homology
TOM	translocase of the outer membrane
TPAF	*Tetrahymena* proliferation-activating factor
TRADD	TNFR1-associated DD protein
TRAIL	TNF-related apoptosis-inducing ligand
UPR	unfolded protein response
VDAC	voltage-dependent anion channel

Contributors

Alpi, Arno Max-Planck-Institute for Biochemistry, Am Klopferspitz 18a, 82152 Martinsried, Germany

Antonsson, Bruno Serono Pharmaceutical Research Institute, 14 Chemin des Aulx, CH-1228 Plan-les-Ouates, Geneva, Switzerland

Bedi, Atul The Johns Hopkins University School of Medicine, 1650 Orleans Street, Baltimore, Maryland 21231-1000, U.S.A.

Böttger, Angelika Department Biologie II, Ludwig-Maximilians-Universität München, Luisenstraße 14, D-80333 München, Germany

Breckenridge, David G. Department of Biochemistry, McIntyre Medical Sciences Building, McGill University, 3655 Drummond St., Montreal, Quebec, Canada H3G 1Y6

Cakouros, Dimitrios Hanson Institute, Frome Road, Adelaide, SA 5000, Australia

David, Charles Department Biologie II, Ludwig-Maximilians-Universität München, Luisenstraße 14, D-80333 München, Germany

Fröhlich, Kai-Uwe IMBM, Universitätsplatz 2, A-8010 Graz, Austria

Gartner, Anton Max-Planck-Institute for Biochemistry, Am Klopferspitz 18a, 82152 Martinsried, Germany

Grimm, Stefan Max-Planck-Institute for Biochemistry, Am Klopferspitz 18a, 82152 Martinsried, Germany

Haughn, Loralee Dana-Farber Cancer Institute, Smith 73b, Department of Cancer, Immunology and AIDS, 44 Binney Street, Boston, MA, 02115, U.S.A.

Hockenbery, David Fred Hutchinson Cancer Research Center, Divisions of Cancer Research and Human Biology, 1100 Fairview Avenue North, Seattle, Washington 98109, U.S.A.

Hofmann, Kay MEMOREC Stoffel GmbH, Stöckheimer Weg 1, 50829 Köln, Germany

Jäättelä, Marja Institute of Cancer Biology, Danish Cancer Society, Strandboulevarden 49, DK-2100 Copenhagen, Denmark

Kumar, Sharad Hanson Institute, Frome Road, Adelaide, SA 5000, Australia

Leist, Marcel Lundbeck A/S, Ottiliavej 9, DK-2500 Valby, Denmark

Madeo, Frank Physiologisch-chemisches Institut, Universität Tübingen, Hoppe-Seyler-Straße 4, D-72076 Tübingen, Germany

Ravi, Rajani The Johns Hopkins University School of Medicine, 1650 Orleans Street, Baltimore, Maryland 21231-1000, U.S.A.

Schumacher, Björn Max-Planck-Institute for Biochemistry, Am Klopferspitz 18a, 82152 Martinsried, Germany

Shore, Gordon C. Department of Biochemistry, McIntyre Medical Sciences Building, McGill University, 3655 Drummond St., Montreal, Quebec, Canada H3G 1Y6

White, Kristin Massachusetts General Hospital, Harvard Medical School, Building 149, 13th Street, Charlestown, MA 02129, U.S.A.

Preface

Who would have thought so? Scientists publish paper after paper, gather in meetings, file patents, found companies, establish web sites – and all that about one phenomenon: apoptosis, the cellular suicide program. The quest to understand this process constitutes a field that remained for a long time as quiet as its subject of investigation but is now the most competitive area of molecular biology.

This frenzied activity reflects a basic understanding that has made its way into the mainstream of molecular biology: cell death is an active process that can be regulated. The cell therefore decides its own life and death and consequently its most fundamental fate.

Many hopes are focused on an understanding of this activity, as it could provide the means to deal with many diseases stemming from a disregulation of apoptosis such as cancer, stroke, autoimmune diseases, and neurodegenerative disorders. But there is also a purely intellectual motive in studying cell death. Having lingered in the backyard of research for so long, apoptosis constitutes a novel concept in biology that has already made its impact on many diverse areas. And there is still plenty to find out.

However, as the reader will notice, many exciting findings about cell death have already set the stage to tell fascinating stories, some of which are contained in this book. Through these, this compendium tries to convey the fascination that the field of cell death is exuding.

My thanks to all who made this book possible: Inga Overkamp and Rita Gernert, who helped enormously with the references; the people at BIOS, especially Dr. Nigel Farrar, Debora Bertasi, and Eleanor Hooker; the scientists who contributed lucid descriptions of their subjects; and their families for their patience.

Stefan Grimm

Dedicated to my parents

Introduction

Stefan Grimm

Given the gravity of the decision between life and death, it is surprising that scientists have only recently begun to unravel the mysteries of apoptosis, the cellular suicide program. Two aspects contributed to the continued neglect of apoptosis as a fundamental decision of cells.

It is a fast process. While many cells undergo apoptosis (estimates range up to 10^9 a day), they are rapidly swallowed by neighboring phagocytotic cells and are therefore hard to detect. The other reason was a psychological hurdle. Many scientists wondered why Nature would want cells to be able to kill themselves. Cell death might just be a direct consequence of an insult and therefore an inevitable fate of the cell.

Now, however, scientists have realized that all cells are able to receive distinct signals, interpret these signals as suicide stimuli, and induce a discrete genetic program that leads to cell death. Rather than a passive fate, apoptosis is therefore an active process that is initiated by the cell and is governed by gene activities that induce or inhibit apoptosis. It must therefore be distinguished from necrosis, which is unregulated, passive cell death. From this fundamental insight it follows that Nature must 'want' cells to perish upon a given stimulus; it must be a biologically meaningful response. But this also requires that the signal is known and deemed grave enough to warrant the demise of a cell.

Proapoptotic stimuli come in a bewildering array of different forms; they can either be generated by other cells or originate within the doomed cells. They can constitute positive signals, such as the activation of membrane receptors (Chapter 1), but might also arise from neglect, as in the case of an absent survival factor. In any event, the original signal activates a cellular 'sensor' that transmits the signal to downstream proteins that are part of signal transduction pathways. This is accomplished by genes that form a highly interconnected lattice of checks and balances of (mostly) protein–protein interactions (Chapters 3, 4) that evaluate the stimulus in relation to other signals that the cell is receiving and compute the appropriate output. If a decision for cell death is reached, the endpoint of such a cascade is the activation of a certain class of ubiquitous proteases: caspases (Chapter 2). These are cysteine proteases that reside in every cell in an inactive form and are activated for cell death. They can then cleave specific substrates in the cell and generate the typical morphology of an apoptotic cell. The fact that apoptotic cells demonstrate a steroetyped appearance was one of the first hallmarks of apoptosis to be discovered (Kerr *et al.*, 1972). The cells shrink dramatically, display membrane blebbing and DNA condensation, and are subsequently digested by neighboring cells.

Apoptosis, in its narrowest interpretation as physiologically intended or 'programmed cell death' – a term often used interchangeably for apoptosis – plays an important role in embryonic development and in tissue homeostasis. But apoptosis is also induced when the cell experiences accidents such as DNA damage. The difference from 'programmed cell death' is therefore that the triggering signal is not genetically encoded. Specifically, apoptosis activation can be interpreted in many cases as an altruistic behavior of the cell to benefit the organism as a whole; an example would be cell death induced in virus-infected cells or in cells that are in cancer progression. The multitude of different instances of apoptosis induction is reflected by the vast array of proapoptotic stimuli mentioned above. While the causes might be different, 'programmed cell death' and accidental apoptosis are in most cases mediated by a similar – or at least an overlapping – group of genes, many of which come in gene families (Chapter 5).

As versatile as this program is, it is no surprise that many diseases can arise when apoptosis fails (Thompson, 1995). Cancer is fostered if apoptosis is repressed, allowing unrestrained proliferation; autoimmune diseases can arise if autoreactive immune cells fail to die and attack the organism. However, degenerative diseases can originate if too much apoptosis is initiated. This is especially serious for cells, such as neurons, that do not renew themselves.

While the above picture reflects the basic understanding of apoptosis, it has been modified and rendered more precise in recent years: Distinct signal transduction pathways have been elucidated that decide when caspases are activated: the so-called extrinsic pathway (Green, 1998) is initiated by receptors sitting on the cytoplasmic membrane (Chapter 1). The 'intrinsic' pathway is activated within the cell and involves mitochondria as important organelles, both as cell-death sensors and for the amplification of the proapoptotic signal (Chapter 7). The involvement of such formerly inconspicuous cell compartments led to speculation that every organelle has its own sensor for apoptosis. The endoplasmic reticulum (ER) is meanwhile an established organelle in cell death that harbors important cell-death regulators, some of which are conserved in evolution (Chapter 6). The phylogenetic origin of apoptosis is attested by the finding that even primitive life forms contain genes of apoptosis. Therefore, systems for the study of apoptosis range from single-celled yeast (Chapter 8) to simple Metazoa (Chapter 9), mammalian cell cultures (Chapter 12), and classical genetic systems such as those of *C. elegans* (Chapter 10) and *Drosophila* (Chapter 11).

This book is subdivided into three main sections, the first of which starts with the description of several genes regulating apoptosis to give the reader a basic understanding of the signaling molecules involved. The narrative then progresses to more complex cellular processes that involve whole organelles in apoptosis (the second section). The third and final section is devoted to organisms that are used to study the biology of apoptosis. It is concluded by a perspective on the subject of caspase-independent cell death (Chapter 13), which could well be the topic of the next paradigm-shift in apoptosis research, and which reminds us how much this is a work in progress. We can remain curious as to what Nature still has in store for us to learn about apoptosis.

1. Death receptors in apoptosis

Rajani Ravi and Atul Bedi

1. Introduction

The requirement of cell death to preserve life is no paradox for multicellular animals. Programmed cell death, or apoptosis, enables the physiologic culling of excess cells during embryonic development and tissue remodeling or regeneration in adult animals. In addition to maintaining homeostasis by controlling cell number in proliferating tissues, apoptosis plays an instrumental role in the selective attrition of neurons that fail to establish functional synaptic connections during the development of vertebrate nervous systems. The vertebrate immune system uses apoptosis to delete lymphocytes with inoperative or autoreactive receptors from its repertoire, and to reverse clonal expansion at the end of an immune response. Cytotoxic T cells and natural killer cells induce apoptosis of target cells to effect innate and adaptive immune responses to intracellular pathogens, cancer cells, or transplanted tissues. The altruistic demise of cells in response to cellular stress or injury, or genetic errors, serves to preserve genomic integrity and constitutes an important mechanism of tumor surveillance.

Given the crucial role of apoptosis in such a diverse array of physiologic functions, it is no surprise that aberrations of this process underlie a host of developmental, immune, degenerative, and neoplastic disorders. This appreciation has fueled a furious investigation of the molecular determinants and mechanisms of apoptosis and a search for the key regulators of this process (Hengartner, 2000). The molecular execution of cell death involves activation of members of a family of cysteine-dependent aspartate-specific proteases (caspases) by two major mechanisms. One mechanism, termed the 'intrinsic' pathway, signals the release of prodeath factors from the mitochondria via the action of pro-apoptotic members of the Bcl-2 family (Green and Reed, 1998; Gross *et al.*, 1999a; Green, 2000). An alternative mechanism of activating caspases and mitochondrial disruption, termed the 'extrinsic' pathway, is triggered by engagement of cell-surface 'death receptors' by their specific ligands (Ashkenazi and Dixit, 1998).

In this chapter, we review our current understanding of the molecular determinants and mechanisms of death receptor-induced apoptosis, and identify the key regulators of these death-signaling pathways. We also review the physiologic role of death receptors and describe pathologic conditions that result from a failure or deregulation of their activity. Finally, we highlight the enormous promise of targeting death receptors or their regulatory circuits for treatment of human disease.

2. Death receptors and ligands

Death receptors are cell-surface receptors that trigger death signals following engagement with their cognate 'death ligands' (Ashkenazi and Dixit, 1998). Death receptor-transduced signals play an instrumental role in 'instructive apoptosis', a mechanism that has evolved to enable the deletion of cells in higher metazoans. Death receptors belong to the tumor necrosis factor receptor (TNFR) gene superfamily, whose members have cysteine-rich extracellular domains (CRDs) in their amino terminal region (Smith *et al.*, 1994). The death receptors constitute a subgroup of this family that also possess a homologous cytoplasmic sequence termed the 'death domain' (DD) (*Figure 1*) (Brakebusch *et al.*, 1992; Itoh and Nagata, 1993). The best-characterized death receptors are TNFR1 (also termed p55 or CD120a) (Smith *et al.*, 1994), CD95 (also called Fas or Apo1) (Nagata, 1997), avian CAR1, death receptor 3 (DR3; also called Apo3, WSL-1, TRAMP, or LARD) (Chinnaiyan *et al.*, 1996b; Kitson *et al.*, 1996; Marsters *et al.*, 1996; Bodmer *et al.*, 1997; Screaton *et al.*, 1997b), TRAIL-R1 (also called DR4) (Pan *et al.*, 1997b), TRAIL-R2 (also called DR5, Apo2, TRICK2, or KILLER) (Pan *et al.*, 1997a; Screaton *et al.*, 1997a; Sheridan *et al.*, 1997; Walczak *et al.*, 1997; Wu, G.S. *et al.*, 1997), and DR6 (Pan *et al.*, 1998a). These receptors are activated by ligands of the TNF gene superfamily; TNFR1 is ligated by TNF and lymphotoxin α, CD95 is bound by CD95L (FasL) (Nagata, 1997), DR3 interacts with Apo3 ligand (Apo3L, also called TWEAK) (Chicheportiche *et al.*, 1997; Marsters *et al.*, 1998), and TRAIL-R1 and TRAIL-R2 are engaged by Apo2 ligand (Apo2L, which is also called TNF-related apoptosis-inducing ligand [TRAIL]) (Wiley *et al.*, 1995; Pitti *et al.*, 1996).

3. Induction of apoptosis by death receptors

3.1 The molecular machinery of cell death – caspases

The molecular machinery of cell death comprises an evolutionarily conserved family of cysteine aspartate proteases (caspases) that execute cell disassembly via cleavage of critical substrates that maintain cytoskeletal and DNA integrity (Earnshaw *et al.*, 1999). Caspases recognize specific tetrapeptide motifs in their target proteins and cleave their substrates at Asp-Xxx bonds (after aspartic acid residues) (Thornberry *et al.*, 1997). At least 14 caspases and more than 100 substrates have been identified (Earnshaw *et al.*, 1999). Caspases are divided into distinct subfamilies according to their structural and sequence identities, and their substrate specificity is determined by the specific pattern of four residues amino-terminal to the cleavage site (the P2–P4 positions). While caspases typically function by inactivat-

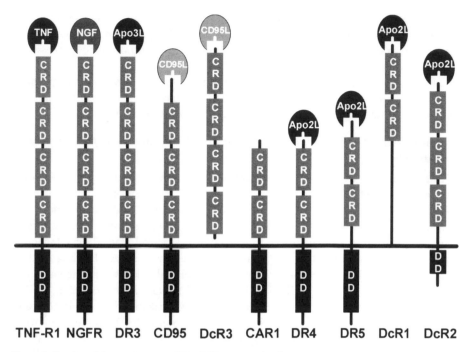

Figure 1. Death and decoy receptors of the TNF receptor family.
Death receptors have variable numbers of cysteine-rich extracellular domains (CRDs) in their extracellular ligand-binding amino terminal regions, and a homologous cytoplasmic sequence, termed the 'death domain' (DD), which is essential for apoptosis signaling. Death receptors are activated by ligands of the TNF gene superfamily; TNFR1 is ligated by TNFα, CD95/Fas is bound by CD95L, DR3 interacts with Apo3L, and TRAIL-R1 and TRAIL-R2 are engaged by Apo2L/TRAIL. DcR1 (TRAIL-R3) is structurally related to DR4 and DR5, but lacks a cytoplasmic tail. DcR2 (TRAIL-R4) also resembles DR4 and DR5, but has a truncated cytoplasmic DD. The extracellular domains of DcR1 and DcR2 compete with DR4 and DR5 for binding to Apo2L/TRAIL, but cannot initiate death signals in response to ligand-engagement. DcR3 binds to CD95L and inhibits CD95-CD95L interactions.

ing proteins by proteolytic cleavage, they can also, in some cases, activate the target by cleaving off a negative regulatory domain. Caspase-mediated cleavage of key substrates underlies many of the characteristic features of apoptosis such as nuclear shrinkage (Rao *et al.*, 1996; Buendia *et al.*, 1999), plasma membrane blebbing (Rudel and Bokoch, 1997; Coleman *et al.*, 2001; Sebbagh *et al.*, 2001), and internucleosomal DNA fragmentation (Liu, X. *et al.*, 1997; Enari *et al.*, 1998; Sakahira *et al.*, 1998). Deficiency of specific caspases or their inhibition can prevent the induction of apoptosis in response to diverse death stimuli (Earnshaw *et al.*, 1999). As such, caspases represent the fundamental executioners of cell death.

3.2 Activation of caspases by death receptors

Caspases are synthesized as inactive zymogens that comprise an N-terminal prodomain and two other domains, p20 and p10, which form the active mature enzyme upon cleavage between the p20 and p10 domains as well as between the p20 domain and the prodomain (Earnshaw *et al.*, 1999). Since these Asp-X cleavage sites correspond to caspase substrate motifs, procaspases can be activated by either previously activated upstream caspases or by autocatalytic processing (Thornberry *et al.*, 1997).

Members of the death receptor family share the same fundamental mechanism(s) of activating caspases and amplifying this enzymatic cascade (*Figure 2*). These sequential steps and signaling pathways are dissected below:

Formation of the death-inducing signaling complex – activation of the initiator caspase-8/FLICE

Death receptors are type I transmembrane proteins containing cytoplasmic sequences (DDs) that are essential for transduction of the apoptotic signal (Itoh and Nagata, 1993; Tartaglia *et al.*, 1993). The oligomerization of death receptors by engagement of their cognate ligands results in the rapid assembly of a membrane-bound death-inducing signaling complex (DISC) (Kischkel *et al.*, 1995). Ligand-induced trimerization of the CD95/Fas/Apo1 receptor facilitates binding of the adapter protein, FADD (Fas-associated DD protein; also known as mediator of receptor-induced toxicity [MORT1]) (Boldin *et al.*, 1995; Chinnaiyan *et al.*, 1995) through homotypic interactions between their DD (Brakebusch *et al.*, 1992; Itoh and Nagata, 1993; Tartaglia *et al.*, 1993). Receptor-bound FADD molecules form higher-order oligomers, or 'fibers' (Siegel *et al.*, 1998). FADD also carries a so-called death-effector domain (DED), which, in turn, interacts with the analogous DED motifs found in the N-terminal region of the zymogen form of caspase-8 (procaspase-8; also called FLICE or MACH) (Boldin *et al.*, 1996; Muzio *et al.*, 1996; Medema *et al.*, 1997). FADD-dependent recruitment and aggregation of multiple procaspase-8 molecules to the receptor/FADD scaffold results in autocatalytic cleavage and cross-activation by induced proximity, thereby releasing active caspase-8 into the cytoplasm (Martin *et al.*, 1998; Muzio *et al.*, 1998; Yang *et al.*, 1998a).

Other death receptors activate caspase-8 in a fashion analogous to that of CD95. While death receptors for Apo2L/TRAIL (TRAIL-R1/DR4 and TRAIL-R2/DR5) directly bind FADD, TNF-α-bound TNFR1 binds the adapter molecule TRADD (TNFR1-associated DD protein), which, in turn, recruits FADD to the receptor complex (Hsu *et al.*, 1996). Experiments with FADD gene-knockout mice (Zhang, L. *et al.*, 1998) or transgenic mice expressing a dominant-negative mutant of FADD (FADD-DN) in T cells (Newton *et al.*, 1998; Zornig *et al.*, 1998) have demonstrated that FADD is essential for induction of apoptosis by CD95/Fas, TNFR1, and DR3. However, a similar obligatory role of FADD in apoptosis signaling by Apo2L/TRAIL or its death receptors has not been uniformly observed. Cells from FADD-deficient mice remain susceptible to DR4-induced apoptosis (Yeh *et al.*, 1998), and ectopic expression of FADD-DN failed to block induction of apoptosis by either Apo2L/TRAIL or overexpression of either DR4 or DR5 (Pan *et al.*, 1997a; Sheridan *et al.*, 1997). While these studies suggest the existence of a FADD-independent mechanism by which Apo2L/TRAIL activates caspase-8 (perhaps through another adapter), other conflicting observations indicate that DR4- or DR5-induced apoptosis can be inhibited by transfection of FADD-DN (Walczak *et al.*, 1997). The physiologic role of FADD is further supported by Apo2L/TRAIL

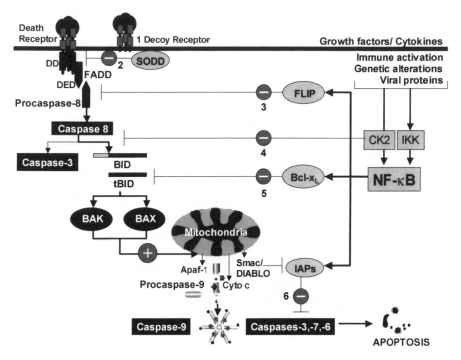

Figure 2. Schematic representation of the molecular mechanisms and regulation of death receptor-induced apoptosis.

Ligand-induced trimerization of death receptors facilitates binding of the adapter protein, FADD/MORT1, through homotypic interactions between their DDs. FADD carries a death-effector domain (DED), which interacts with the analogous DED motifs found in procaspase-8/FLICE. FADD-dependent recruitment of multiple procaspase-8 molecules to the receptor:FADD scaffold results in autocatalytic cleavage and cross-activation. Caspase-8 triggers proteolytic activation of caspase-3. Caspase-8 also cleaves and activates BID. The active truncated form of BID (tBID) translocates to the outer mitochondrial membrane, where it binds BAX or BAK. tBID-induced homoligomerization of BAX or BAK results leads to mitochondrial disruption and release of prodeath cofactors (cytochrome c, Smac/DIABLO). The interaction of cytochrome c with Apaf-1 results in a nucleotide-dependent conformational change that allows binding of procaspase-9. The formation of the procaspase-9/Apaf-1/cytochrome c complex promotes the transactivation of caspase-9. Caspase-9 activates downstream caspases (caspase-3 and caspase-7), thereby amplifying the caspase cascade and promoting apoptosis.

The death receptor-induced signaling pathway is regulated at multiple levels: (1) Decoy receptors interfere with the interaction of death ligands with their cognate death receptors. DcR1/TRAIL-R3 and DcR2/TRAIL-R4 compete with TRAIL-R1/TRAIL-R2 for Apo2L/TRAIL; DcR3 binds to CD95L and competitively inhibits the interaction of CD95 with CD95L. (2) Silencer of death domains (SODD) inhibit the intrinsic self-aggregation of the death domain of TNFR1. (3) The recruitment and activation of caspase-8 are inhibited by FLICE-inhibitory protein (FLIPs). (4) Phosphorylation of BID by casein kinases (CKI and CKII) renders BID resistant to caspase-8-mediated cleavage. (5) Sequestration of tBID by the Bcl-2 homolog, Bcl-x_L, curtails its ability to promote the allosteric activation of BAX or BAK. (6) Inhibitor-of-apoptosis proteins (IAPs) inhibit effector caspases (caspase-3, caspase-7, caspase-9, and caspase-6), until they are themselves sequestered by Smac/DIABLO. NF-κB promotes the expression of c-FLIP, Bcl-x_L, and members of the IAP family (c-IAP1, c-IAP2, XIAP). By inducing the concurrent expression of multiple antiapoptotic proteins that interrupt different steps along the death receptor-signaling pathway, NF-κB exerts a multipronged inhibition of death-receptor signals.

dependent recruitment of FADD and caspase-8 to DR4 and DR5 (Kischkel *et al.*, 2000). Regardless of the specific mechanism employed to activate caspase-8, experiments with embryonic fibroblasts from caspase-8-deficient mice confirm that caspase-8 is essential for initiation of apoptosis by CD95/Fas, TNFR1, DR3, DR4, and DR5 (Varfolomeev *et al.*, 1998).

Caspase-8-mediated activation of downstream effector caspases

Caspase-8-mediated activation of caspase-3. In some cell types (termed 'type I'), robust activation of caspase-8 by formation of the DISC results in the direct cleavage and activation of the downstream effector caspase-3, which in turn, cleaves other caspases (such as caspase-6) and vital substrates, leading to the terminal events of apoptosis (Scaffidi *et al.*, 1998). Such cell types can undergo apoptosis via death receptor-induced activation of the caspase cascade independently of the mitochondria. However, experiments with caspase-3 knockout mice indicate that while caspase-3 may serve an important role in internucleosomal DNA fragmentation, it is not required for CD95- or TNF-induced apoptosis (Woo *et al.*, 1998).

Caspase-8-mediated cleavage of BID – mitochondrial disruption by BAX or BAK. Caspase-8 cleaves and activates BID (p22), a 'BH-3 domain only' prodeath member of the Bcl-2 family (Wang, C.Y. *et al.*, 1996; Li, J.H. *et al.*, 1998; Luo *et al.*, 1998; Gross *et al.*, 1999b; Roy and Nicholson, 2000). The active truncated form of BID (tBID; p15) translocates to the outer mitochondrial membrane, where it binds and homooligomerizes BAX or BAK, two multidomain proapoptotic members of the Bcl-2 family (Eskes *et al.*, 2000; Wei, M.C. *et al.*, 2000). tBID-induced homooligomerization of BAX or BAK results in an allosteric conformational change that leads to mitochondrial disruption and release of a cocktail of prodeath cofactors (such as cytochrome c and Smac/DIABLO) into the cytoplasm (Li, H. *et al.*, 1998; Gross *et al.*, 1999b). The interaction of the released cytochrome c with Apaf-1 results in a nucleotide-dependent conformational change that allows binding of procaspase-9 through N-terminal caspase recruitment domains (CARD) present on both molecules (Vaux, 1997). The formation of the procaspase-9/Apaf-1/cytochrome c complex (also called the 'apoptosome') promotes the transcatalytic cleavage and scaffold mediated transactivation of caspase-9 (Srinivasula *et al.*, 1998; Stennicke *et al.*, 1999). Caspase-9 activates further downstream caspases such as caspase-3 and caspase-7, thereby amplifying the caspase cascade and promoting apoptosis (Deveraux *et al.*, 1998). Therefore, BID represents a mechanistic link between death receptor-induced activation of caspase-8 (the 'extrinsic' pathway) and the mitochondrial activation of caspases-9 and -3 (the 'intrinsic' or 'mitochondrial' pathway) (Roy and Nicholson, 2000).

Caspase-8-induced cleavage of BID is instrumental for death receptor-induced apoptosis in certain cell types (termed 'type II') which show weak DISC formation and therefore depend upon mitochondrial activation of caspase-9 to amplify the caspase cascade (Scaffidi *et al.*, 1998). Studies with BID-deficient mice indicate that BID is required for CD95-induced apoptosis in hepatocytes, but not in thymocytes or fibroblasts (Yin *et al.*, 1999). Studies of BAX-deficient, BAK-deficient, or BAX/BAK-deficient mice suggest that BAX and BAK play essential, yet mutually redundant, roles in death receptor-mediated apoptosis in hepatocytes, but are not required for CD95-induced death of thymocytes or fibroblasts (Lindsten *et al.*, 2000; Wei *et al.*, 2001). While BAX$^{-/-}$/BAK$^{-/-}$ hepatocytes resist CD95-induced apoptosis, hepatocytes from either BAX$^{-/-}$ or BAK$^{-/-}$ mice remain sensitive to death receptor-induced apoptosis. In contrast, recent studies using colon carcinoma lines that have wild-type BAX and their isogenic BAX-deficient sister clones have demonstrated that BAX may be required for death receptor-induced apoptosis in cancer cells (Burns and el Deiry, 2001; Deng *et al.*, 2002; LeBlanc *et al.*, 2002; Ravi and Bedi, 2002). These reports indicate that although BAX is dispensable for apical death receptor signals, including activation of caspase-8 and cleavage of BID, it is necessary for mitochondrial activation of caspase-9 and induction of apoptosis in response to Apo2L/TRAIL, CD95/Fas, or TNF-α (LeBlanc *et al.*, 2002; Ravi and Bedi, 2002). These data suggest that the basal expression of BAK in these cells cannot substitute for BAX in mediating death receptor-induced apoptosis of tumor cells. However, upregulation of BAK expression by exposure to the chemotherapeutic agents, etoposide or irinotecan, is associated with sensitization of BAX$^{-/-}$ cancer cells to death receptor-induced apoptosis (LeBlanc *et al.*, 2002). Therefore, tBID may employ BAX or BAK for mitochondrial activation of apoptosis in a cell-type- and death signal-specific manner.

Since loss of BAX and BAK confer long-term resistance to death receptor-induced apoptosis, mitochondrial disruption appears to be critical for induction of apoptosis in type II cells. However, activation of the caspase-9/Apaf1 complex and caspase-3 is not the only mechanism by which mitochondrial disruption results in apoptosis (Cheng *et al.*, 2001). Absence of these downstream effectors provides only transient protection from tBID-induced apoptosis. Cells from caspase-9$^{-/-}$, Apaf$^{-/-}$, or caspase-3$^{-/-}$ mice remain viable for 24 h after retroviral expression of tBID, but are killed after 48 h (Cheng *et al.*, 2001). In line with these observations, cells from caspase-9$^{-/-}$ (Hakem *et al.*, 1998; Kuida *et al.*, 1998), Apaf$^{-/-}$ (Cecconi *et al.*, 1998; Yoshida *et al.*, 1998), or caspase-3$^{-/-}$ (Kuida *et al.*, 1996; Woo *et al.*, 1998) mice remain susceptible to CD95-induced apoptosis. One possible mechanism by which tBID-mediated mitochondrial depolarization can promote death in the absence of either cyto c/Apaf-1/caspase-9 or caspase-3 may involve the activation of redundant effector caspases via release of Smac/DIABLO (second mitochondria-derived activator of caspase). Smac/DIABLO promotes caspase activation by binding and antagonizing members of the IAP (inhibitor of apoptosis) family of proteins (Du *et al.*, 2000; Verhagen *et al.*, 2000). The human IAP family comprises six members which inhibit the effector caspases-9, -3, and -7 (Deveraux *et al.*, 1997; Deveraux *et al.*, 1998; Goyal, 2001). The mitochondrial release of Smac/DIABLO into the cytoplasm via the caspase-8-BID-BAX/BAK pathway may sequester IAPs and allow activation of multiple effector caspases. The simultaneous activation of multiple redundant effector caspases may explain why deficiency of any single caspase (caspase-9 or caspase-3) is insufficient to block death receptor-induced apoptosis. It is also possible that irreversible damage to the mitochondria may itself be sufficient to induce apoptosis in a caspase-independent manner.

While these observations indicate that mitochondrial disruption via the BID-BAX/BAK pathway is essential for death receptor-induced apoptosis of certain cell types (hepatocytes, cancer cells), embryonic fibroblasts and thymocytes from

BID-deficient or BAX/BAK double-knockout (DKO) mice remain sensitive to CD95/Fas-induced apoptosis (Lindsten *et al.*, 2000; Wei *et al.*, 2001). The precise reasons for the differential requirement of the cross-talk between the extrinsic and intrinsic pathways in type I and type II cells have yet to be elucidated. These may involve biochemical differences at the receptor level and/or differences in the expression of initiator caspases and/or antiapoptotic proteins that determine the threshold that must be crossed for death receptors to activate downstream caspases.

Role of caspase-10 in death receptor-induced apoptosis

In addition to caspase-8, death receptors can also induce recruitment and activation of the structurally related protein, caspase-10. In cells with endogenous expression of both caspase-8 and caspase-10, CD95L and Apo2L/TRAIL can recruit either protein to their DISC, where both enzymes are proteolytically activated with similar kinetics (Kischkel *et al.*, 2001). Caspase-10 recruitment and cleavage requires the adapter FADD/Mort1 and DISC assembly. Cells expressing either caspase-8 or caspase-10 can undergo ligand-induced apoptosis, indicating that each caspase can initiate apoptosis independently of the other (Kischkel *et al.*, 2001). Thus, apoptosis signaling by death receptors involves not only caspase-8 but also caspase-10, and both caspases may have equally important roles in apoptosis initiation. However, caspase-8 plays an obligatory role in death receptor-induced apoptosis of cell types that do not express caspase-10, such as many cancer cells (Kischkel *et al.*, 2001).

Role of apoptosis signal regulating kinase (ASK1) and c-Jun amino terminal kinase (JNK)

While there is overwhelming evidence that confirms the requirement of FADD/caspase-8-mediated signaling pathways in death receptor-induced apoptosis, other putative death receptor signaling pathways have also been described. Apoptosis signal regulating kinase (ASK)1 is a mammalian MAPKKK that activates SEK1 (or MKK4), which, in turn, activates stress-activated protein kinase (SAPK, also known as JNK: c-Jun N-terminal kinase). Overexpression of ASK1 induces apoptotic cell death, and ASK1 is activated in cells treated with TNF-α; TNF-α-induced apoptosis is inhibited by a catalytically inactive form of ASK1 (Ichijo *et al.*, 1997). These observations suggest that ASK1 may be a key element in the mechanism of stress- and cytokine-induced apoptosis. ASK1 leads to activation of c-Jun N-terminal protein kinase (JNK). However, the role of JNK activation in TNF-α-induced apoptosis is less clear. Experiments using JNK activators or dominant negative forms of the JNK substrate c-Jun suggest a proapoptotic role of JNK in TNF-α-induced death (Verheij *et al.*, 1996). However, JNK-deficient mouse fibroblasts remain sensitive to TNF-α- or CD95L-induced apoptosis (Tournier *et al.*, 2000). Therefore JNK activation may potentiate death receptor-induced apoptosis, but is not obligatory for this process.

4. Regulation of death receptor-induced apoptosis

Death receptors play an instrumental role in the physiologic induction of apoptosis during development and tissue turnover in adult animals. However, the unscheduled activation of death receptor-induced signals could lead to inadvertent caspase activation with devastating consequences for the organism. In order to direct the 'instructive' apoptosis of cells without sustaining uncontrolled cell death, death receptor-induced signaling is tightly regulated at multiple levels (*Figure 2*). These regulatory mechanisms are described below.

4.1 Expression of death and decoy receptors

At the most apical level, death receptor-ligand interactions may be regulated by the tissue-specific or inducible expression of death receptors or their respective ligands. TNFR1 is expressed ubiquitously, while its ligand (TNF) is expressed mainly by activated T cells and macrophages (Smith *et al.*, 1994). Likewise, CD95/Fas is widely expressed and its cell-surface expression is elevated by immune activation of lymphocytes or in response to cytokines such as interferon-γ, TNF, and CD40 ligand (CD40L) (Leithauser *et al.*, 1993; Krammer, 2000). Expression of CD95 ligand (CD95L) is, however, restricted to cytotoxic T cells, NK cells, and antigen presenting cells (APCs) (Li, J.H. *et al.*, 1998).

Akin to TNFR1 and CD95, the death receptors for Apo2L/TRAIL (DR4/TRAIL-R1 and DR5/TRAIL-R2) are broadly expressed in most organ systems (Golstein, 1997; Ashkenazi and Dixit, 1998). However, unlike the restricted pattern of TNF and CD95L expression in immune activated T cells and APCs, Apo2L/TRAIL mRNA is expressed constitutively in many tissues, and transcript levels increase upon stimulation in peripheral blood T cells (Screaton *et al.*, 1997a; Jeremias *et al.*, 1998; Martinez-Lorenzo *et al.*, 1998). Since several tissues constitutively express both Apo2L and its death receptors, normal cells must employ mechanisms to protect themselves from autocrine or paracrine Apo2L-DR4/DR5 interactions. One such line of defense is provided by expression of a set of decoy receptors (DcRs). DcR1 (also called TRAIL-R3, TRID, or LIT) is a glycosyl phosphatidylinositol (GPI)-anchored cell-surface protein that is structurally related to DR4 and DR5, but lacks a cytoplasmic tail (Degli-Esposti *et al.*, 1997b; Pan *et al.*, 1997b; MacFarlane *et al.*, 1997; Mongkolsapaya *et al.*, 1998; Schneider *et al.*, 1997; Sheridan *et al.*, 1997). DcR2 (also called TRAIL-R4 or TRUNDD) also resembles DR4 and DR5, but has a truncated cytoplasmic DD that is only a third as long as that of functional DDs that are capable of transducing apoptotic signals (Degli-Esposti *et al.*, 1997a; Marsters *et al.*, 1997; Pan *et al.*, 1998b). The extracellular domains of DcR1 and DcR2 compete with DR4 and DR5 for binding to Apo2L/TRAIL, but cannot initiate death signals in response to ligand engagement. Transfection of Apo2L-sensitive cells with either DcR1 or DcR2 substantially reduces their sensitivity to Apo2L-induced apoptosis. Deletion of the truncated cytoplasmic region of DcR2 does not affect its ability to protect cells from Apo2L-induced death. Enzymatic cleavage of the GPI anchor also results in the sensitization of DcR1-expressing cells to Apo2L-induced apoptosis. These observations indicate that DcR1 and DcR2 may protect normal cells from Apo2L by acting as 'decoys' that compete with TRAIL-R1/TRAIL-R2 for their shared ligand. The expression of decoy receptors provides a potential molecular basis for the relative resistance of normal cells to TRAIL/Apo2L-induced death. In support of this notion, resting peripheral T cells (that resist Apo2L) exhibit an elevation of DR5

(Screaton *et al.*, 1997a) and concomitant reduction of DcR1 levels (Mongkolsapaya *et al.*, 1998) when they acquire an Apo2L-sensitive phenotype upon activation by interleukin-2 (Martinez-Lorenzo *et al.*, 1998). However, many tumor cell lines express high levels of decoy receptors, yet remain susceptible to Apo2L-induced death (Griffith and Lynch, 1998). Therefore, it is likely that the susceptibility of cells to Apo2L-induced apoptosis must involve additional regulatory mechanisms beyond the ratio of death and decoy receptors.

A third decoy receptor, DcR3, is a secreted soluble protein the binds to CD95L (Pitti *et al.*, 1998). DcR3 competitively inhibits the interaction of CD95 with CD95L, and overexpression of DcR3 inhibits CD95-induced apoptosis (Pitti *et al.*, 1998). DcR3 mRNA is expressed in the spleen, colon, and lung. While its physiologic role remains unclear, the frequent amplification of the DcR3 gene in primary lung and colon cancers may protect tumor cells from CD95L-induced death.

4.2 Inhibition of DD signaling by silencer of death domains (SODD)

The DDs of death receptors (TNF-R1, CD95, DR3, DR4, and DR5) can self-associate and bind other DD-containing proteins. Overexpression of DD receptors may lead to ligand-independent receptor aggregation and cell death. However, cells are protected from such spontaneous ligand-independent signaling by death receptors via expression of a ~60-kDa protein termed silencer of death domains (SODD) (Jiang *et al.*, 1999). SODD associates with the DDs of TNF-R1 and inhibits the intrinsic self-aggregation of the DD (Jiang *et al.*, 1999). This inhibition is lost by triggering the release of SODD from the DD in response to cross-linkage of TNF-R1 with TNF-α. This allows ligand-dependent recruitment of adapter proteins to form an active signaling complex. However, the duration of TNF signaling is controlled by the rapid dissociation of signaling proteins from TNF-R1 and reformation of the TNF-R1-SODD complex (Jiang *et al.*, 1999). While SODD interacts with TNF-R1 and DR3, other SODD-related proteins may play a similar role in preventing ligand-independent signaling by CD95, DR4, or DR5.

4.3 Regulation of caspase-8 by FLICE-inhibitory proteins (FLIPs)

The recruitment and activation of caspases by death receptor engagement can be inhibited by FLICE-inhibitory protein (FLIP; also called I-FLICE, CASH, CLARP, MRIT, or usurpin) (Bertin *et al.*, 1997; Goltsev *et al.*, 1997; Han *et al.*, 1997; Hu *et al.*, 1997; Inohara *et al.*, 1997; Irmler *et al.*, 1997; Srinivasula *et al.*, 1997; Thome *et al.*, 1997; Rasper *et al.*, 1998; Chai, J. *et al.*, 2001). FLIP contains death-effector domains (DEDs) that bind to the DED of FADD and the prodomains of procaspases-8 and -10, thereby inhibiting their recruitment to the CD95-FADD- or TNFR1-induced activation complex (Goltsev *et al.*, 1997; Han *et al.*, 1997; Goltsev *et al.*, 1997; Han *et al.*, 1997; Irmler *et al.*, 1997; Srinivasula *et al.*, 1997). Enforced expression of v-FLIP (found in γ-herpes and pox viruses) inhibits apoptosis induced by CD95, TNFR1, DR3, and DR4/TRAIL-R1 (Thome *et al.*, 1997). The equine herpes II virus E8 protein and molluscum contagiosum MC159 and MC160 also contain DEDs homologs to those of procaspase-8, and inhibit its recruitment to the death receptor signaling complex (Bertin *et al.*, 1997).

Multiple splice variants of the human homologs of FLIP have been identified.

The longer, more abundant form, FLIP$_L$, has two DEDs and a caspase-like C-terminal domain, but lacks the catalytic cysteine and histidine residues that contribute to substrate binding. The shorter splice variant, FLIP$_S$, comprises only the two DEDs. The effects of cellular FLIPs appear to vary depending on the cellular context. Enforced expression of FLIP (the long splice variant) has an apparently paradoxical pro-apoptotic effect, possibly mediated by the aggregation of procaspase-8. However, experiments with c-FLIP-deficient mice support an antiapoptotic role of c-FLIP that is analogous to its viral counterparts. Embryonic fibroblasts from c-FLIP-deficient mice are hypersensitive to TNF-α- or CD95/Fas-induced apoptosis (Yeh *et al.*, 2000). Akin to v-FLIP, c-FLIP appears to function as a physiologic inhibitor of death receptor-induced apoptosis via homotypic DED-mediated interactions with FADD and procaspase-8 (Irmler *et al.*, 1997; Srinivasula *et al.*, 1997). c-FLIP also interacts with TRAF2 and receptor-interacting protein (RIP), which are responsible for TNFR1-induced activation of NF-κB and JNK. However, c-FLIP-deficient cells do not exhibit any change in TNF-α-induced activation of NF-κB (Yeh *et al.*, 2000). Conversely, NF-κB activation is required for TNF-α-induced expression of c-FLIP (Kreuz *et al.*, 2001). These observations indicate that NF-κB-induced expression of c-FLIP protects cells from death receptor-induced apoptosis by preventing initiation of the caspase cascade.

4.4 Regulation of BID cleavage and function

Caspase-8-induced cleavage of BID is required for mitochondrial amplification of downstream caspases in response to death receptor engagement in type II cells. Therefore, regulation of BID cleavage or activity is a mechanism of controlling death receptor-induced apoptosis in type II cells. These regulatory mechanisms are described below.

Regulation of BID phosphorylation and cleavage by casein kinases I and II

BID is cleaved by caspase-8 at Asp 59, which resides in a large flexible loop between the second and third α helices (Li. H. *et al.*, 1998; Luo *et al.*, 1998). This cleavage site is located between the Thr and Ser residues which are phosphorylated by casein kinases I and II (CKI and CKII) (Desagher *et al.*, 2001). CKI exists as monomers of seven isoforms encoded by distinct genes (α, β, γ1, γ2, γ3, δ, and ε) (Tuazon and Traugh, 1991). CKII is an evolutionarily conserved holoenzyme composed of two catalytic α (and/or α′) subunits and two regulatory β subunits (Tuazon and Traugh, 1991). Phosphorylation of BID by CKI and CKII has been reported to render BID resistant to caspase-8-mediated cleavage (Desagher *et al.*, 2001). Conversely, a mutant of BID that cannot be phosphorylated at these residues is apparently more sensitive to caspase-8-induced cleavage and more effective than wild-type BID in promoting apoptosis. Consistent with these observations, activation of CKI and CKII reportedly delays CD95/Fas-induced apoptosis, whereas CK inhibitors potentiate death receptor-induced apoptosis (Desagher *et al.*, 2001).

While these observations suggest that the protection conferred by CKs is mediated by phosphorylation of BID, CKs may also target other proteins, such as NF-κB, that regulate the caspase-8-BID-BAX/BAK death pathway. It is also possible that other kinases, such as protein kinase C (PKC), may be involved in the phosphorylation of BID. Activation of PKC by phorbol esters prevents CD95-induced cleavage of BID and apoptosis, and this protective effect is reversed by inhibition of PKC (Holmstrom *et al.*, 2000). It remains to be determined whether CKs cooperate with PKC or other kinases in the phosphorylation of BID. The Ser/Thr phosphatase that dephosphorylates BID and renders it susceptible to caspase-8-mediated cleavage in response to death receptor engagement has yet to be identified.

Sequestration of BID by Bcl-x$_L$ – competition with BAX or BAK

BH-3 domain-only members of the Bcl-2 family, such as BID, absolutely require multidomain members of the Bcl-2 family (BAX, BAK) to induce apoptosis. Antiapoptotic members of the Bcl-2 family, such as Bcl-x$_L$, heterodimerize with BAX or BAK, as well as BID (Cheng *et al.*, 2001). While mutants of Bcl-x$_L$ that cannot bind either BAX or BAK (bearing F131V or D133A substitutions) remain capable of protecting cells from death receptor-induced apoptosis, Bcl-x$_L$ mutants that fail to bind BID (bearing G138E, R139L, and I140N substitutions) are unable to inhibit apoptosis (Cheng *et al.*, 2001). These observations support a model in which antiapoptotic Bcl-2 family members, such as Bcl-x$_L$, sequester tBID in stable mitochondrial complexes, thereby curtailing its ability to promote the allosteric activation of BAX or BAK. In this scenario, proapoptotic multidomain members (BAX, BAK) compete with antiapoptotic members (Bcl-2, Bcl-x$_L$) for binding to tBID to regulate the mitochondrial disruption and efflux required for the terminal events of apoptosis.

4.5 Inhibitor-of-apoptosis proteins (IAPs) – sequestration of Smac/DIABLO and inhibition of caspases

Caspases are directly regulated by interactions with inhibitor-of-apoptosis (IAP) proteins. At least five mammalian homologs of the baculovirus IAP have been identified. Four of these (cIAP-1, cIAP-2, XIAP, and NAIP) consist of an N-terminal domain containing multiple copies of a so-called baculovirus IAP repeat (BIR) motif (Birnbaum *et al.*, 1994), and a C-terminal zinc-containing protein–protein interaction domain (RING finger) (Lovering *et al.*, 1993). The fifth member (survivin) contains only the BIR domain. XIAP (X chromosome-linked IAP; also known as hILP), cIAP-1, and c-IAP2 (but not NAIP) directly bind and inhibit effector caspases, such as caspase-3 and caspase-7 (Deveraux *et al.*, 1997; Roy *et al.*, 1997). In addition, they also prevent activation of procaspase-9 and procaspase-6 by upstream signals (Deveraux *et al.*, 1998). XIAP inhibits caspase-3 and caspase-7 via its second BIR domain and BH2-terminal linker (Takahashi *et al.*, 1998), and prevents activation of procaspase-9 through a region containing its third BIR domain (BIR3) (Deveraux *et al.*, 1999). The BIR2 region facilitates caspase-binding, and the NH2-terminal linker directly blocks the catalytic cleft of caspase-3 and caspase-7 (Chai, J. *et al.*, 2001; Huang *et al.*, 2001; Riedl *et al.*, 2001). Consistent with its ability to inhibit multiple effector caspases, overexpression of XIAP can inhibit TNF-α-induced apoptosis (Stehlik *et al.*, 1998). While these direct interactions with caspases may be responsible for the antiapoptotic effects of IAPs, cIAP-1 and c-IAP2 also interact with the TNFR1-associated proteins, TRAF-1 and

TRAF-2, via their BIR domains (Rothe *et al.*, 1995). Therefore, although c-IAPs do not directly interact with caspase-8, it is possible their recruitment to the TNFR1 signaling complex via an interaction with TRAF-2 may regulate caspase-8 activation and/or TRAF-dependent signaling. Consistent with this scenario, expression of c-IAP1 or cIAP-2 alone was not sufficient to reduce cellular sensitivity to TNF-α-induced death; however, the expression of both c-IAP1 and c-IAP2, coupled with TRAF1 and TRAF2, suppressed TNF-α-induced apoptosis (Wang, C.Y. *et al.*, 1998).

The basal expression of mammalian IAPs varies in different cell types in response to cytokines, such as TNFα. As shall be discussed below, TNF-α-induced expression of c-IAP1, cIAP-2, and XIAP is dependent upon the NF-κB transcription factor (Chu *et al.*, 1997; Wang, C.Y. *et al.*, 1998). In these physiologic situations, IAPs may serve to keep caspases in check until they are themselves sequestered and antagonized by the mitochondrial efflux of Smac/DIABLO in response to death signals. However, the constitutively high expression of IAPs, such as survivin, in many different types of tumor cells may render such cells abnormally resistant to death receptor-induced apoptosis (Ambrosini *et al.*, 1997).

4.6 NF-κB – a master regulator of death receptor-induced apoptosis

While each of the regulatory mechanism(s) described above serves to interrupt specific steps along the death receptor-induced signaling pathway, the master regulator responsible for orchestrating the coordinated control of death receptor-induced apoptosis is NF-κB, a family of heterodimeric transcription factors (Rel proteins) that plays an important role in determining lymphocyte survival during immune, inflammatory, and stress responses (Sha *et al.*, 1995; Beg and Baltimore, 1996; Liu, Z.G. *et al.*, 1996; Van Antwerp *et al.*, 1996; Wang, C.Y. *et al.*, 1996; Attar *et al.*, 1997; Franzoso *et al.*, 1998; Alcamo *et al.*, 2001; Senftleben *et al.*, 2001b; Karin and Lin, 2002). Mammals express five Rel proteins that belong to two classes (Grimm and Baeuerle, 1993; Karin and Ben Neriah, 2000). Members of one group (RelA, c-Rel, and RelB) are synthesized as mature proteins, while the other (encoded by *NFkb1* and *NFkb2*) includes precursor proteins (p105 and p100, respectively) that undergo proteolysis to yield their mature products (p50 and p52 NF-κB proteins).

NF-κB dimers containing RelA or c-Rel are held in an inactive cytoplasmic complex with inhibitory proteins, the IκBs. Phosphorylation of IκBs at two critical serine residues (Ser^{32} and Ser^{36} in IκBα, Ser^{19} and Ser^{23} in IκBβ) in their N-terminal regulatory domain by the IκB kinase (IKK) complex targets them for rapid ubiquitin-mediated proteasomal degradation (Karin and Ben Neriah, 2000). IKK is a multisubunit protein kinase consisting of two catalytic subunits, IKKα and IKKβ, which phosphorylate IκB, and a regulatory subunit, IKKγ (also called NEMO, NF-κB essential modifier/modulator or IKKAP1), which is required for activation of IKKα/IKKβ heterodimers in response to proinflammatory cytokines, such as TNF-α and interleukin-1 (IL-1). The C-terminus of IKKγ subunit serves as a docking site for upstream signals, and the N-terminal half of IKKγ (minus the first 100 amino acids) binds to IKKβ. This results in phosphorylation of specific conserved serine residues (S177 and S181) within the T-loop (activation domain) in the catalytic domain of IKKβ. Activation of the canonical NF-κB pathway involving degradation of IκB is mostly dependent on the IKKβ subunit, and is essential for innate immunity (Li *et al.*, 1999a; Delhase *et al.*, 1999; Li *et al.*, 1999b; Senftleben *et al.*, 2001b). A second pathway is involved in activation of the NF-κB dimer between RelB and p52 (Solan *et al.*, 2002). RelB is held in an inactive cytoplasmic complex by NF-κB2p100 until IKKα-dependent degradation of the IκB-like COOH-terminus of p100 allows the release and nuclear translocation of the active RelB/p52 dimer (Solan *et al.*, 2002). The activation of the RelB/p52 dimer by proteolytic processing is important for lymphoid organ development and the adaptive immune response (Senftleben *et al.*, 2001a).

In addition to the release and nuclear translocation of the dimer, transcriptional induction of target genes by NF-κB requires phosphorylation of Rel proteins by serine/threonine kinases, such as casein kinase II and Akt (Zhong *et al.*, 1997; Sizemore *et al.*, 1999; Madrid *et al.*, 2000; Wang, D. *et al.*, 2000).

Role of NF-κB in protection of cells from death receptor-induced apoptosis

Targeted disruption of the RelA subunit of NF-κB or either IKKβ or IKKγ/NEMO results in embryonic death of mice as a result of massive hepatic (liver) apoptosis (Beg *et al.*, 1995; Li *et al.*, 1999a; Rudolph *et al.*, 2000). RelA$^{-/-}$ fibroblasts, unlike their wild-type (RelA$^{+/+}$) counterparts, exhibit a profound sensitivity to TNF-α-induced apoptosis (Beg and Baltimore, 1996). Likewise, IKKβ$^{-/-}$ fibroblasts, or cells stably transfected with phosphorylation mutants of IκBα, fail to activate NF-κB and display increased sensitivity to TNF-α-induced death (Van Antwerp *et al.*, 1996; Li *et al.*, 1999b; Senftleben *et al.*, 2001b). These observations demonstrate an important role of NF-κB in protecting cells from death receptor-induced apoptosis.

Engagement of TNFR1 by TNF leads to the recruitment of the adapter protein TRADD to the clustered DDs of the trimerized receptors. TRADD, in turn, serves as a platform for the docking of multiple signaling molecules to the activated receptor complex. As discussed earlier, TNF-induced apoptosis is triggered by recruitment of the adapter molecule FADD to the TNFR1-TRADD complex. Therefore, the apoptotic signaling pathways triggered by different members of the death receptor family (TNFR1, CD95/Fas, and TRAIL-R1/R-2) are all initiated by ligand-induced recruitment of FADD and FADD-mediated activation of caspase-8. While they share a common death-signaling pathway, these receptors exhibit a differential ability to activate NF-κB. While FasL is unable to activate NF-κB, TNF-α induces activation of the transcription factors, NF-κB and JNK/AP1, via recruitment of receptor-interacting protein (RIP) and TNFR-associated factor-2 (TRAF-2) to the receptor complex. TRAF-2 and RIP activate the NF-κB-inducing kinase (NIK), which, in turn, activates the IκB kinase (IKK) and IKKβ-dependent activation of NF-κB (Malinin *et al.*, 1997; Kelliher *et al.*, 1998; Scheidereit, 1998). TRAF2 and RIP also stimulate JNK/AP-1 via activation of apoptosis signal regulating kinase (ASK)1 (Nishitoh *et al.*, 1998). Cells from TRAF2 gene knockout mice or transgenic mice expressing a dominant negative TRAF2 mutant fail to activate JNK in response to TNF, but have only slight defects in TNF-induced activation of NF-κB (Lee *et al.*, 1997; Yeh *et al.*, 1997). In contrast, RIP-deficient cells remain capable of activating JNK but lack the ability to activate NF-κB in response to TNF (Kelliher *et al.*, 1998). Therefore, RIP is essential for TNF-induced activation of NF-κB, while TRAF2 is required for signaling the activation of JNK. In addition to inducing expression of diverse proinflammatory and immunomodulatory genes, NF-κB promotes the expression of genes that protect cells from TNF-induced apoptosis (discussed below). Since the proapoptotic activity of TNF-α is opposed by the concurrent expression of antiapoptotic NF-κB target genes, the ability of TNF-α to induce apoptosis requires the inhibition of NF-κB. The differential ability to activate NF-κB may explain why TNF-α, unlike FasL, rarely triggers apoptosis unless new protein synthesis is simultaneously blocked (Baud and Karin, 2001).

In addition to protecting cells from the latent death-signaling arm of TNFR1, TNF-α-induced activation of NF-κB promotes the expression of a host of proinflammatory and immunomodulatory genes that mediate the biologic function of this cytokine. In the absence of the protection conferred by NF-κB, TNF-α loses its native function in the immune response and, instead, acquires a proapoptotic role. The mid-gestational lethality of RelA $^{-/-}$, IKKβ $^{-/-}$, or IKKγ $^{-/-}$ mice results from the extensive hepatocyte apoptosis induced by the production of TNF-α by hematopoietic progenitors that are resident in the fetal liver. The massive liver apoptosis resulting from embryonic deficiency of RelA is completely reversed by the concurrent deficiency of TNFR1 or TNF-α in DKO mice (Doi *et al.*, 1999; Rosenfeld *et al.*, 2000; Alcamo *et al.*, 2001). NF-κB also protects lymphoid cells from death receptor-induced apoptosis during the immune response (Van Parijs *et al.*, 1996a). Activation of NF-κB by co-stimulation of lymphocytes mediates cell survival and clonal proliferation, while inhibition of NF-κB by IκB mutants promotes activation-induced apoptosis of T cells, and loss of CD8$^+$ T cells in the thymus. As shall be discussed later, NF-κB-mediated protection of cells from death receptor-induced apoptosis plays an instrumental role in regulating the immune response.

Molecular mechanisms by which NF-κB regulates death receptor-induced apoptosis

NF-κB is a critical determinant of the expression of genes that modulate death receptor-induced apoptosis. NF-κB promotes the expression of a number of survival factors, including the caspase-8/FLICE inhibitor (c-FLIP), members of the inhibitor of apoptosis (IAP) family (c-IAP1, c-IAP2, XIAP), TNFR-associated factors (TRAF1 and TRAF2), and the Bcl-2 homologs, A1 (also known as Bfl-1) and Bcl-x$_L$. As discussed above, these proteins serve to interrupt different steps along the death receptor-signaling pathway. By inducing the concurrent expression of multiple antiapoptotic proteins, NF-κB exerts a multipronged inhibition of death-receptor signals (*Figure 2*).

c-FLIP, is an NF-κB-inducible protein (encoded by *Cflar*) that prevents death receptor-induced activation of the initiator procaspase-8 (Irmler *et al.*, 1997; Yeh *et al.*, 2000). NF-κB activation is required for TNF-α-induced expression of c-FLIP (Kreuz *et al.*, 2001). Although TNF-α-induced induction of c-FLIP is repressed by a degradation-resistant mutant of IκBα, c-FLIP can still be induced by TNF-α in RelA$^{-/-}$ cells (Yeh *et al.*, 2000). Therefore, the identity of the NF-κB dimer responsible for promoting c-FLIP expression remains unknown.

The promoter of the *cIap2* gene contains two functional κB sites (Hong *et al.*, 2000). Induction of *c-Iap2* by TNF-α is blocked by introduction of a phosphorylation mutant form of IκBα that resists IKK-induced degradation (Wang, C.Y. *et al.*, 1998). Akin to c-IAP2, c-IAP1 and XIAP are also NF-κB-induced proteins which block the activation of caspases (-3, -7, and -9) by death receptors (Liston *et al.*, 1996).

c-IAPs cannot directly interact with procaspase-8, and expression of either protein alone is not sufficient to protect cells from TNF-α-induced death (Wang, C.Y. *et al.*, 1998). Therefore, the recruitment of c-IAPs to the TNFR1 signaling complex via interactions with TRAF2 may be required for inhibition of caspase-8 and proximal blockade of the death signal (Shu *et al.*, 1996; Wang, C.Y. *et al.*, 1998). TRAF1 and TRAF2 are also NF-κB-inducible genes that, along with c-IAP1 and c-IAP2, suppress TNF-α-induced activation of caspase-8. Since TRAF2 also serves as an adapter that augments TNF-α-induced activation of NF-κB, TRAF2 and NF-κB may participate in a positive feedback loop to prevent TNF-α-induced apoptosis. Consistent with this notion, cells from TRAF2$^{-/-}$ mice are partially defective in TNF-α-induced NF-κB activation and exhibit exaggerated sensitivity to TNF-α-induced apoptosis (Lee *et al.*, 1997; Yeh *et al.*, 1997).

The prototypic antiapoptotic member of the Bcl-2 family, Bcl-x$_L$, contains a κB DNA site (TTTACTGCCC; 298/+22) in its promoter (Chen, C. *et al.*, 2000). The Rel-dependent induction of Bcl-x$_L$ in response to TNF-α is sufficient to protect cells carrying a degradation-resistant form of IκB from TNF-α-induced death (Chen, C. *et al.*, 2000). Likewise, NF-κB-dependent expression of Bcl-x$_L$ in response to co-stimulatory signals (CD40-CD40L or CD28-B7 interactions) serves to protects B or T cells from CD95/Fas- or Apo2L/TRAIL-induced apoptosis (Chen, C. *et al.*, 2000; Ravi *et al.*, 2001) .

Bfl-1/ A1 is a hematopoietic-specific Bcl-2 homolog which contains a functional κB site in its promoter and is induced in an NF-κB-dependent fashion in response to TNF-α (Zong *et al.*, 1999). Overexpression of A1 partially protects Rel-deficient cells from TNF-α-induced apoptosis (Zong *et al.*, 1999). Since A1$^{-/-}$ mice exhibit only increased neutrophil apoptosis, it is likely that A1 is dispensable for the antiapoptotic function of NF-κB in all other tissue types (Hamasaki *et al.*, 1998).

NF-κB may also protect cells from death receptor-induced apoptosis by attenuating expression of the proapoptotic protein, BAX (Bentires-Alj *et al.*, 2001). In certain cell types, inhibition of NF-κB by a degradation-resistant mutant of IκBα results in increased *Bax* promoter activity and expression of BAX. Although the *Bax* promoter has a κB site that binds Rel proteins, it is not required for NF-κB-mediated inhibition of BAX expression. Therefore, NF-κB may increase BAX expression via an indirect mechanism. One possibility is that NF-κB may repress stimulation of the *Bax* promoter by interfering with the function of the *p53* tumor suppressor gene. While the precise molecular mechanism remains unclear, the reduced expression of BAX may contribute to the resistance of cells with constitutive activation of NF-κB to death receptor-induced apoptosis. As discussed earlier, targeted loss of the *Bax* gene renders cancer cells resistant to CD95L, TNF-α, and Apo2L/ TRAIL-induced apoptosis Deng *et al.*, 2002; LeBlanc *et al.*, 2002; Ravi and Bedi, 2002). Since NF-κB is constitutively activated by diverse genetic aberrations in human cancers, NF-κB-mediated repression of BAX may play a role in protecting tumor cells from death receptor-ligand interactions (Ravi and Bedi, 2002).

Finally, NF-κB also induces expression of a JNK inhibitor (De Smaele et al., 2001; Javelaud and Besancon, 2001; Tang et al., 2001a). TNF-α-induced activation of JNK occurs transiently in normal cells but is increased and prolonged in IKKβ- or RelA-deficient cells (Tang et al., 2001a). Although the identity of the NF-κB-induced JNK inhibitor remains unclear, the identified candidates include XIAP and the GADD45β protein (De Smaele et al., 2001; Tang et al., 2001a). Ectopic expression of GADD45β abrogates TNF-α-induced activation of JNK and rescues RelA$^{-/-}$ cells from TNF-α-induced apoptosis. Although the suppression of JNK may contribute to the antiapoptotic function of NF-κB, JNK is not an essential mediator of death receptor-induced apoptosis.

These observations suggest that NF-κB inhibits death receptor-induced apoptosis by concomitant induction of multiple survival genes as well as repression or inactivation of proapoptotic genes.

4.7 The dynamic balance between death receptors and apoptosis-inhibitors – a tug-of-war that determines cell fate

The activation of caspases by death receptor-induced signals is held in check by the antiapoptotic proteins described above. While this serves to prevent unscheduled or uncontrolled cell death, the induction of apoptosis in response to physiologic death signals requires mechanisms to circumvent or overcome these antiapoptotic proteins. Without such mechanisms, the failure of death receptor-induced apoptosis would disrupt homeostasis and result in immune and neoplastic disorders. The dynamic balance between the antagonistic functions of death receptors and antiapoptotic proteins ensures that death receptor-induced apoptosis is allowed to proceed in a signal-dependent, scheduled, and controlled fashion. The molecular mechanisms by which this balance is tipped in favor of instructive apoptosis involves signal-induced expression of death receptors and caspase-mediated proteolysis of many of the key proteins that inhibit death receptor-induced apoptosis.

Expression of death receptors

Cell-surface expression of CD95/Fas is elevated by immune activation of lymphocytes or in response to cytokines such as interferon-γ, TNF, and CD40 ligand (CD40L) (Leithauser et al., 1993; Krammer, 2000a)[Q4]. Likewise, immune activation of lymphocytes results in an elevation of DR5 (Screaton et al., 1997a) and concomitant reduction of DcR1 levels (Mongkolsapaya et al., 1998). Since immune activation and clonal lymphoid expansion must be followed by cellular demise to preserve homeostasis, the immune system must employ molecular mechanisms to couple immune activation (and expression of survival proteins) with the induction of death receptors that mediate the decay of the immune response. One mechanism by which the immune system accomplishes this task is by employing NF-κB. While activation of NF-κB by costimulation of lymphocytes induces death receptors (CD95, DR5) (Ravi et al., 2001; Zheng et al., 2001), the activity of the induced receptors is held in check by the concurrent NF-κB-mediated induction of antiapoptotic proteins (described earlier) (Ravi et al., 2001). Engagement of the induced receptors by their cognate ligands (expressed by activated T cells or DCs) may become capable of inducing apoptosis following decay of NF-κB activity at the termination of immune stimulation or by caspase-mediated inactivation of NF-κB (see below).

Caspase-mediated cleavage of antiapoptotic proteins

Once death receptors are induced and engaged, the activation of caspases is amplified by caspase-induced proteolytic cleavage of several key antiapoptotic proteins, including NF-κB and NF-κB-induced survival proteins. The proteins targeted by caspases and the functional effects of such cleavage are described below.

Inactivation of NF-κB by caspase-mediated proteolysis – loss of survival gene expression. The proteins responsible for mediating TNF-α-induced activation of NF-κB are themselves substrates of caspases. Caspase-8-mediated cleavage of RIP at Asp^{32} destroys its ability to activate IKK (Lin *et al.*, 1999). In addition to inhibiting NF-κB activation, the NH_2-terminal-deficient fragment (RIPc) generated by such cleavage promotes the assembly of TNFR1-TRADD-FADD complex and potentiates TNF-α-induced apoptosis (Lin *et al.*, 1999). Proteolysis of TRAF1 and TRAF2 also results in increased sensitivity to death receptor-induced apoptosis (Duckett and Thompson, 1997; Arch *et al.*, 2000; Leo *et al.*, 2001). Caspase-8-mediated cleavage of TRAF1 at Asp^{163} during TNF-α- or CD95L-induced apoptosis generates a COOH-terminal fragment that inhibits TRAF2- or TNFR1-mediated activation of NF-κB (Leo *et al.*, 2001). Since the truncated protein contains a TRAF domain, it may act as a dominant negative inhibitor of interactions of TRAF1 with either TRAF2 or c-IAPs (Schwenzer *et al.*, 1999).

IKKβ, the catalytic subunit responsible for the canonical pathway of NF-κB activation, is itself inactivated by caspase-3-mediated proteolysis (at Asp^{78}, Asp^{214}, Asp^{373}, and Asp^{546}) during TNF-α- or CD95-induced apoptosis (Tang *et al.*, 2001b). Expression of the IKKβ(1-546) fragment inhibits endogenous IKK and sensitizes cells to TNF-α-induced apoptosis (Tang *et al.*, 2001b). Conversely, overexpression of a caspase-resistant mutant of IKKβ promotes the sustained activation of NF-κB and prevents TNF-α-induced apoptosis. Therefore, caspase-mediated cleavage of IKKβ may be a mechanism by which caspases terminate the activation of NF-κB and remove the key obstacle to their own activity. Caspases may also achieve this end by proteolytic removal of the NH_2-terminal domain (containing the Ser^{32} and Ser^{36} phosphorylation residues) of IκBα, thereby generating a IκB fragment that is resistant to TNF-α-induced degradation and functions as a super-repressor of NF-κB activation (Barkett *et al.*, 1997; Reuther and Baldwin, 1999).

While caspases may prevent activation of NF-κB via proteolytic inactivation of the upstream signals (described above), ligation of death receptors can directly induce caspase-mediated cleavage of RelA (Ravi *et al.*, 1998a). The truncation of the transactivation domain generates a transcriptionally inactive dominant-negative fragment of RelA that serves as an efficient proapoptotic feedback mechanism between caspase activation and NF-κB inactivation (Levkau *et al.*, 1999).

These observations suggest that the protection conferred by NF-κB against death receptor-induced apoptosis may be eliminated by caspase-mediated proteolysis of the RIP/TRAF-IKK-IκBα-RelA pathway, thereby tilting the dynamic balance between death receptors and NF-κB-induced survival proteins in favor of cell death. The direct cleavage of RelA ensures the irreversible loss of NF-κB activity, resulting in the rapid amplification of caspase activity and inevitable cell death.

Caspase-mediated cleavage of Bcl-2, Bcl-x$_L$, and IAPs – amplification of caspase activity. Many of the key antiapoptotic proteins that inhibit caspases are themselves targets of caspases. Antiapoptotic members of the Bcl-2 family (Bcl-2 and Bcl-x$_L$) compete with proapoptotic multidomain members (BAX, BAK) for binding to tBID. As such, Bcl-2 and Bcl-x$_L$ inhibit tBID-mediated activation of BAX/BAK, thereby limiting mitochondrial disruption. However, both Bcl-2 and Bcl-x$_L$ are themselves substrates for caspases. The loop domain of Bcl-2 is cleaved at Asp[34] by caspase-3 *in vitro*, in cells overexpressing caspase-3, and during induction of apoptosis by death receptors (CD95/Fas, TRAIL-R1/R2) (Cheng *et al.*, 1997; Ravi *et al.*, 2001). Death receptor-induced caspase-mediated proteolytic cleavage of Bcl-2 inactivates its survival function by removal of the BH4 domain. The carboxyl-terminal Bcl-2 cleavage product, which retains the BH3 homology and transmembrane regions, behaves as a BAX-like death effector and potentiates apoptosis (Cheng *et al.*, 1997). Cleavage of Bcl-2 contributes to amplification of the caspase cascade, and cleavage-resistant mutants of Bcl-2 confer increased protection against apoptosis. Akin to Bcl-2, Bcl-x$_L$ is cleaved by caspases during induction of apoptosis by diverse stressful stimuli (Clem *et al.*, 1998; Fujita *et al.*, 1998). Likewise, proteolytic cleavage converts Bcl-x$_L$ into two prodeath fragments. However, it is not yet known whether Bcl-x$_L$ is cleaved during death receptor-induced apoptosis or whether cleavage-resistant mutants of Bcl-x$_L$ offer better protection against death receptor-induced apoptosis. Akin to Bcl-2 family members, both c-IAP1 and XIAP are also caspase substrates (Deveraux *et al.*, 1999; Clem *et al.*, 2001). Overexpression of the caspase-induced cleavage product of c-IAP1 induces apoptosis (Clem *et al.*, 2001). Likewise, caspase-induced cleavage of XIAP at Asp[24] generates a COOH-terminal fragment that potentiates CD95-induced apoptosis (Deveraux *et al.*, 1999).

These observations suggest that caspase-induced proteolysis of these survival proteins may serve to amplify rapidly the caspase cascade, thereby potentiating death receptor-induced apoptosis.

5. Role of death receptor-induced apoptosis in the immune system

5.1 Regulation of the immune response

Apoptosis plays an essential role in the immune system. The molecular regulation of cell survival is a fundamental determinant of lymphocyte maturation, receptor repertoire selection, homeostasis, and the cellular response to stressful stimuli, such as DNA damage (Krammer, 2000). Increased apoptosis is involved in the pathogenesis of diverse immune disorders. Conversely, genetic aberrations that render cells incapable of executing their suicide program result in autoimmune disorders and tumorigenesis. The physiologic role played by death receptors in the immune response is summarized below.

Activation of immune and inflammatory responses

TNF-α is produced by activated T cells and macrophages. TNF-α or TNFR knockout mice exhibit a deficient inflammatory response to bacterial endotoxin and have an exaggerated susceptibility to microbial infections (Smith *et al.*, 1994). While these findings demonstrate a role of TNF-α in immune and inflammatory responses, TNF-α can also induce apoptosis in certain cell types or contexts. The molecular mechanisms that underlie these divergent effects of TNF-α have been described earlier.

Maintenance of lymphocyte homeostasis

The elimination of lymphocytes during development, receptor repertoire selection, and the decay phase of the immune response involve death receptor-ligand interactions.

Deletion of thymocytes and peripheral T cells. Pre-T cells undergo maturation and rearrangement of T-cell antigen receptor (TCR) genes in the thymus. T cells that fail to undergo TCR rearrangement are incapable of stimulation by self major histocompatibility (MHC) antigen-peptide complexes and suffer death by neglect. Since this process is impaired in transgenic mice carrying a dominant-negative form of the adapter FADD, death receptors may be involved in the induction of apoptosis at this pre-TCR stage of development (Newton *et al.*, 2000). Thymocytes that survive pre-TCR selection mature into $CD4^+ CD8^+$ T cells and undergo further positive and negative selection depending on the affinity of their TCRs for self MHC antigens. T cells with high affinity for self MHC molecules and peptide are eliminated, and the surviving mature $CD4^+$ MHC class-II-restricted and $CD8^+$ MHC class-I-restricted T cells leave the thymus and enter the peripheral T-cell pool in secondary lymphoid organs. The negative and positive selection that occurs during T-cell development is instrumental for self-MHC restriction and prevention of autoimmunity. Studies of T-cell receptor transgenic mice indicate that CD95 is involved with peripheral, but not thymic, deletion of T cells (Singer and Abbas, 1994; Van Parijs *et al.*, 1996). Although the TCR repertoire is not altered in mice that have genetic deficiencies of the CD95-CD95L system (*lpr* and *gld* mice), CD95-induced apoptosis may be involved in the negative selection of thymocytes that encounter high antigen concentrations (Kishimoto *et al.*, 1998; Newton *et al.*, 1998). The role of other death receptor-ligand interactions (such as DR4/DR5-Apo2L/TRAIL) in thymic deletion is as yet unknown.

Peripheral T cells are activated by cross-linkage of the TCR by the MHC-peptide complex together with costimulatory signals delivered by engagement of the CD28 receptor on the T cell by members of the B7 family expressed on APCs. Costimulation of T cells protects cells from TCR-induced apoptosis via activation of NF-κB- and NF-κB-dependent expression of antiapoptotic proteins, such as Bcl-x_L and c-FLIP (Boise *et al.*, 1995; Khoshnan *et al.*, 2000). TCR ligation in the absence of costimulatory signals result in T-cell apoptosis via mechanism(s) that do not require CD95-CD95L interactions (Van Parijs *et al.*, 1996). Following clonal expansion, antigen-specific T cells acquire an apoptosis-sensitive phenotype that enables their eventual demise during the decay phase of the immune response. This form of instructive apoptosis (termed 'activation-induced cell death' [AICD]) is required to return the expanded pool of cells to baseline levels. Unlike TCR-induced death of resting T cells, AICD is mediated by CD95–CD95L interactions at its initiation (Singer and Abbas, 1994; Alderson *et al.*, 1995; Brunner *et al.*, 1995; Dhein *et al.*, 1995; Ju *et al.*, 1995) and the TNF-α-TNFR system at a later phase of execution (Zheng *et al.*, 1995). However, AICD may also involve other death receptors-ligands (such as Apo2L/TRAIL) or death-inducing mechanism(s) that activate caspases and/or disrupt mitochondria independent of death receptor–ligand interactions (Martinez-Lorenzo *et al.*, 1998; Hildeman *et al.*, 1999; Spaner *et al.*, 1999).

Instructive apoptosis of B cells. B cells express cell-membrane receptors (antibodies) with single antigen specificity. The specificity of the immune response is achieved by antigen binding and selection of pre-existing clones of antigen-specific lymphocytes. Autoreactive B cells are deleted in the bone marrow. Mature B cells that populate secondary lymphoid organs undergo somatic hypermutation with elimination of low-affinity or autoreactive B-cell mutants (Lam and Rajewsky, 1998). Although B cells do not express CD95L, they express CD95 and may be susceptible to elimination by CD95L expressed by T cells and APCs (Scott *et al.*, 1996). It is possible that such fratricidal death receptor–ligand interactions may play a role in elimination of low-affinity or autoreactive B cells.

Ligation of the antigen-specific B-cell receptor (BCR) alone is an insufficient signal to activate a B cell and instead results in apoptosis or functional elimination. Activation of a B cell to effector function requires delivery of a second signal by activated helper T cells. Activation of the CD4+ T cell (by engagement of the TCR and ligation of CD28) induces expression by the T cell of a 39-kDa glycoprotein called gp39 or CD40 ligand (CD40L), whose receptor, CD40, is present on B cells (Foy *et al.*, 1994; Renshaw *et al.*, 1994; Xu *et al.*, 1994; Foy *et al.*, 1996). In addition to CD40-CD40L, two other receptor-ligand pairs within the TNF-R/TNF superfamilies regulate B-cell survival. Ligation of the B-cell receptors, TACI and BCMA, by BLyS (TALL-1, THANK, BAFF, zTNF4) and APRIL, respectively, promotes B-cell survival, proliferation, and immunoglobulin production (Mackay *et al.*, 1999; Moore *et al.*, 1999; Mukhopadhyay *et al.*, 1999; Khare *et al.*, 2000; Yan *et al.*, 2000). There are striking similarities between the BLyS-TACI and CD40L-CD40 systems; both ligands are TNF family members expressed on activated T cells and DCs, and both receptors are TNFR homologs expressed on B cells. Upon engagement by their respective ligands, both TACI and CD40 induce activation of NF-κB- and NF-κB-dependent expression of survival proteins (Bcl-x$_L$) that rescue B cells from BCR- or CD95-induced apoptosis (Lagresle *et al.*, 1996). In the absence of such protection, cell death may be triggered via CD95–CD95L interactions as well as CD95-independent BCR-induced signals that lead to mitochondrial disruption and caspase activation (Berard *et al.*, 1999; Craxton *et al.*, 1999).

After clonal expansion, antigen-reactive lymphocytes are titrated down until the lymphoid cell pool is restored to its basal level. The decay of the immune response is achieved by elimination of lymphocytes via instructive apoptosis involving death receptor/ligand systems. While CD40 engagement is required for B-cell proliferation, it also promotes the expression of CD95 on activated B cells.

Activated APCs and T cells synthesize the adversarial death ligands that kill B cells at the end of an immune response (Scott *et al.*, 1996).

Induction of cell death by cytotoxic T cells and NK cells

Mature CD8[+] T cells (cytotoxic T lymphocytes [CTLs]) and natural killer (NK) cells are effectors of innate and adaptive immune responses to intracellular pathogens, cancer cells, or transplanted tissues. CTLs and NK cells induce apoptosis of these targets by two major mechanisms. One mechanism involves ligation of CD95 on target cells by FasL expressed on CTLs (Li, J.H. *et al.*, 1998). The second mechanism involves calcium-dependent exocytosis of the CTL-derived granule proteins, perforin and granzymes (Heusel *et al.*, 1994; Shresta *et al.*, 1995). Perforin facilitates the delivery of granzyme B into target cells via an as yet obscure mechanism that does not require plasma membrane pore formation (Shi *et al.*, 1997; Metkar *et al.*, 2002). Granzyme B, the prototypic member of this family of serine proteases, induces cleavage and activation of multiple caspases, including caspase-3, -6, -7, -8, -9, and -10 (MacDonald *et al.*, 1999). Granzyme B also cleaves BID at a site distinct from that targeted by caspase-8 (Alimonti *et al.*, 2001). Akin to tBID (generated by caspase-8), the truncated BID generated by granzyme B (gtBID) translocates to the mitochondrial membrane and promotes the release of mitochondrial death factors via BAX or BAK (Heibein *et al.*, 2000; Alimonti *et al.*, 2001; Wang, N.S. *et al.*, 2001). Since granzyme B and CD95L can both activate the BID-BAX/BAK death-signaling pathway, they provide independent mechanisms of inducing target cell apoptosis. Accordingly, cells deficient in CD95 or overexpressing c-FLIP remain susceptible to CTL-induced death (Kataoka *et al.*, 1998). It will be important to determine whether interruption of a distal step of the death-signaling pathway shared by CD95L and granzyme B (such as loss of BAX/BAK) reduces CTL-induced death of type II target cells that require cross-talk between the extrinsic and intrinsic pathways to undergo apoptosis. Such genetic impediments to CTL-induced death may be an important mechanism by which tumor cells evade immune surveillance.

Establishment of zones of immune privilege

Immune-privileged sites such as the eye, brain, and the testes may evade damage by constitutively expressing CD95L to counterattack and eliminate CD95-expressing infiltrating lymphocytes (Green and Ferguson, 2001). Some reports suggest that expression of CD95L by certain types of cancer cells may protect such tumors from immune surveillance. While ectopic expression of FasL by gene transfer can confer immune privilege on some tissues, it can also induce a granulocytic infiltrate and increased rejection in tissue transplants. The role of CD95L–CD95 interactions in the creation of zones of immune privilege in tumors or tissue allografts *in vivo* remains debatable (Green and Ferguson, 2001).

5.2 Disorders of the immune system resulting from deregulation of death receptor-induced apoptosis

The functional importance of the normal physiologic role of death receptors in the immune system is evident from the occurrence of various disorders resulting from dysfunctional death receptors/death ligands or their signaling pathways.

Autoimmune disorders from genetic defects in CD95-induced apoptosis

Mutations that result in defects of the CD95-CD95L system result in immune disorders that feature lymphadenopathy and autoimmunity. The recessive *lpr* (lymphoproliferation) mutation results in a splicing defect that reduces expression of CD95, while the *lpr*[cg] is a point mutation in the DD of CD95 that makes it incapable of transducing a death signal (Nagata, 1997). The *gld* (generalized lymphoproliferative disease) mutation in the carboxy domain of CD95L interferes with its interaction with CD95 (Nagata, 1997). The immune disorders manifest in mice carrying these mutations result from the loss of CD95-CD95L-induced apoptosis. Defects of the CD95-CD95L system also result in similar immune disorders in humans (Fisher *et al.*, 1995; Rieux-Laucat *et al.*, 1995; Nagata, 1998). Children with type Ia autoimmune lymphoproliferative syndrome (ALPS; Canale Smith syndrome) frequently have mutations in the DD of CD95 and exhibit lymphadenopathy, aberrant accumulation of T cells, and autoimmunity. While type Ib ALPS is associated with defects in CD95L, type II ALPS involves defects in CD95 signaling, such as mutations in caspase-10 (Wang, H.G. *et al.*, 1999).

Augmentation of death receptor-induced apoptosis in AIDS

Acquired immunodeficiency syndrome (AIDS) is characterized by an excessive depletion of CD4[+] T helper cells via apoptosis. Several different mechanisms underlie the increased apoptosis of CD4[+] T cells in response to infection with the human immunodeficiency virus (HIV). HIV-encoded gene products (such as HIV-1 Tat) increase expression of CD95L, which promotes TCR-induced CD95-mediated apoptosis as well as fratricidal deletion of uninfected T cells (Debatin *et al.*, 1994; Finkel *et al.*, 1995; Li, C.J. *et al.*, 1995; Westendorp *et al.*, 1995). T cells from HIV-infected individuals also show increased sensitivity to Apo2L/TRAIL (Jeremias *et al.*, 1998). In addition to CD95L- and Apo2L/TRAIL-induced apoptosis, HIV-binding of CD4 and the chemokine receptor CXCR4 also contributes to the rapid depletion of CD4[+] T cells in AIDS (Gougeon and Montagnier, 1999).

6. Role of death receptor-induced apoptosis in development

Physiologic cell death is essential for animal development, maintenance of adult tissue homeostasis, tissue remodeling and regeneration, and elimination of cells with genetic or stochastic developmental errors (Meier *et al.*, 2000). The apoptotic culling of overproduced, unnecessary, misplaced, or damaged cells occurs via different mechanisms. Such cell death may occur autonomously via withdrawal of the appropriate trophic survival signals or transcriptional induction of genes that promote apoptosis (Raff, 1992). The survival of different cell types depends on specific survival signals within their specialized microenvironments. These survival signals include cytokines, hormones, interactions with neighboring cells, and the extracellular matrix, as well as more specialized signals such as neuronal synaptic connections or productive assembly of an appropriate immune receptor. Withdrawal of such survival signals results in cell death by default or neglect. Interference with the autonomous deletion of cells is manifested by not only developmental abnormalities, but also a wide variety of adult pathologies, such as neoplastic disorders. Spontaneous or induced mutations in the mitochondrial cell death machinery manifest as phenotypic defects that are especially evident in the immune and nervous systems. Mice lacking caspase-9, Apaf-1, or caspase-3 exhibit gross neuronal hyperproliferation and abnormalities in brain development (Kuida *et al.*, 1996; Cecconi *et al.*, 1998; Hakem *et al.*, 1998; Kuida *et al.*, 1998; Yoshida *et al.*, 1998). Transgenic mice with tissue-specific overexpression of Bcl-2 or Bcl-x_L exhibit extended cell survival and pathologic cellular accumulation in the targeted tissues (Strasser *et al.*, 2000). Conversely, mice lacking Bcl-2 show premature demise of mature lymphocytes, while Bcl-x-deficiency results in increased neuronal apoptosis and embryonic lethality (Motoyama *et al.*, 1995; Middleton *et al.*, 2000).

In addition to death by neglect from loss of trophic survival signals, developmental cell death may also require activation of death receptor-induced caspase-8-mediated signaling pathways. The principal effects of the loss of TNF, CD95 or their cognate ligands are on the immune system (Yeh *et al.*, 1999). However, death receptor-induced apoptosis may also be important in other systems. Apoptosis of spinal motor neurons in response to trophic factor withdrawal is inhibited by an antagonistic anti-CD95 antibody (Raoul *et al.*, 1999). Caspase-8$^{-/-}$ mice suffer embryonic lethality after day 11.5, with poorly developed heart musculature and abnormal accumulation of erythrocytes in the liver, lung, lens, and mesenchyma (Varfolomeev *et al.*, 1998). FADD$^{-/-}$ embryos die at the same stage of development with similar cardiac abnormalities, and FADD$^{-/-}$ T cells exhibit impaired antigen-induced proliferation (Zhang *et al.*, 1998). The absence of obvious developmental defects in most tissues in mice lacking specific TNF-R or TNF family members, and the focal failure of specific tissues in FADD$^{-/-}$ or caspase-8$^{-/-}$ mice are indicative of functional overlap and redundancy within the TNF-R/TNF-family and between the mitochondrial and death receptor-induced signaling pathways.

7. Death receptor-induced apoptosis of tumor cells

7.1 Role of death receptors/ligands in tumor surveillance

Death receptor-ligand interactions may serve a critical physiologic function in tumor surveillance (Kashii *et al.*, 1999). NK cells play a pivotal role in the control of tumor metastasis (Talmadge *et al.*, 1980; Karre *et al.*, 1986). Freshly isolated murine liver NK cells, but not natural killer T cells or ordinary T cells, constitutively express cell-surface Apo2L/TRAIL, which, together with perforin and Fas ligand (FasL), mediate NK cell–dependent suppression of experimental liver metastasis of tumor cells (Takeda *et al.*, 2001). Administration of neutralizing monoclonal antibodies against either Apo2L/TRAIL or FasL significantly increases hepatic metastases of several tumor cell lines. While inhibition of perforin-mediated killing also inhibits NK-mediated cytotoxicity (Smyth *et al.*, 1999), complete inhibition is achieved only with the combination of anti-TRAIL and anti-FasL antagonistic antibodies (Takeda *et al.*, 2001). Endogenously produced interferon-γ plays a critical role in inducing Apo2L/TRAIL expression on NK cells and T cells

(Kayagaki *et al.*, 1999). These findings suggest that Apo2L/TRAIL and FasL may contribute to the natural suppression of tumors by NK cells. Expression of FasL on cells other than NK cells might also contribute to tumor suppression (Owen-Schaub *et al.*, 1998).

7.2 Involvement of death receptors in response of tumor cells to anticancer therapy

The response of cancer cells to chemotherapeutic agents and γ-radiation involves induction of apoptosis in response to the inflicted cellular damage (Rich *et al.*, 2000). The ability of anticancer agents to induce tumor-cell apoptosis is influenced by a host of oncogenes and tumor suppressor genes that regulate cell-cycle checkpoints and death-signaling pathways. The *p53* tumor suppressor gene is a key determinant of these responses (el Deiry, 1998; Kirsch and Kastan, 1998). Phosphorylation-induced stabilization of p53 in response to cellular damage plays a pivotal role in mediating cell-cycle arrest as well as apoptosis. The particular response elicited by p53 depends on the cell type and context, as well as the presence of coexisting genetic aberrations. The induction of apoptosis by p53 involves multiple and apparently redundant mechanisms (Schuler and Green, 2001). p53 can directly activate the mitochondrial death pathway by inducing the expression of specific target genes, such as *Noxa* (Oda *et al.*, 2000), *PUMA* (Nakano and Vousden, 2001), or *Bax* (Miyashita and Reed, 1995). p53 may also promote cell death by inhibiting the transcriptional activity of NF-κB (Ravi *et al.*, 1998b; Wadgaonkar *et al.*, 1999; Webster and Perkins, 1999), thereby repressing NF-κB-dependent expression of a host of survival genes. These observations indicate that p53-induced death is mediated by multiple redundant pathways leading to mitochondrial activation of Apaf-1/caspase-9. In support of this notion, cells from caspase-9$^{-/-}$ mice are highly resistant to chemotherapeutic drugs and irradiation (Kuida *et al.*, 1998). Accordingly, inhibition of this mitochondrial pathway by overexpression of Bcl-2 (Strasser *et al.*, 1994) or inactivation of *Bax* (Zhang, L. *et al.*, 2000) can also render tumor cells resistant to anticancer drugs or γ-radiation.

DNA damage also promotes the expression of the death receptors, CD95 (Muller *et al.*, 1998) and DR5/TRAIL-R2 (Wu, G.S. *et al.*, 1997). Although p53 promotes their DNA damage-induced expression, death receptors are not essential for DNA damage-induced apoptosis. Cells from CD95-deficient (*lpr*), CD95L-deficient (*gld*), FADD$^{-/-}$, or caspase-8$^{-/-}$ mice are resistant to death receptor-induced signals, but remain sensitive to chemotherapy and irradiation-induced apoptosis (Eischen *et al.*, 1997; Fuchs *et al.*, 1997; Newton and Strasser, 2000). Likewise, overexpression of FLIP prevents tumor-cell apoptosis by death receptors, but not by chemotherapeutic agents or γ-radiation (Kataoka *et al.*, 1998). Conversely, cells from *p53*$^{-/-}$ mice resist DNA damage-induced apoptosis, but remain normally susceptible to CD95-induced death, and p53-deficient tumor cells can be killed with CD95L or Apo2L/TRAIL (Fuchs *et al.*, 1997; Ravi *et al.*, 2001). These findings indicate that the death receptor and DNA damage/stress-induced signaling pathways operate largely independently until they converge at the level of mitochondrial disruption (*Figure 3*).

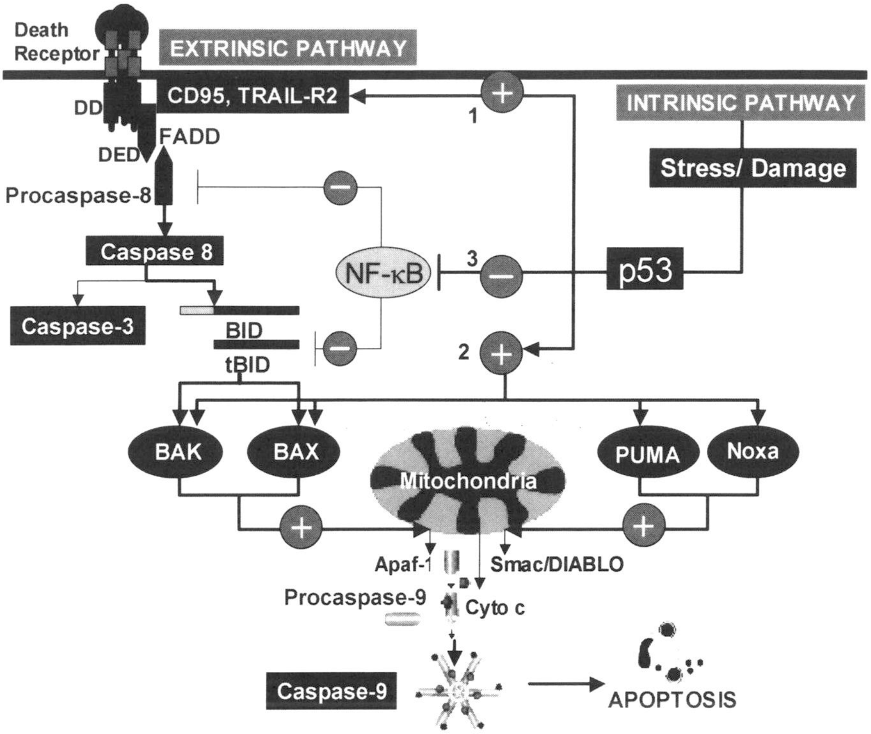

Figure 3. Cross-talk between the death receptor (extrinsic) and stress-induced (intrinsic/mitochondrial) death signaling pathways.
Cellular stress or DNA damage results in stabilization of p53. p53 can promote mitochondrial activation of caspase-9/Apaf-1 by inducing the expression of specific target genes, such as *Noxa*, *PUMA*, or *Bax*. The stress/DNA damage and death receptor signaling pathways operate largely independently until they converge at the level of mitochondrial disruption. The following mechanistic links between the intrinsic and extrinsic pathways may account for the synergistic cytotoxicity of chemotherapeutic agents/irradiation and death ligands: (1) DNA damage promotes expression of the death receptors (CD95 and DR5/TRAIL-R2). (2) Cellular damage inflicted by chemotherapeutic agents can promote expression of BAK. (3) p53 may inhibit the transcriptional activity of NF-κB.

Although death receptors play contributory, yet dispensable, roles in the response to conventional chemotherapy or irradiation, death receptor–ligand interactions may be instrumental for the action of cancer immunotherapy or specific anticancer agents. One example of how anticancer agents may employ death receptors-ligands in their actions is all-*trans*-retinoic acid (ATRA), a retinoid that induces complete remissions in approximately 10% of patients with acute promyelocytic leukemia. ATRA induces APL-cell differentiation followed by postmaturation apoptosis through induction of Apo2L/TRAIL (Altucci *et al.*, 2001). ATRA-induced expression of Apo2L/TRAIL is associated with activation of NF-κB-induced antiapoptotic target genes (Altucci *et al.*, 2001). Since the majority of patients with APL are resistant to ATRA, and complete remission induced by

ATRA is followed by the emergence of ATRA-resistant disease, it will be interesting to determine whether ATRA-resistant APL cells can be treated by inhibition of NF-κB and/or administration of Apo2L/TRAIL.

7.3 Targeting death receptors for treatment of cancers

Genetic aberrations that render cells resistant to diverse chemotherapeutic agents or ionizing radiation, such as loss of the *p53* tumor suppressor gene or overexpression of Bcl-2, underlie the observed resistance of human cancers to conventional anticancer therapy (Kaufmann and Earnshaw, 2000). Identifying approaches to induce apoptosis in tumors that harbor such genetic impediments could lead to effective therapeutic interventions against resistant human cancers. Since death receptors provide an alternative mechanism of activating caspases and triggering cell death, ligand- or antibody-induced engagement of death receptors may be an attractive strategy for anticancer therapy. However, the clinical utility and therapeutic ratio of this approach depends on the differential sensitivity of tumor cells and normal tissues to each agent and/or the ability to target delivery of death ligands/antibodies to tumor cells. Although TNF-α and CD95L can induce apoptosis of several types of tumor cells *in vitro*, their clinical application in cancer therapy is hindered by the serious toxicity of these ligands *in vivo*. Systemic administration of TNF-α causes a serious inflammatory septic shock-like syndrome that is induced by NF-κB-mediated expression of proinflammatory genes in macrophages and T cells. Systemic administration of FasL or agonistic antibodies against CD95 causes lethal hepatic apoptosis. In contrast to these ligands, Apo2L/TRAIL holds enormous promise for anticancer therapy.

A broad spectrum of human cancer cell lines express death receptors for Apo2L/TRAIL (TRAIL-R1/DR4 and TRAIL-R2/DR5) and exhibit variable sensitivity to Apo2L/TRAIL-induced apoptosis (Ashkenazi et al., 1999). Although Bcl-2 protects cells from diverse cytotoxic insults, Bcl-2 overexpression does not confer significant protection against induction of apoptosis by Apo2L/TRAIL (Ravi et al., 2001). Likewise, tumor cells that resist DNA damage-induced apoptosis by virtue of loss of p53 also remain susceptible to induction of apoptosis by TRAIL/Apo2L. Cancer cells with wild-type p53 (p53$^{+/+}$) and their isogenic p53$^{-/-}$ derivatives generated by deletion of p53 via targeted homologous recombination are equally sensitive to TRAIL/Apo2L-induced apoptosis (Ravi and Bedi, 2002). The tumoricidal activity of Apo2L/TRAIL *in vivo* has been confirmed in preclinical animal models (athymic nude or SCID mice) carrying human tumor xenografts without any evidence of toxicity to normal tissues (Ashkenazi et al., 1999; Roth et al., 1999; Walczak et al., 1999). Apo2L/TRAIL prevented the growth of evolving breast or colon or glial cancers after xenotransplantation, and decreased the size of established tumors. More importantly, systemic treatment with Apo2L/TRAIL significantly improved the survival of tumor-bearing animals without any deleterious effects on normal tissues. Likewise, monoclonal antibodies that engage the human Apo2L/TRAIL receptors, DR4 and DR5, also demonstrate potent antitumor activity without any evidence of toxicity (Chuntharapai et al., 2001; Ichikawa et al., 2001).

An especially encouraging feature of Apo2L/TRAIL or agonistic antibodies against death receptors for Apo2L/TRAIL is that they induce apoptosis of tumor cells while sparing normal tissues. While TRAIL-R1 and TRAIL-R2 are broadly expressed in most organ systems, normal cells frequently express two additional TRAIL receptors, TRAIL-R3 (TRID or DcR1) and TRAIL-R4 (TRUNDD or DcR2) that serve as decoys and confer protection against Apo2L/TRAIL-induced death. The expression of decoy receptors provides a potential molecular basis for the relative resistance of normal cells to Apo2L/TRAIL-induced death (Marsters *et al.*, 1999). Other studies indicate that high levels of FLIP may protect normal cells from death receptor-induced apoptosis (Kim, K. *et al.*, 2000). Regardless of the specific mechanism(s) involved, the differential sensitivity of tumor cells and normal tissues to Apo2L/TRAIL-induced cytotoxicity makes this ligand a promising investigational anticancer agent. One potential concern was raised by the reported ability of a polyhistidine-tagged recombinant version of human Apo2L/TRAIL (Apo2L/TRAIL.His) to induce apoptosis *in vitro* in isolated human hepatocytes (Jo *et al.*, 2000). This concern was alleviated by subsequent studies using a Zn-bound homotrimeric version of human Apo2L/TRAIL that lacks exogenous sequence tags (Apo2L/TRAIL.0), which has been developed as a candidate for human clinical trials (Lawrence *et al.*, 2001). These studies demonstrated that Apo2L/TRAIL retains potent antitumor activity but is non-toxic to human or nonhuman primate hepatocytes *in vitro*. Moreover, intravenous administration of Apo2L/TRAIL in cynomolgus monkeys or chimpanzees was well-tolerated with no evidence of changes in liver enzyme activities, bilirubin, serum albumin, coagulation parameters, or liver histology (Lawrence *et al.*, 2001).

Although Apo2L/TRAIL can induce apoptosis independently of p53 or Bcl-2, cancer cell lines exhibit a wide heterogeneity in their sensitivity to Apo2L/TRAIL *in vitro*, and certain lines are resistant to this ligand (Ashkenazi *et al.*, 1999). As might be appreciated from the earlier discussion of the molecular determinants and regulators of death receptor-induced apoptosis, cancer cells may evade Apo2L/TRAIL-mediated apoptosis by mutational inactivation of death-signaling genes or aberrant expression of proteins that block death-signaling pathways. *Bax* is a frequent target of mutational inactivation in human cancers that harbor mutations in genes that govern DNA mismatch repair (MMR) (approximately 15% of human colon, endometrial, and gastric carcinomas). More than 50% of MMR-deficient colon adenocarcinomas contain somatic frame-shift mutations in an unstable tract of eight deoxyguanosines in the third coding exon (spanning codons 38–41)[(G)$_8$] within Bax (Rampino *et al.*, 1997). In addition, a similar frame-shift mutation results from loss of a G residue from a repetitive sequence in the second exon (LeBlanc *et al.*, 2002). MMR-deficient human colon carcinoma cells are rendered completely resistant to Apo2L/TRAIL by inactivation of *Bax* (Deng *et al.*, 2002; LeBlanc *et al.*, 2002; Ravi and Bedi, 2002). While both Bax$^{+/-}$ and Bax$^{-/-}$ sister clones activated apical death receptor signals, including activation of caspase-8 and cleavage of BID, Apo2L/TRAIL could not induce mitochondrial disruption or cell death in Bax$^{-/-}$ tumor cells. These data suggest that the basal expression of BAK in tumor cells may not be sufficient to substitute for BAX in mediating death receptor-induced apoptosis. In addition to loss of BAX, cancer cells may also evade

Apo2L/TRAIL-induced death by overexpression of the caspase-8 inhibitor, FLIP, or upregulation of Bcl-x_L (Burns and el Deiry, 2001). As both FLIP and Bcl-x_L are NF-κB-inducible proteins, this mechanism of resistance may operate in cancers that have constitutively high NF-κB activity (Ravi et al., 2001). Since NF-κB is frequently activated by diverse genetic aberrations, growth factors, cytokines, viral proteins, costimulatory interactions, and stressful stimuli in diverse cancer types, it may be a common denominator of the resistance of many human cancers to Apo2L/TRAIL. Conversely, such cancers may be sensitized to Apo2L/TRAIL-induced death by inhibitors of NF-κB (Ravi and Bedi, 2002). Tumor cells can also be sensitized to Apo2L/TRAIL-induced apoptosis by various chemotherapeutic agents or ionizing radiation. Since death-receptor and DNA damage/stress-induced pathways operate largely independently until they converge at the level of mitochondrial disruption, the simultaneous delivery of both signals may have synergistic cytotoxicity. Moreover, conventional chemotherapeutic agents may also potentiate Apo2L/TRAIL-induced tumor cell death by upregulating p53, DR5/TRAIL-R2, and BAK (LeBlanc et al., 2002). Although the combination of Apo2L/TRAIL with either NF-κB inhibitors or conventional anticancer agents may exert synergistic antitumor effects, additional studies are required to evaluate and optimize the safety and therapeutic ratio of such regimens in vivo.

8. Conclusion

The last decade has witnessed breathtaking progress in our understanding of cell death and its fundamental physiologic importance in multicellular animals. Enormous strides have been made in identifying the molecular assassins and the mechanisms by which they direct cell death. As our knowledge of death receptors and their signaling pathways has grown, so too has our appreciation of the key survival signals that keep them in check. It is now evident that evolution has designed an intricate molecular circuitry that maintains a dynamic balance between death receptors and antiapoptotic proteins. The stringent regulation of death receptor-induced apoptosis enables signal-dependent induction of physiologic cell death while protecting the organism from the devastating consequences of unscheduled or uncontrolled apoptosis. These insights into the molecular regulation of cell death have opened exciting avenues for therapeutic interventions against diseases that involve too much apoptosis or the failure of physiologic cell death. The challenge before us is to design innovative therapeutic strategies that counteract these defects by targeting death receptors or their regulatory pathways. Apo2L/TRAIL may prove that we have already embarked on a journey from death receptors to successful anticancer therapy. The next decade will test whether our investment in defining the basic mechanisms of cell death will harvest rich dividends in diverse fields of clinical medicine.

Acknowledgments

Work in our laboratories is supported by the National Institutes of Health (National Cancer Institute), the US Army Medical Research and Materiel Command – Department of Defense, the Passano Foundation, the Valvano

Foundation for Cancer Research, the Mary Kay Ash Charitable Foundation, the Susan G. Komen Breast Cancer Foundation, and the Virginia and D. K. Ludwig Fund for Cancer Research. We thank our colleagues for their insights, stimulating discussions and contributions, and apologize to those scientists whose work we have either inadvertently failed to mention or cited indirectly through reviews.

2. | The role of caspases in apoptosis

Sharad Kumar and Dimitrios Cakouros

1. What are caspases?

As the name suggests, caspases (<u>c</u>ysteinyl <u>asp</u>artate protein<u>ases</u>) are cysteine proteases that cleave their substrates following an Asp residue. The prototypic member
of the caspase family is CED-3, a *Caenorhabditis elegans* protein essential for programmed cell death during development (Chapter 10). Mutations in the *ced-3* gene
result in the survival of 131 cells that normally die during development of the worm
(Ellis and Horvitz, 1986). Cloning of *ced-3* indicated that its gene product is similar to a mammalian cysteine protease termed interleukin-1β-converting enzyme
(now called caspase-1) and a developmentally regulated protein Nedd2 (now called
caspase-2) (Yuan *et al.*, 1993; Kumar *et al.* 1994; Wang *et al.*, 1994). Subsequently,
several caspases have been cloned from various mammalian and nonmammalian
species. A total of 11 caspases have been described in man, 10 in mouse, four in
chicken, four in zebrafish, seven in *Drosophila melanogaster,* and three in *C. elegans*
(reviewed in Lamkanfi *et al.*, 2002). Most work so far has been done with mammalian caspases, although a significant body of information is now also available
on *Drosophila* caspases. Not all known caspases play a role in apoptosis, and caspase gene knockout (k/o) in mice suggests that some caspases have tissue- and signal-specific functions (reviewed in Colussi and Kumar, 1999; Zheng *et al.*, 1999).
This chapter focuses mainly on caspases that play a role in cell death pathways.

2. Types of caspases

Most living cells contain several caspases, and the inactive proteins exist constitutively as unprocessed zymogen. Upon receiving an apoptotic signal, the proforms
of caspases undergo proteolytic processing to generate two subunits that comprise
the active enzyme. However, recent studies have shown that the cleavage of zymogen is not always an obligatory requirement for caspase activation (see later). The
structural studies predict that the mature caspase is a heterotetramer, composed of

two heterodimers derived from two precursor molecules (reviewed in Shi, 2001). In addition to the regions that give rise to two subunits, procaspases contain N- terminal prodomains of varying lengths. By the length of prodomain, caspases can be divided into two groups: those containing a relatively long prodomain, and those containing a short prodomain (*Figure 1*). The long prodomains in many caspases consist of specific protein–protein interaction domains that play a crucial role in caspase activation (reviewed in Kumar, 1999; Kumar and Colussi, 1999; Shi, 2002). These prodomains seem to mediate recruitment of the procaspase molecules to specific death complexes. This recruitment of procaspase molecules, mediated by specific adaptor molecules, results in caspase activation by autocatalysis. Thus, the caspases that get activated via recruitment to specific signaling complexes are also known as the initiator caspases. The main initiator caspases are caspases-8, -9, and -10 in mammals and DRONC in *Drosophila* (reviewed in Nicholson, 1999; Kumar and Doumanis, 2000). Caspases that lack a long amino-terminal prodomain also lack the ability to self-activate and require cleavage by activated initiator caspases. As most of the cellular caspase substrates are cleaved by these downstream caspases, these caspases are often referred to as the effector caspases. The key effector caspases in mammals include caspases-3, -6, and -7, and in *Drosophila* DCP-1 and DRICE (reviewed in Nicholson, 1999; Kumar and Doumanis, 2000).

Two well-defined protein–protein interaction domains have been found in initiator caspases. *C. elegans* CED-3, *Drosophila* DRONC, and mammalian caspases-1, -2, -4, -5, -9, -11, and -12 contain a caspase recruitment domain (CARD) in their prodomain, whereas *Drosophila* DREDD and mammalian caspases -8 and -10 con-

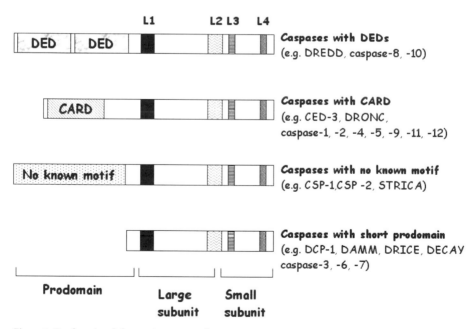

Figure 1. Prodomains define various types of caspases.

tain a pair of death effector domains (DEDs) in their prodomain regions (reviewed in Kumar, 1999; Kumar and Colussi, 1999; Shi, 2002) (*Figure 1*). A number of other caspases, including *C. elegans* CSP-1 and CSP-2, and *Drosophila* STRICA, have long prodomains, lacking any CARD or DED (Shaham, 1998; Doumanis *et al.*, 2001). While the CARDs and DEDs in some caspases clearly play a role in caspase activation, the function of long prodomains in CSP-1, CSP-2, and STRICA is currently unclear.

3. Caspase structure

Structures of several active caspases, including caspases-1, -3, -8, and -9, and caspase-7 zymogen, have been determined (Walker *et al.*, 1994; Wilson *et al.*, 1994; Mittl *et al.*, 1997; Rotonda *et al.*, 1996; Blanchard *et al.*, 1999; Watt *et al.*, 1999; Renatus *et al.*, 2001; Chai *et al.*, 2001b; Riedl *et al.*, 2001b; reviewed in Shi, 2001). The functional caspase unit is a homodimer, with each monomer consisting of a large and a small subunit. The active site Cys is located within the large subunit, whereas the residues that form the S1 subsite are derived from both large and small subunits. Homodimerization between the two caspase molecules is mediated by hydrophobic interactions with six antiparallel β strands from each catalytic subunit, forming a contiguous 12-stranded β sheet. The recently determined procaspase-7 structure shows that the overall fold of the homodimeric procaspase-7 is similar to the active tetrameric caspase-7 in that each monomer is organized in two structured subdomains connected by partially flexible linkers (Chai *et al.*, 2001b; Riedl *et al.*, 2001a). However, in the procaspase, the linker asymmetrically occupies and blocks the central cavity that is normally present in the active caspases. This leads to the inability of the procaspase molecule to interact with the substrate or inhibitor. Following the processing of the zymogen, the cleavage within the linkers results in conformational changes that govern the formation of an active site (Chai *et al.*, 2001b; Riedl *et al.*, 2001b). Thus, in the case of effector caspases, such as caspase-7 (and perhaps also caspase-3), the cleavage of the zymogen into two subunits is necessary for the activation of the enzyme. The zymogen structure of any of the initiator caspases has not been solved.

4. Substrate specificity of caspases

The substrate binding site of active caspases is formed by four loops L1–L4, which determine the substrate sequence specificity of various caspases (*Figure 1*). Of these loops, L2 and L4 are highly variable among different caspases, whereas L1 and L3 are of relatively similar length. The binding pockets for the P4–P1 residues in the substrates (termed S4–S1 subsites) are mostly located between L1, L3, and L4 (reviewed in Shi, 2001). The minimum substrate requirement for caspase-mediated cleavage is a tetrapeptide with an Asp residue in P1 position, with the P4–P2 residues detemining caspase target specificities. However, there are exceptions to this case. For example, mammalian caspase-2, and *Drosophila* DRONC and DAMM prefer a minimum of five residues in the substrates, and tetrapeptide substrates are poorly cleaved by these caspases (Talanian *et al.*, 1997a; Dorstyn *et al.* 1999a; Hawkins *et al.*, 2000; Harvey *et al.*, 2001). By the P4–P2 specificities, the

Table 1. Substrate specificities of various mammalian caspases[a]

Group	Caspase	Preferred peptide substrate(s)
Group I	Caspase-1	YEVD, WEHD
	Caspase-4	LEVD, (W/L)EHD
	Caspase-5	(W/L)EHD
Group II	Caspase-2	VDVAD, DEHD
	Caspase-3	DEVD, DEHD
	Caspase-7	DEVD
	Caspase-6	VEID, VEHD
Group III	Caspase-8	IETD, LETD
	Caspase-9	LEHD
	Caspase-10	IEAD

[a]P4–P1 residues are shown (except for caspase-2 substrate VDVAD). In all cases, caspase-mediated cleavage occurs following the Asp residue in P1 position. See text for details and references.

mammalian caspases have been divided into three groups (Thornberry *et al.*, 1997). The group I caspases (caspase-1, -4, and -5) prefer a hydrophobic residue in the P4 position, group II caspases (caspase-2, -3, and -7) have a strict requirement for an Asp in the P4 position, and group III caspases (caspase-6, -8, -9, and -10) prefer a branched chain aliphatic amino acid in P4 (*Table 1*). Group II caspases, caspase-3 and -7 in particular, with a P4–P1 specificity of DxxD are considered to be activated by upstream initiator caspases such as caspase-8 and caspase-9. The preferred P3 residue for most mammalian caspases is Glu (Thornberry *et al.*, 1997). Although all caspases require an Asp in the P1 position, *Drosophila* DRONC has been shown to possess some cleavage activity on peptide substrates carrying Glu in P1 (Hawkins *et al.*, 2000). DRONC also shows a strong preference for Thr, Ile, or Val at P2 (Hawkins *et al.*, 2000). The *Drosophila* effector caspases DCP-1, DRICE, and DECAY share a substrate specificity identical to that of mammalian group II caspases (Fraser and Evan, 1997; Dorstyn *et al.*, 1999b; Song *et al.*, 2000).

5. Caspase activation

As mentioned above, specific adaptor molecules, via protein–protein interaction domains, mediate the recruitment of procaspase molecules to the death-signaling complexes. Among the three types of domains mediating interactions between molecules of the death effector apparatus, death domains (DDs) are found in upstream components of the apoptotic pathways, such as death receptors (e.g., CD95, TNFR1, DR3, DR4, and DR5) and the adaptor molecules that are recruited to these receptors (e.g., FADD and TRADD) (reviewed in Ashkenazi and Dixit, 1998). However, DEDs and CARDs are generally responsible for recruiting class I caspase precursors to death effector complexes through specific adaptor molecules (*Table 2*). DD-mediated association of receptors and adaptors usually occurs as a consequence of ligand-dependent activation of death receptors. NMR structures of the DD, DED, and CARD show that the three domains share a very similar struc-

Table 2. Adaptors for various apoptotic initiator caspases and the domains that mediate caspase/adaptor interactions

Caspase	Adaptor	Domain mediating interaction
CED-3	CED-4	CARD
DRONC	DARK	CARD
DREDD	dFADD (?)	DED
Caspase-8	FADD	DED
Caspase-9	Apaf-1	CARD
Caspase-10	FADD	DED

ture consisting of six or seven antiparallel, amphipathic α-helices (Huang *et al.*, 1996; Eberstadt *et al.*, 1998; Chou *et al.*, 1998; Qin *et al.*, 1999).

5.1 Caspase activation by oligomerization

Various studies have shown that procaspase oligomerization can induce caspase activation (reviewed in Kumar, 1999; Kumar and Colussi, 1999). Both initiator caspases, such as CED-3, and caspase-8 and -9, and effector caspases, such as caspase-3, can undergo proteolytic processing when induced to oligomerize by artificial means (Butt *et al.*, 1998; Colussi *et al.*, 1998b; MacCorkle *et al.*, 1998; Martin *et al.*, 1998; Muzio *et al.*, 1998; Yang *et al.*, 1998a). On the basis of these studies, the events leading to the activation of initiator caspases were proposed to involve the following steps: first, conformational changes due to upstream signaling events result in the recruitment of adaptor molecules to a death complex; second, this is followed by further changes that promote the recruitment of caspases through specific domains; and, finally, close proximity of procaspase molecules allows inter- or intramolecular catalysis and activation of the zymogen. The exact mechanism of processing of caspase zymogen is not fully understood, but zymogen forms of initiator caspases are predicted to contain low-level protease activity that mediates their autoactivation (reviewed in Stennicke and Savesen, 1999).

5.2 Activation of caspase-8

Mammalian caspase-8 and -10, and the fly caspase DREDD contain two DEDs in their prodomain region (Boldin *et al.*, 1996; Muzio *et al.*, 1996; Chen *et al.*, 1998). Most of the studies have focused on caspase-8, but given the structural similarities between DED-containing caspases, it is likely that caspase-10 is activated by a mechanism similar to that of caspase-8 activation. As DD-containing death receptors have not been found in the fly, it is unlikely that DREDD activation occurs at the level of membrane recruitment, although activation may still require dFADD-mediated oligomerization (Hu and Yang, 2000). Upon binding of the Fas/CD95 ligand, the CD95 receptor recruits a cytosolic adaptor protein, FADD (Fas-associated death domain), through a homotypic interaction between the CD95 intracellular DD and the FADD C-terminal DD (Boldin *et al.*, 1995; Chinnaiyan *et al*, 1995). FADD contains a DED in its N-terminal region that interacts with the DEDs in the prodomain of procaspase-8 and recruits procaspase-8 to CD95 (Boldin *et al.*, 1996; Muzio *et al.*, 1996). Therefore, it seems likely that the aggregation of CD95 by its ligand brings the DDs in the CD95 intracellular C-termini in close proximity to create high-affinity binding sites for the FADD DD. The complex formed by CD95, FADD, procaspase-8, and possibly other proteins is known as DISC (death-inducing signaling complex) (see Chapter 1). Recruitment of procaspase-8 to the DISC leads to proximity-induced autocatalytic activation (Muzio *et al.*, 1998).

5.3 The Apaf-1/caspase-9 apoptosome

The activation of caspase-9 requires Apaf-1, dATP, and cytochrome *c*, which is released from the mitochondria in cells committed to apoptosis (Li *et al.*, 1997; Srinivasula *et al.*, 1998). Apaf-1 contains multiple WD40 motifs in the carboxyl terminus, which appear to act as a negative regulatory domain by preventing Apaf-1 oligomerization and Apaf-1/caspase-9 interaction (Srinivasula *et al.*, 1998; Zou *et al.*, 1997, 1999). The deletion of the Apaf-1 WD-40 repeats makes Apaf-1 constitutively active and capable of activating procaspase-9 independently of cytochrome *c* and dATP. Thus, binding of cytochrome *c* and dATP to Apaf-1 causes conformational changes in the Apaf-1 molecule that expose the amino-terminal CARD, enabling it to interact with the CARD in procaspase-9. Reconstitution studies using purified components found that Apaf-1 binds dATP. This is followed by binding of cytochrome *c*, an effect which promotes multimerization of Apaf-1 molecules (Zhou *et al.*, 1997, 1999). The heptameric Apaf-1 complex, often called apoptosome, then recruits and activates procaspase-9. The Apaf-1/caspase-9 complex acts as a holoenzyme that cleaves and activates procaspase-3 (Rodriguez and Lazebnik, 1999). As procaspase-9 processing is not required for its activation (Stennicke *et al.*, 1999), and since processed caspase-9 is only weakly active in the absence of Apaf-1 (Rodriguez and Lazebnik, 1999), it is likely that the apoptosome primarily functions to enhance allosterically caspase-9 activity, rather than mediate its cleavage. The recombinant procaspase-9, containing mutations that abolish processing sites of the zymogen, can activate downstream caspases, but only in the presence of cytosolic factors, presumably cytochrome *c* and Apaf-1 (Stennicke *et al.*, 1999).

The three-dimensional structure of the apoptosome has been determined recently. It shows that the apoptosome is a wheel-like heptameric particle with the Apaf-1 CARDs located at the central 'hub' and the WD40 repeats at the extended arms (Acehan *et al.*, 2002). In this structure procaspase-9 is localized to the central region of the apoptosome, whereas cytochrome *c* associates with the extended WD40 arms (*Figure 2*). This model defines a role for cytochrome *c* in apoptosome assembly.

As in the case of caspase-9, the activation of CED-3 is mediated by its adaptor CED-4 (Chinnaiyan *et al.*, 1997); however, there is no evidence for the requirement of cytochrome *c* in caspase activation in *C. elegans*. The role of cytochrome *c* in caspase activation in *Drosophila* is also currently unclear. Recent data suggest that while DARK, the Apaf-1 homolog in fly, is clearly necessary for the activation of the apical caspase DRONC, cytochrome *c* is not released from mitochondria during apoptosis of *Drosophila* cells (Dorstyn *et al.*, 2002). Although an apoptosome-like complex containing DRONC can form *in vitro*, the actual composition of this complex and its role in caspase activation remains unknown (Dorstyn *et al.*, 2002).

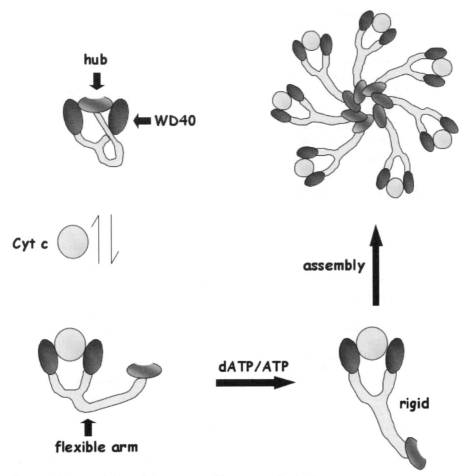

Figure 2. **The assembly of Apaf-1 apoptosome.** Please see text for details.

5.4 Other mechanisms

In vitro immunodepletion studies have shown that activated initiator caspases (caspase-8 and -9 in particular) are necessary for the activation of effector caspases (Slee *et al.*, 1999). Although many effector caspases, including caspases-3, -6, and -7, are capable of undergoing self-activation *in vitro* and when expressed in *E. coli* (reviewed in Kumar and Lavin, 1996; Kumar, 1999), there is currently no evidence for autoactivation under physiologic conditions. Among the apoptotic caspases, caspases-2 is an unusual caspase, and its mechanism of activation is still a matter of some debate (Kumar, 1999). Caspase-2 contains a CARD that has been shown to homodimerize (Butt *et al.*, 1998) and to interact with a CARD in the putative adapter RAIDD (Duan and Dixit, 1997; Shearwin-Whyatt *et al.*, 2000). However, definitive evidence for a function of RAIDD in caspase-2 activation is still lacking

(Shearwin-Whyatt *et al.*, 2000). Interestingly, *in vitro* cleavage and immunodepletion studies suggest that caspase-3 mediates caspase-2 activation (Harvey *et al.*, 1996; Slee *et al.*, 1999), and in the thymocytes derived from Apaf-1 and caspase-9 k/o mice, caspase-2 processing is severely impaired (O'Reilly *et al.*, 2002), suggesting that caspase-2 may be a downstream effector caspase. As caspase-2 is rapidly activated during apoptosis induced by a variety of different signals (Harvey *et al.*, 1997), and it has been shown to act upstream of mitochondria to cause the release of cytochrome *c* and DIABLO (Guo *et al.*, 2002), it is possible that while caspase-2 can be activated by caspase-3-mediated processing in the downstream caspase amplification loop, its primary role may be as an initiator caspase.

6. Regulation of caspase activation

As caspase activation is a key step in the initiation of apoptosis, it is tightly controlled at various steps by a variety of proteins, including FLIP, the Bcl-2 family of proteins, and the inhibitor of apoptosis proteins (IAP). Although low levels of caspase zymogens are present in most cells, transcriptional regulation of caspases, particularly during developmentally programmed cell death, is likely to play an important role. Furthermore, caspase activation may also be regulated by compartmentalization, as some caspases, such as caspase-2 and -12, are localized to specific cellular compartments (Colussi *et al.*, 1998a; Nakagawa *et al.*, 2000).

6.1 Regulation of caspase-8 activation by FLIPs

Many viruses encode apoptosis inhibitors to avoid the host's apoptotic response. These include baculovirus caspase inhibitors p35 and IAPs, and gamma-herpesvirus encoded v-FLIPs (for viral FLICE-inhibitory proteins), which interfere with apoptosis signaled through death receptors (reviewed in Ekert *et al.*, 1999). Although cellular homologs of p35 have not been found, IAP and v-FLIP-like proteins are present in mammals. Both viral and cellular FLIPs contain two DEDs that interact with the DED in FADD, thus inhibiting the recruitment and activation of caspase-8 by TNFR family members (Irmler *et al.*, 1997; Thome *et al.*, 1997). The cellular FLIP can be upregulated in some cell lines under critical involvement of the NF-kB pathway (Michaeu *et al.*, 2001). *flip*[−/−] embryos do not survive past day 10.5 of embryogenesis, and they exhibit impaired heart development, a phenotype similar to that reported for *fadd*[−/−] and *casp8*[−/−] embryos (Yeh *et al.*, 2000; Varfolomeev *et al.*, 1998; Zhang, J. *et al.*, 1998). However, unlike *fadd*[−/−] and *casp8*[−/−] cells, *flip*[−/−] embryonic fibroblasts are highly sensitive to FasL- or TNF-induced apoptosis and show rapid induction of caspase activities, suggesting that while FLIP cooperates with FADD and caspase-8 during embryonic development, it is also necessary for protecting cells against apoptosis mediated by TNFR family members (Yeh *et al.*, 2000).

6.2 Regulation of caspases by IAPs

As mentioned above, genes encoding IAP proteins were originally cloned from baculoviruses through their ability to inhibit virus-induced apoptosis in lepidopteran cells (Crook *et al.*, 1993). Subsequently, IAP-like proteins have been found in yeast, *C. elegans*, *Drosophila*, and mammals (see Chapter 3 for details). There are two IAP-like proteins in *Drosophila* (DIAP-1 and -2), and several in mammals, including XIAP, cIAP-1, cIAP-2, NAIP, and survivin (Deveraux and Reed, 1999). Each IAP protein contains in its N-terminal region one to three copies of the baculovirus IAP repeats (BIRs), which are required for the function of all IAPs (Deveraux and Reed, 1999; Hay, 2000). In addition, some IAPs, including cIAP-1 and XIAP, have a ring-finger domain that act as ubiquitin-protein ligases. Although not all IAPs act as inhibitors of apoptosis, XIAP, and cIAP-1 and -2 can physically interact with caspases and inhibit mature caspase-3, -7, and -9 (Chapter 3; Deveraux and Reed, 1999). The recent structural studies suggest that the N-terminal BIR2 linker region of XIAP occupies the catalytic site in the active caspase-3 and -7, whereas the adjacent BIR2 may help stabilize the interaction (Chai *et al.*, 2001b; Huang *et al.*, 2001; Riedl *et al.*, 2001b). In contrast, the XIAP BIR3 domain is required for inhibiting caspase-9 (Sun *et al.*, 2000). These data suggest that distinct mechanisms govern IAP-mediated inhibition of initiator and effector caspases.

In *Drosophila*, the cell death activators reaper (RPR), head involution defective (HID), and GRIM physically interact with DIAP1 (Hay, 2000). Genetic and biochemical data suggest that RPR, HID, and GRIM promote cell death by disrupting DIAP1–caspase interactions and that DIAP1 is required to block apoptosis-inducing caspase activity (Wang, S.L. *et al.*, 1999; Goyal *et al.*, 2000). RPR, HID, and GRIM contain a short region of homology in their N-termini called the RHG motif, which allows binding to DIAP1 and DIAP2. Loss of DIAP1 function results in global early embryonic cell death and a large increase in DIAP1-inhibitable caspase activity, and DIAP1 is still required for cell survival when expression of *rpr*, *hid*, and *grim* is eliminated (Wang, S.L. *et al.*, 1999). DIAP1 and DIAP2 interact with a number of long and short prodomain *Drosophila* caspases, including DRONC, DRICE, DCP-1, and STRICA, but not DAMM (Fraser *et al.*, 1997; Hawkins *et al.*, 2000; Meier *et al.*, 2000; Quinn *et al.*, 2000; Doumanis *et al.*, 2001; Harvey *et al.*, 2001). In mammals, the proteins SMAC/DIABLO and HtrA2 are released during apoptosis from the mitochondria into the cytoplasm, where they bind to IAPs and prevent their ability to inhibit caspase activity (Verhagen *et al.*, 2000, 2002; Du *et al.*, 2000; Suzuki *et al.*, 2001; Hedge *et al.*, 2002; Martin *et al.*, 2002; van Loo *et al.*, 2002). SMAC and HtrA2 seem to function similarly to RPR, HID, and GRIM, and contain the characteristic tetrapeptide RHG motifs in the mature protein that binds to the BIR domain of IAPs. Structural data show that the binding of SMAC to the BIR2 and BIR3 domains of XIAP disrupts its ability to inhibit the caspases by steric hindrance (Chai *et al.*, 2001b).

6.3 Regulation by Bcl-2 family

In *C. elegans*, the CED-4 adapter is required for the activation of CED-3, whereas CED-9 negatively regulates this process by directly interacting with CED-4, and perhaps sequestering it (see Chapter 10). However, EGL-1 binds CED-9, thereby acting as a trigger for CED-4-dependent CED-3 activation (Conradt and Horvitz, 1998; del Peso *et al.*, 1998). Both EGL-1 and CED-9 homologs, BH3-only proteins and Bcl-2-like proteins, respectively, also play pivotal roles in the regulation of caspase-9 activation in mammals (see Chapter 4 for details of the Bcl-2 gene family). However, unlike the *C. elegans* CED-9, mammalian Bcl-2 family members do not interact with Apaf-1; rather, they seem to regulate the release of cytochrome *c*. In addition to the EGL-1 and CED-9 homologs, mammals also have proapoptotic Bcl-2-like proteins (Bax, Bak, and Bok) that contain three BH (Bcl-2 homology) domains (Chapter 4). The antiapoptotic members of this family, such as Bcl-2 and Bcl-x_L, prevent cytochrome *c* release, while the proapoptotic members such, as Bax, and BH3-only proteins (e.g., Bid, Bim, Bmf, Bad, and others) promote it (Green and Reed, 1998; Adams and Cory, 2001). Following apoptotic signaling, many of the Bcl-2 family proteins undergo conformational changes and translocate to the outer membrane of the mitochondria. The exact mechanisms by which Bcl-2 family proteins regulate cytochrome *c* release are still a matter of considerable debate; these mechanisms are discussed in detail in Chapter 4.

The BH3-only proteins share sequence homology with Bcl-2 only in the BH3 domain, which is required to interact with other Bcl-2 family members (Huang and Strasser, 2000). These proteins appear to act as key molecules that link the upstream signaling to the apoptotic machinery. Like EGL-1, mammalian BH3 proteins interact with the prosurvival members of the Bcl-2 family, and perhaps inhibit their function. These proteins are regulated in a number of different ways. For example, Noxa and Puma are transcriptionally regulated by p53 (Oda *et al.*, 2000; Nakano *et al.*, 2001; Yu *et al.*, 2001), Bid is regulated by caspase-8-mediated cleavage (Li *et al.*, 1998a; Luo *et al.*, 1998), Bad is regulated by phosphorylation-dependent binding to 14-3-3 (Datta *et al.*, 1997), whereas Bim and Bmf are regulated by sequestration to the cellular motor proteins (Putalakath *et al.*, 1999; 2001). In all cases, the activation and interaction of the BH3-only proteins with other Bcl-2 family members initiates the caspase activation events (Chapter 4).

6.4 Regulation by transcription

Apoptosis has been generally considered as a process whereby pre-existing proteins involved in cell death are activated by proteolytic cleavage or protein modification in response to an apoptotic signal, resulting in the activation of the caspase cascade. In the last few years, however, transcriptional regulation of many proteins of the apoptotic machinery, including caspases and Bcl-2 family members, has been demonstrated both in mammals and in *Drosophila*. Many mammalian caspases are transcriptionally upregulated under certain conditions (e.g., Kinoshita *et al.*, 1996; Chen, M. *et al.*, 2000; von Mering *et al.*, 2001). However, the clearest example of transcriptional regulation of caspase activation comes from studies in *Drosophila*.

During the larval/pupal metamorphosis in *Drosophila*, obsolete larval organs undergo stage-specific, programmed cell death in response to steroid hormone ecdysone (reviewed in Riddiford, 1993; Thummel, 1996; Baehrecke, 2000). At the end of the third larval instar, a high-titer ecdysone pulse triggers puparium formation, initiating metamorphosis and the prepupal stage of development. This is followed by a second pulse of ecdysone, approximately 10 h after puparium formation, which signals eversion of the adult head and defines the prepupal to pupal transition. In response to the ecdysone pulse at the late third instar stage, apoptosis is initiated in the larval midgut, and the larval salivary glands die in early pupae immediately after the prepupal pulse of ecdysone. Recent studies have shown that ecdysone mediates cell death by transcriptionally upregulating a number of death effectors, including the caspase *dronc* and the Apaf-1 homologs *dark*, *rpr*, and *hid*, and downregulating the apoptosis inhibitors, *diap1* and *diap2* (reviewed in Baehrecke, 2000).

A network of transcriptional regulators spatially and temporally control ecdysone-mediated cell death. Ecdysone binds to its heterodimeric EcR/UsP receptor and transcriptionally regulates a set of transcription factors. These include the Broad-Complex (BR-C), which are zinc finger transcription factors; an Ets-like transcription factor E74; and the orphan nuclear receptor, E75 (Burtis *et al.*, 1990; Segraves and Hogness, 1990; DiBello *et al.*, 1991). These transcription factors then regulate secondary response genes. Ten hours after puparium formation, a second pulse of ecdysone triggers salivary gland cell death, which is also accompanied by the expression of BR-C, E74, and the additional transcription factors E93 and βFTZ-F1 (reviewed in Thummel, 1996; Baehrecke, 2000). These transcription factors have been shown to play a role in ecdysone-mediated cell death in the salivary gland and midgut. For example, *rpr* transcription in salivary glands is directly regulated by the EcR/Usp complex (Jiang *et al.*, 2000). *BR-C* is required for maximal *rpr* and *dronc* expression (Cakouros *et al.*, 2002; Jiang *et al.*, 2000). *BR-C* and *E74* are needed for optimal induction of *hid* in salivary glands, whereas βFTZ-F1 is required for the induction of *diap2*, and E75 represses this death inhibitor immediately before the induction of *rpr* and *hid* (Jiang *et al.*, 2000). In the salivary glands of E93 mutants, *rpr*, *hid*, *dark*, and *dronc* mRNA levels are severely reduced (Lee *et al.*, 2000). Thus, during development and metamorphosis, upregulation of caspase RNA, along with the regulation of other death effectors and inhibitors, may play an important role in the synchronous removal of unwanted cells.

7. Caspase targets

Over 100 cellular proteins are now known to be cleaved by effector caspases in cells undergoing apoptosis. These proteins include several functional classes, such as cytoskeletal proteins, proteins involved in DNA and RNA metabolism, protein kinases, proinflammatory cytokines, cell-cycle regulators, apoptosis regulators, signal transducers, transcription factors, and proteins involved in neurodegenerative disorders (see reviews by Nicholson, 1999; Chang and Yan, 2000). Depending on the target, the cleavage by caspases can functionally inactivate or activate a protein. Although the functional significance of all caspase cleavage events is not known, many of the proteins inactivated by caspase-mediated cleavage play a role in cellular homeostasis, and their cleavage may simply result in shutting down cell-cycle and repair machinery, while the cleavage of others is more directly associated with morphologic changes characteristic of apoptosis. A comprehensive description of the growing number of caspase substrates is beyond the scope of this chapter, but a few need special mention here. Apoptosis has long been known to be associated with cleavage of DNA in characteristic nucleosomal length fragments (Wyllie, 1980). The DNase responsible for this fragmentation has been identified and named CAD (caspase-activated DNase) (Enari *et al.*, 1998). CAD normally remains

bound to a cytoplasmic inhibitor (ICAD), which is degraded by caspase-3 in apoptotic cells (Liu *et al.*, 1997; Sakahira *et al.*, 1998). The free CAD then enters the
nucleus and acts upon chromatin (Sakahira *et al.*, 1998). Cells that are devoid of
caspase-3 fail to show characteristic DNA fragmentation, although they are still
able to undergo apoptosis (Jänicke *et al.*, 1998; Woo *et al.*, 1998), suggesting that
caspase-3 plays an essential role in activating CAD, but characteristic DNA fragmentation is not essential for apoptosis to proceed. Cleavage of gelsolin, an actin-
associated protein, generates a fragment that promotes the characteristic
rounding-up of cells (Kothakota *et al.*, 1997). PAK2, a serine/threonine kinase, is
activated by caspase cleavage (Rudel and Bokoch, 1997). As PAK2 is involved in the
regulation of the actin cytoskeleton, its activation seems to help formation of apoptotic bodies. Thus, cleavage of specific proteins by caspases mediates the characteristic morphologic changes and ultimate death of the cell.

Prosurvival members of the Bcl-2 family, including Bcl-2 and Bcl-x$_L$ proteins,
are also cleaved by caspases during apoptosis, and the fragments generated by the
cleavage appear to enhance apoptotic changes in the dying cells (Cheng *et al.*, 1997;
Clem *et al.*, 1998). Although initiator caspases mainly target effector caspases for
cleavage and activation, there are some examples where a cellular protein is directly
cleaved by activated initiator caspase. For example, caspase-2 and caspase-8 cleave
Bid, a BH3-only protein (Li *et al.*, 1998a; Luo *et al.*, 1998; Guo *et al.*, 2002). As discussed above, the cleaved Bid translocates to mitochondria and initiates
cytochrome *c* release, thereby linking the receptor pathway to the mitochondrial
pathway of apoptosis. As the intrinsic and extrinsic pathways of apoptosis are independent of each other, the bridging via Bid probably serves as an amplification loop
for caspase activation, rather than an initiating event.

8. Function of individual caspases

8.1 Caspase mutations and gene knockouts

In *C. elegans*, the loss of function mutations results in a complete inhibition of
developmental cell death (Chapter 10). In *Drosophila melanogaster*, specific
mutants are only available for *dcp-1* and *dredd*. Most *dcp-1* mutant embryos die
during larval stages, and the larvae that survive lack imaginal disks and gonads, and
have melanotic tumors (Song *et al.*, 1997). The *dcp-1* mutation also causes female
sterility by defective transfer of the cytoplasmic contents from nurse cells to developing oocytes, a process which requires apoptosis (McCall and Steller, 1998). The
loss of *dredd* function suggests a role for this caspase primarily in the regulation of
innate immune response in the fly (Leulier *et al.*, 2000; Stoven *et al.*, 2000).
Although *dronc* mutants are currently not available, RNA interference studies suggest that *dronc* is essential for programmed cell death in the embryos and cell lines
(Quinn *et al.*, 2000; Cakouros *et al.*, 2002).

Caspase gene k/o data in the mouse emphasizes the complexity of the caspase-
activation pathways. The phenotypes of the *casp9*$^{-/-}$ and *casp3*$^{-/-}$ mice are very similar, suggesting that these caspases are in the same pathway. Both *casp9*$^{-/-}$ and
casp3$^{-/-}$ mutations are lethal during embryogenesis or shortly after birth (Kuida *et
al.*, 1996, 1998; Hakem *et al.*, 1998; Woo *et al.*, 1998). The most profound defect of

these mice is hyperplasia in the brain resulting from decreased apoptosis. Defective neural tube closure is seen in the hindbrain region at E10.5 of *casp9*$^{-/-}$ embryos. These mutants also have enlarged proliferation zones in the fore- and midbrain, resulting from reduced apoptosis, and causing expansion and protrusion of cranial tissues. *casp3* mutant mice show a protrusion of brain tissue consisting of ectopic cell masses with similarities to the cerebral cortex (Kuida *et al.*, 1996; Woo *et al.*, 1998). Increased cell number and density of the brain stem in *casp3*$^{-/-}$ embryos, resulting from reduced apoptosis, becomes apparent by E12. Protrusions of the neuroepithelium in the retina are also seen as a result of reduced cell death in this tissue. These phenotypes indicate that caspase-3 and -9 play nonredundant roles in brain development and that caspase-3 is also required for normal eye development. In addition, cells from *casp9*- and *casp3*-null mice show similar sensitivities to various apoptotic stimuli. For example, embryonic stem (ES) cells from *casp9* and *casp3* mutants are resistant to apoptosis induced by a range of apoptotic stimuli, including treatment by adriamycin, etoposide, sorbitol, cisplatin, anisomycin, and UV irradiation (Hakem *et al.*, 1998; Kuida *et al.*, 1998; Woo *et al.*, 1998). ES cells from *casp9* mutants are also resistant to apoptosis induced by γ irradiation whereas *casp3*-null ES cells are not. MEFs derived from *casp9*$^{-/-}$ and *casp3*$^{-/-}$ mice are resistant to adriamycin and cytotoxic lymphocyte-mediated apoptosis, and *casp3* null MEFs are also resistant to TNF-α-induced apoptosis while *casp9* MEFs are not. A direct role for caspase-9 in activation of caspase-3 has been demonstrated in *casp9*$^{-/-}$ mice (Hakem *et al.*, 1998; Kuida *et al.*, 1998). Caspase-3 processing is deficient in *casp9* mutant brain tissue, and lysates from caspase-9-deficient thymocytes or embryonic brains are unable to process caspase-3 even in the presence of cytochrome *c* and dATP. This activity can be restored by the addition of *in vitro* translated caspase-9 to the lysates.

casp8 k/o in mice is embryonically lethal (Varfolomeev *et al.*, 1998). *casp8*$^{-/-}$ embryos are smaller than normal, display impaired heart muscle development and an accumulation of erythrocytes in the abdomen and in blood vessels in the trunk area at E11.5 and E12.5. Excessive erythrocytosis is also seen in the liver of mutants, although the mutants contain reduced hemopoietic stem-cell numbers (Varfolomeev *et al.*, 1998). *casp8*-null fibroblasts do not undergo apoptosis in response to CD95 and DR3 ligation, although cell death induced by UV, cytotoxic drugs, vesicular stomatitis virus, and serum deprivation is unaffected (Varfolomeev *et al.*, 1998). These results suggest that while caspase-8 plays nonredundant roles in normal embryonic development and apoptosis mediated by the TNFR family members, it is dispensable for other apoptotic pathways.

Evolutionarily, caspase-2 is one of the most conserved caspases and is highly related to CED-3 and DRONC (Kumar *et al.*, 1994; 1997). Caspase-2 is expressed in most tissues and cell types and undergoes rapid activation in response to a variety of apoptotic stimuli (Kumar *et al.*, 1994; Harvey *et al.*, 1997; O'Reilly *et al.*, 2002). Despite this, *casp2*$^{-/-}$ mice are grossly normal and survive to adulthood (Bergeron *et al.*, 1998; O'Reilly *et al.*, 2002). *casp2*$^{-/-}$ females have an apparently increased number of female germ cells, but these mice show normal fertility (Bergeron *et al.*, 1998). The thymocytes derived from *casp2*-null mice are normally sensitive to CD95-mediated cell death, dexamethasone treatment, and γ-irradiation (Bergeron *et al.*, 1998; O'Reilly *et al.*, 2002). Furthermore, DRG neurons from these mice undergo factor-withdrawal-induced cell death as wild-type mice (O'Reilly *et al.*, 2002). These results suggest that caspase-2 is not essential for normally occurring cell death, or plays a redundant role in apoptosis. Some support for a compensatory pathway in *casp2*-deficient cells comes from a recent report where the authors found that in *casp2*-null neurons, the NGF-deprivation-induced, caspase-2-dependent cell death becomes dependent on the caspase-9 pathway, and that *casp2$^{-/-}$* neurons show a threefold compensatory elevation of caspase-9 expression (Troy *et al.*, 2001).

Caspase-12 is unique in that it localizes primarily to the endoplasmic reticulum (ER) (Nakagawa *et al.*, 2000). Caspase-12 is activated by ER stress, such as accumulation of excess proteins in ER, and in response to the disruption of ER calcium homeostasis, but not by TNFR or mitochrondrially mediated death signals (Nakagawa *et al.*, 2000). *Casp12*-null mice are resistant to ER stress-induced apoptosis, but their cells undergo apoptosis in response to other death stimuli. Interestingly, caspase-12-deficient cortical neurons are defective in apoptosis induced by amyloid-beta protein, but not by staurosporine or trophic factor deprivation (Nakagawa *et al.*, 2000). Thus, caspase-12 appears to be specifically involved in certain stress response pathways, independently of the TNFR and mitochondrial pathways of apoptosis.

Unpublished data on *casp6* k/o suggest that ablation of this caspase has no deleterious effect in mice, whereas *casp7* mutants are embryonically lethal (cited in Zheng *et al.*, 1999).

8.2 Knockout of the caspase adaptors

In *C. elegans*, loss-of-function mutants of *ced-4* show complete abrogation of developmentally programmed cell death (Yuan and Horvitz, 1990). In *Drosophila*, the hypomorphic alleles of *dark*, the APAF-1 homolog, attenuate programmed cell deaths during development, causing hyperplasia of the central nervous system, ectopic melanotic tumors, and defective wings (Kanuka *et al.*, 1999; Rodriguez *et al.*, 1999; Zhou *et al.*, 1999). Furthermore, in *dark* mutants, caspase activation is suppressed, and killing by RPR, GRIM, and HID, is substantially suppressed (Kanuka *et al.*, 1999; Rodriguez *et al.*, 1999). In the mouse, the *apaf-1*-null mutation is embryonically or perinatally lethal (Cecconi *et al.*, 1998; Yoshida *et al.*, 1998b). The *apaf-1*-deficient embryos show a range of craniofacial abnormalities resulting from aberrant midline fusion of craniofacial structures, and brain abnormalities including ectopic forebrain cell masses, similar to those seen in *casp3* and *casp9* mutant animals. In E12.5, the retina of *apaf1$^{-/-}$* mutants is thicker than normal and occupies the optic cup, and the lenses are smaller and incorrectly polarized. Delayed removal of interdigital webbing in *apaf1$^{-/-}$* embryos is also seen (Cecconi *et al.*, 1998; Yoshida *et al.*, 1998b). Caspase-3 activation is reduced in *apaf1*-deficient cells, and mutant cells show resistance to a variety of apoptosis stimuli (Cecconi *et al.*, 1998; Yoshida *et al.*, 1998b). These data indicate that the Apaf-1 pathway plays a nonredundant role in controlling cell number during brain and eye development and the morphogenetic cell death in the brain, skull, face, eye, and limbs. Additional defects seen in *apaf1$^{-/-}$* mutants not seen in *casp9$^{-/-}$* mutants indicate that Apaf-1 has roles in developmental cell death that are not dependent on caspase-9 activity and suggest that Apaf-1 may mediate the activation of other caspases.

The *fadd* k/o in mice is embryonically lethal by day 12.5 (Zhang, J. *et al.*, 1998; Yeh *et al.*, 1998). *fadd*$^{-/-}$ embryos show a delayed and underdeveloped phenotype, somewhat similar to the *casp8* k/o phenotype. At E10.5, the ventricular myocardium is thinner than normal, and the inner trabeculation is poorly developed, indicating that FADD plays a nonredundant role in heart development. The mutants also show abdominal hemorrhage that is not caused by abnormal blood-vessel development. *fadd*$^{-/-}$ MEFs are resistant to apoptosis induced by CD95, TNFR-1, or DR3, but not DR4; however, apoptosis induced by overexpression of oncogenes or by cytotoxic drugs is not impaired in these cells. FADD also appears to be required for T-cell proliferation. The similarity in the phenotypes of *casp8* and *fadd* mutant mice suggests that FADD and caspase-8 are essential for apoptosis mediated by many of the TNFRs, but dispensable for death pathways induced by other stimuli.

The caspase mutant and k/o data suggest that while some caspases play essential roles in specific cell-death pathways, others are either redundant or compensated for in the k/o situations. Nevertheless, the results show that individual caspases may participate in apoptosis execution in a cell- and signal-specific manner.

3. Making sense of the Bcl-2 family of apoptosis regulators

Loralee Haughn and David Hockenbery

1. Introduction

Although they were the first identified regulators of mammalian programmed cell death/apoptosis and have been the focus of intensive study, we have more unanswered questions about Bcl-2 and its relatives than other apoptotic factors with more recent histories. Conflicting results have appeared in the literature concerning the subcellular localizations, interactions with self and other binding partners, intracellular site of action, and primary function(s) of this enigmatic protein family. Several of these areas of confusion can be attributed to the physicochemical properties of these proteins: targeting to various intracellular membranes, mitochondrial residence, pore-forming activity, and existence of different conformational states. Other issues stem from the still-evolving conceptions of apoptotic cell death, such as the opposing death and survival functions of Bcl-2 homology (BH) proteins, the suitability of linear substrate-product pathways as a model of mammalian apoptotic programs, and the lack of models to explore the functions of these proteins in healthy cells. In this review, we will examine several as yet unresolved issues relating to the BH protein family as regulators of apoptosis.

2. Internecine warfare or private agendas

One of the most intriguing aspects of the BH family of proteins is its division into anti- and proapoptotic branches. The founding member, Bcl-2, was discovered as the predominant oncogene in follicular lymphomas, located at one reciprocal breakpoint of the t(14;18) (q32;q21) chromosomal translocation (Tsujimoto *et al.*, 1984; Cleary *et al.*, 1986). Although lacking in transforming activity, cells transduced with Bcl-2 remained viable for extended periods in the absence of growth factors and, in combination with the c-myc oncogene, made murine B-cell precursors tumorigenic (Vaux *et al.*, 1988). The first proapoptotic BH protein to be identified, Bax, was co-immunoprecipitated with Bcl-2 in stoichiometric amounts

(Oltvai *et al.*, 1993). Bax-transfected cells died faster in the absence of growth factor than control cells and, later, expression of Bax was shown to be capable of directly triggering apoptosis (Xiang *et al.*, 1996). Since the discovery of Bcl-2 and Bax, the BH family in mammalian cells has expanded by 18 members, with six acting principally as survival factors, and 12 hastening cell death in various experimental systems (Antonsson and Martinou, 2000) (*Table 1*). Further clouding our understanding of these fundamentally opposed activities are the rare observations of a BH protein changing its stripes. For example, introduction of proapoptotic Bax into embryonic neurons promoted survival in the absence of neurotrophic factors (Middleton *et al.*, 1996). In some cases, too much survival protein is a bad thing, resulting in apoptotic death. Homologs of BH proteins exist in all metazoans and several animal DNA viruses (Barry and McFadden, 1998; Wiens *et al.*, 2000). *C. elegans* has both a cell-survival and a death-promoting homolog (Liu and Hengartner, 1999), while only proapoptotic versions have been identified to date in *D. melanogaster* (Vernooy *et al.*, 2000).

2.1 Rheostat model of dimerization

Early experiments showed that the relative steady-state levels of antiapoptotic BH members and death-promoting BH members correlated with the cellular sensitivity to a death stimulus, such as withdrawal of growth factors (Oltvai *et al.*, 1993). Moreover, the relative amounts of pro- and antiapoptotic members were manifested by differential associations between these factors (Oltvai and Korsmeyer, 1994). In the original rheostat model, both anti- and proapoptotic BH proteins

Table 1. Chronology of Bcl-2 homology family

Gene	Apoptotic function
Bcl-2 (Cleary *et al.*, 1986)	anti
Bax (Oltvai *et al.*, 1993)	pro
Bcl-xL (Boise *et al.*, 1993)	anti
Mcl-1 (Kozopas *et al.*, 1993)	anti
Bad (Yang *et al.*, 1995)	pro
Bak (Farrow *et al.*, 1995) (Chittenden *et al.*, 1995b) (Kiefer *et al.*, 1995)	pro
Bfl-1/A1 (Choi *et al.*, 1995)	anti
Nr13/Boo/Diva (Gillet *et al.*, 1995)	anti/pro
Bik/Nbk/Bip1 (Boyd *et al.*, 1995)	pro
Bcl-w (Gibson *et al.*, 1996)	anti
Bid (Wang, K. *et al.*, 1996)	pro
Bok/Mtd (Hsu *et al.*, 1997a)	pro
Hrk/Dp5 (Inohara *et al.*, 1997)	pro
Bim/Bod (O'Connor *et al.*, 1998), (Hsu *et al.*, 1998)	pro
Bcl-rambo (Kataoka *et al.*, 2001)	pro
Noxa (Oda *et al.*, 2000)	pro
Bcl-B (Ke *et al.*, 2001)	anti
Bmf (Puthalakath *et al.*, 2001)	pro
Bcl-G (Guo *et al.*, 2001)	pro
PUMA/Bbc3 (Nakano and Vousden, 2001), (Yu *et al.*, 2001) (Han *et al.*, 2001)	pro

form homodimers or associate as mixed heterodimers (Korsmeyer *et al.*, 1993). As the initial biologic activity of Bax was shown to inhibit Bcl-2-mediated survival, Bcl-2:Bax heterodimers were postulated as the molecular basis for the Bax death-promoting activity, as a means of inhibiting an innate Bcl-2 survival function. This ordering of action/reaction has fallen out of fashion, primarily due to two developments. Recent experimental models are more focused on the direct killing activities of proapoptotic BH proteins, studied with inducible or transient expression systems (Korsmeyer *et al.*, 1999; Antonsson, 2001). This conception of Bax as a direct mediator of cell death, rather than a regulator, has also been reinforced by the demonstration that endogenous Bax is a predominantly cytosolic protein that translocates to a mitochondrial site of action once an apoptotic stimulus is delivered (Hsu *et al.*, 1997b; Wolter *et al.*, 1997). Finally, the discovery that BH proteins have pore-forming activities, predicted by the solution of the Bcl-x_L x-ray crystallographic structure, has been more easily integrated with an inherent death-inducing activity of the proapoptotic BH members than a survival function (Muchmore *et al.*, 1996).

Whichever direction the action/reaction axis is pointing, mutational studies have tended to support the functional importance of heterodimeric associations of pro- and antiapoptotic proteins. The BH family is recognized by the conserved homology domains BH1–4. Certain mutations in the antiapoptotic BH members, such as the BH1 domain Gly^{138}Ala in Bcl-2 and Gly^{159}Ala in Bcl-x_L, disrupt associations with Bax and result in strong loss of survival function (Yin *et al.*, 1994; Yang *et al.*, 1995). Similarly, substitution of hydrophobic amino acids in the Bax BH3 domain with charged amino acids, including the Gly^{67}Arg mutation identified in a leukemic cell line (Meijerink *et al.*, 1998), eliminate heterodimeric interactions with Bcl-x_L and Bcl-2 and reduce cell-death activity (Wang, K. *et al.*, 1998).

A ratchet was thrown into this model by the demonstration that cytosolic BH proteins could be converted from monomers to dimers with detergents. Youle and colleagues reported that Bax associations with self, Bcl-2, and Bcl-x_L could result from the presence of nonionic detergents, such as Triton X-100, used to prepare cell extracts (Hsu and Youle, 1998). Furthermore, Bax dimers were not detected in this study with the membrane-inserted form of endogenous Bax found in apoptotic cells, using whole-cell cross-linking methods or membrane extraction in 1% CHAPS detergent, which did not induce artifactual Bax dimerizations.

This study runs counter to numerous reports demonstrating the presence of Bax hetero- and homodimers, in isolated mitochondria treated with cross-linkers (Gross *et al.*, 1998; Desagher *et al.*, 1999; Antonsson *et al.*, 2000; Makin *et al.*, 2001), and by fluorescence resonance energy transfer experiments in intact cells (Mahajan *et al.*, 1998; Degterev *et al.*, 2001). As with other features of these enigmatic proteins, while there is little doubt that dimeric and perhaps oligomeric associations occur under some circumstances, it remains unresolved whether the core and physiologically relevant functions of BH proteins depend on partnering. Indicative of the confusion on this point, there have been reports of mutations of both Bcl-x_L and Bax proteins that disrupt heterodimeric interactions without disabling their respective apoptotic functions (Cheng *et al.*, 1996; Simonian *et al.*, 1996; Kelekar *et al.*, 1997).

2.2 BH3 domains

One approach to distinguish death-promoting functions of the Bax side of the BH family from antagonism of Bcl-2-related survival functions is to isolate these activities as independent protein domains. The BH3 homology region was first identified in Bak as a domain required for both Bak cytotoxicity and formation of Bak:Bcl-x_L heterodimers (Chittenden *et al.*, 1995). Synthetic Bak BH3 peptides induced apoptosis and mitochondrial cytochrome *c* release in cell-free systems and when added to cells as a fusion protein with the Antennapedia homeoprotein internalization domain (Cosulich *et al.*, 1997; Holinger *et al.*, 1999). A mutated Bak BH3 peptide (Leu^{78}Ala) with reduced Bcl-x_L affinity is nontoxic, suggesting that these isolated peptides act by inhibiting Bcl-x_L function (Narita *et al.*, 1998). This interpretation is complicated, however, by the observation that BH3 peptides interfere with the binding of Bcl-x_L to full-length partners (Diaz *et al.*, 1997). Thus, BH3 domains removed from the context of a proapoptotic BH protein may act by either directly inhibiting an intrinsic Bcl-x_L function or indirectly stimulating the activity of Bcl-x_L proapoptotic binding partners by promoting their dissociation (Ottilie *et al.*, 1997; Kelekar and Thompson, 1998). Since the structure of dimeric BH proteins is not yet available, the possibility remains that dimerization-dependent and -independent functions can be approached in a cellular system by domain analysis. Short of this, further progress in this area must await a new generation of hypotheses for intrinsic functions of the BH proteins independent of other BH-binding partners. In this vein, it is noteworthy that the intrinsic pore-forming activities of Bcl-x_L and Bcl-2 are suppressed by Bak BH3 peptides (Tzung *et al.*, 2001).

2.3 Mouse genetic models

The study of gene function in genetic backgrounds lacking known binding partners is a classic approach to determine epigenetic relationships. Thus, in the case of BH proteins, an agonist–antagonist relationship would predict loss of Bax-dependent phenotypes in the absence of Bcl-2 or Bcl-x_L, or vice versa. Studies of murine knockout strains for Bcl-2 and Bax do not support this model, but rather suggest that each protein can function independently of the other (Knudson and Korsmeyer, 1997). Bcl-2-deficient mice developed accelerated thymic involution within the first month after birth (Veis *et al.*, 1993). Loss of Bax function in double-knockout mice leads to a nearly complete suppression of thymic hypoplasia and excess thymocyte apoptosis (Knudson and Korsmeyer, 1997). Conversely, Bcl-2 function is not redundant in the absence of Bax, as lck promoter-driven Bcl-2 expression enhances survival of thymocytes and mature T cells in Bax$^{+/+}$ or Bax$^{-/-}$ backgrounds. These conclusions are mildly weakened by the potentially redundant expression of other pro- and antiapoptotic family members capable of partnering with Bcl-2 or Bax. However, attempts to identify substitute binding partners for Bcl-2 in Bax$^{-/-}$ thymocytes were unsuccessful (Knudson and Korsmeyer, 1997).

2.3 Proapoptotic functions: yeast models

Demonstration of direct cytotoxic activities for the proapoptotic Bax-type BH proteins has proved to be substantially easier than assaying direct survival effects of Bcl-

2. One of the more compelling arguments has been the conservation of Bax-induced cytotoxicity with features similar to mammalian cell death in *Saccharomyces cerevisiae*, which lacks Bcl-2 homologs and, arguably, a comparable endogenous death pathway (Ligr *et al.*, 1998; Del Carratore *et al.*, 2002). Reed and coworkers originally reported Bax-induced lethality in budding yeast, which could be rescued by coexpression of Bcl-2 (Hanada *et al.*, 1995). Respiratory-deficient petite strains are resistant to Bax killing, although a growth-arrest phenotype is still present (Greenhalf *et al.*, 1996). Bax is recruited to yeast mitochondria following induction in galactose-containing media (Zha, H. *et al.*, 1996). The mitochondrial consequences of Bax expression in yeast include cytochrome *c* egress (Manon *et al.*, 1997), initial mitochondrial membrane hyperpolarization (Minn *et al.*, 1999; Gross *et al.*, 2000), ROS production (Gross *et al.*, 2000), and outer membrane permeabilization (Priault *et al.*, 1999), recapitulating Bax-induced events in mammalian cells.

2.4 Proapoptotic functions: membrane permeability

Based on the recognition that the x-ray Bcl-x_L fold is related to structures of the known pore-forming proteins, diphtheria toxin T domain and bacterial colicins, Minn *et al.* reported pore-forming activities for Bcl-x_L (Minn *et al.*, 1997). Similar properties have been identified with Bcl-2, Bax, Bcl-x_S, and Bid (Schendel *et al.*, 1997; Schlesinger *et al.*, 1997; Schendel *et al.*, 1999). Escape of mitochondrial proteins enclosed in the intermembrane space, including cytochrome *c* and AIF, through large pores in the outer mitochondrial membrane, is one of several competing theories for the relocalization of these proteins during apoptosis.

Introduction of Bax by transient transfection or inducible expression assays triggers cell death, usually accompanied by mitochondrial cytochrome *c* release. Under these conditions, Bax delivery to mitochondria occurs almost simultaneously with cytochrome *c* release (Pastorino *et al.*, 1998; Rosse *et al.*, 1998). Similar kinetics are observed for galactose or tetracycline-regulated expression in yeast (Manon *et al.*, 1997). Addition of recombinant Bax (0.2–4 µM) to mitochondria in cell-free systems liberates cytochrome *c* within minutes (Jurgensmeier *et al.*, 1998; Eskes *et al.*, 1998; 2000). In many cell types, Bax is predominantly a soluble protein in the cytoplasm that traffics to mitochondria following apoptotic signals. Recruitment of endogenous Bax to mitochondria during apoptosis is accompanied by a significant lag period between Bax translocation and cytochrome *c* release (e.g., 1–5 h) (Saikumar *et al.*, 1998; McGinnis *et al.*, 1999; Putcha *et al.*, 1999; De Giorgi *et al.*, 2002).

Several distinct processes affecting mitochondrial membrane integrity have been reported to take place once Bax is in place. Increased permeability of the outer mitochondrial membrane has been demonstrated by the release of multiple proteins normally retained within the intermembrane space and by accessibility of the inner membrane electron transport complexes to exogenous cytochrome *c* (Vander Heiden *et al.*, 1997; Kluck *et al.*, 1999). There may be an upper limit to Bax-mediated outer membrane permeability, as cytochrome *c* fused to green fluorescent protein (GFP), with a molecular mass of 45 kDa, remains localized to mitochondria following Bax expression in yeast, while the 14-kDa cytochrome *c* is released (Roucou *et al.*, 2000).

2.5 Proapoptotic functions: permeability transition pores

There is an extensive body of evidence to support an interplay between the mitochondrial effects of Bax-type proteins and the mitochondrial permeability transition pore (PTP) (Brenner *et al.*, 2000; Shimizu *et al.*, 2001). The PTP is a calcium-activated high conductance channel in the inner mitochondrial membrane that is implicated in mitochondrial membrane depolarization and osmotic swelling of mitochondria. Bax-induced release of mitochondrial cytochrome *c* is inhibited, according to some, but not all, reports, by cyclosporin A (CsA), an inhibitor of the PTP (Jurgensmeier *et al.*, 1998; Marzo *et al.*, 1998; Narita *et al.*, 1998; Pastorino *et al.*, 1998). Bax-induced lethality and cytochrome *c* release in yeast are also compromised in strains with deletions of porin or the adenine nucleotide translocator (ANT), outer and inner membrane components of the PTP, respectively (Marzo *et al.*, 1998; Shimizu *et al.*, 1999). Bax was reported by different groups to bind VDAC, the mammalian porin, and ANT (Marzo *et al.*, 1998; Narita *et al.*, 1998). Finally, reconstitution of VDAC and the proapoptotic proteins Bax or Bak in synthetic liposomes resulted in formation of pores large enough to pass cytochrome *c* (Shimizu *et al.*, 1999).

Ca^{2+}-induced PTP activation results in osmotic swelling of the matrix space with secondary rupture of the less-expandable outer mitochondrial membrane and release of cytochrome *c* and other proteins in the intermembrane space. Although mitochondrial swelling is described in apoptosis, it is not universally observed and, even when present, may follow rather than coincide with cytochrome *c* release. This mechanism can also be discounted in cell-free systems, in which mitochondrial swelling can be prevented by suspending mitochondria in hypertonic solutions (Scarlett and Murphy, 1997). Observations of inhibition of Bax-induced cytochrome *c* release by CsA in the absence of mitochondrial swelling are possibly explained by heterogeneity in the mitochondrial population (with swelling of a small fraction contributing cytochrome *c*) or a transient PTP activation insufficient for osmotic swelling (Pastorino *et al.*, 1999; Gogvadze *et al.*, 2001). However, the possibility that CsA has mitochondrial effects beyond inhibition of PTP cannot be excluded (Jurgensmeier *et al.*, 1998).

Not all reports agree that PTP activation is necessary for Bax-induced cytochrome *c* release. A Mg^{2+}-sensitive mechanism resistant to the PTP inhibitors CsA, bongkrekic acid, ADP, and EGTA has been reported (Eskes *et al.*, 1998). CsA-resistant induction of cytochrome *c* release has also been described for Bax, Bid, and Bik (Shimizu and Tsujimoto, 2000; von Ahsen *et al.*, 2000a). Furthermore, studies of Bax in mutant yeast backgrounds have disputed that VDAC is required for Bax-mediated translocation of cytochrome *c*, growth arrest, or lethality (Priault *et al.*, 1999; Gross *et al.*, 2000; Harris *et al.*, 2000).

2.6 Proapoptotic functions: homooligomers and pores

Pores formed by homooligomers of Bax are also candidates for the mechanism of cytochrome *c* escape from mitochondria. Recombinant Bax forms high-conductance channels in artificial membranes with predominant conductance of 0.5–1.5 nS (Antonsson *et al.*, 1997; Schlesinger *et al.*, 1997). Pore sizing with dextran molecules of different Stokes diameters indicates Bax pore diameters of up to 22–30 Å, large enough for cytochrome *c* passage (Saito *et al.*, 2000). Analysis of the transport as a function of Bax concentration by Hill plots suggested that four Bax molecules were required for cytochrome *c* transport. Cross-linker studies and CHAPS extraction of endogenous Bax recruited to mitochondria following apoptotic stimuli demonstrate Bax complexes in the range of 41–260 kDa, consistent with homodimers and larger oligomers (Gross *et al.*, 1998; Antonsson *et al.*, 2001; Mikhailov *et al.*, 2001). Recently, patch-clamping studies of mitochondria and proteoliposomes incorporating mitochondrial outer membranes identified a novel, high-conductance channel (mitochondrial apoptosis-induced channel [MAC]) that appeared within 12 h of removal of IL-3 from FL5.12 prolymphoid cells (Pavlov *et al.*, 2001). A channel with similar properties was formed in yeast outer mitochondrial membranes within 16 h of Bax induction. MAC activity was not observed in IL-3-deprived FL5.12 cells previously transfected with Bcl-2. The maximal conductance of MAC in mammalian cells corresponds to a pore diameter of 40 Å, suitable for passage of cytochrome *c*. Bax has also been reported to decrease the stability of planar lipid bilayers in a concentration-dependent manner (Basañez *et al.*, 1999). The mechanism involved a decrease in linear tension, which is expected to result in hydrophilic pores within the lipid membrane itself. This effect was not observed with Bcl-x_L, nor could Bcl-x_L protect membrane stability when added to membranes concurrently or sequentially with Bax.

2.7 Proapoptotic functions: different designs

What conclusions can be drawn from the inconsistent results cited above? Mitochondrial responses, such as permeability transition and swelling, are influenced by the mitochondrial suspension media, the energization state of mitochondria, and even the specific redox donor employed. In *Table 2*, assay conditions for addition of recombinant Bax protein to isolated mitochondria are listed. Methodological differences in osmotic strength (sucrose/mannitol vs. KCl), divalent cations (Mg^{2+} vs. Ca^{2+}), adenine nucleotide (+/– regenerating system), and redox donor (succinate vs. glutamate/malate) have yet to be compared systematically for Bax-mediated cytochrome *c* release (for example, the requirement for Ca^{2+} in Narita *et al.*; 1998). Additional variables are likely to modify the membrane insertion and topology of Bax. As demonstrated by Pastorino *et al.*; 1999, mitochondrial swelling and depolarization are dependent on Bax concentration. Bax concentration effects may reflect the operation of more than one mechanism (such as PTP activation versus direct channel formation versus lipidic pore) with CsA-resistant mechanisms predominant at certain concentration ranges. The study by Eskes *et al.*; 1998 is the only one to employ full-length Bax, and shows a lack of inhibition by the classic PTP inhibitors CsA and bongkrekic acid. Recent reports showing that BH3-only proteins, such as Bid, release mitochondrial cytochrome *c* by a non-PTP mechanism suggest that re-examining the Bax/PTP relationship in the context of full-length Bax versus truncated Bax may be worthwhile (Shimizu and Tsujimoto, 2000; Kim *et al.*, 2000b; Zhai *et al.*, 2001). It is also possible that CsA binding to mitochondrial cyclophilin D affects certain processes in addition to PTP formation. A recent report by Scorrano *et al.* (2002) described tBid-induced

Table 2. Experimental conditions for Bax studies with isolated mitochondria

Reference	Reaction conditions	Buffer	Results
(Jurgensmeier et al., 1998)	0.2–1 µM Bax ΔC20, 1 mg/ml mitochondrial protein, 30°C, 1 h	220 mM mannitol/68 mM sucrose, 2 mM MgCl$_2$, succinate, rotenone, ATP, Cr/PCr	Cytochrome c release No mitochondrial swelling Inhibited by CsA
(Eskes et al., 1998)	0.085–0.17 µM Bax full-length, 0.5 mg/ml mitochondrial protein, 30°C, 1 h	125 mM KCl or 210 mannitol/ 70 mM sucrose + 4 mM MgCl$_2$, succinate, rotenone	Cytochrome c release Not inhibited by CsA or BKA
(Pastorino et al., 1999)	(a) 0.125 µM Bax ΔC19, (b) 0.25–1 µM Bax ΔC19, 0.5 mg/ml mitochondrial protein, 37°C, 15 min	150 mM KCl, 1 mM MgCl$_2$, glutamate/malate	(a) Cytochrome c release No swelling or ↓ΔΨm (b) Cytochrome c release, mitochondrial swelling and ↓ΔΨm Both inhibited by CsA
(Marzo et al., 1998a)	1 µM Bax ΔC21, 1 mg/ml mitochondrial protein, 1 h	220 mM sucrose, 68 mM mannitol, 2 mM MgCl$_2$ succinate, rotenone	Cytochrome c release, ↓ΔΨm Inhibited by CsA
(Narita et al., 1998)	5 µM Bax ΔC21, 1 mg/ml mitochondrial protein, 25°C, 20 min	300 mM mannitol, 40 µM CaC$_2$, succinate	Cytochrome c release, mitochondrial swelling, ↓ΔΨm Inhibited by CsA, BKA
(Finucane et al., 1999)	1.5 µM Bax ΔC19, 0.2 mg/ml mitochondrial protein, 37°C, 30 min	210 mM mannitol, 70 mM sucrose, ATP, Cr/PCr, succinate	Cytochrome c release
(von Ahsen et al., 2000a)	2.5 µM Bax ΔC19, 0.3 mg/ml mitochondrial protein, 22°C, 30 min	125 mM KCl or 210 mM mannitol, 60 mM sucrose, ADP, succinate	Cytochrome c release No mitochondrial swelling, ↓ΔΨm Not inhibited by CsA

Abbrev. BKA bongterekic acid; Cr creatine; PCr phoshocreatine ultrastructural changes to intra-mitochondrial cristae, consistent with opening of junctions between cristae and intermembrane spaces. This reorganization is associated with mitochondrial release of an additional cytochrome c pool that resides in the intercristal spaces and is inhibited by CsA. This CsA-dependent process may thus amplify cytochrome c release initiated by a separate mechanism (such as direct channel formation).

2.8 Antiapoptotic functions: membrane permeability

An intrinsic activity of Bcl-2-type survival proteins has been more difficult to envision. Like Bax, a direct action of Bcl-2/ Bcl-x_L on the PTP has been proposed (Tsujimoto and Shimizu, 2000). Shimizu et al. (1999) demonstrated an association of Bcl-x_L and VDAC, with a Bcl-x_L-dependent inhibition of ^{14}C-sucrose or FITC-cytochrome c uptake in VDAC-containing liposomes. In particular, a peptide comprising the NH_2-proximal BH4 domain of Bcl-2 or Bcl-x_L was sufficient to inhibit the VDAC channel (Shimizu et al., 2000b). The addition of Bax did not interfere with the inhibition of VDAC by Bcl-x_L. Some support for this mechanism is provided by electrophysiologic studies of mitoplasts from MDA-MB-231 breast carcinoma cells transfected with Bcl-2 or empty vector (Murphy et al., 2001). Under basal energized conditions or after addition of micromolar Ca^{2+}, a significantly lower frequency of patches with PTP activity was observed in mitoplasts derived from Bcl-2-expressing cells. However, the single-channel characteristics of the PTP were not altered by Bcl-2. In addition, no novel channel activities were identified in mitochondrial membranes from Bcl-2 expressing cells in this or a second study by this group (Pavlov et al., 2001).

Careful mitochondrial fractionation studies by Thompson and coworkers suggested, counterintuitively, that the primary mitochondrial action of Bcl-2 is maintenance of the permeability of the outer membrane (Vander Heiden et al., 1999; 2000; 2001). Their studies of IL-3-deprived FL5.12 prolymphoid cells demonstrated that Bcl-x_L acted to preserve mitochondrial ATP generation, despite similar O_2 consumption and mitochondrial membrane potentials as in ATP-poor control cells. ATP/ADP exchange in mitochondria was substantially reduced upon deprivation of growth factor, but was restored to basal levels by removal of the outer mitochondrial membrane. Bcl-x_L enhanced mitochondrial ATP/ADP exchange in growth factor-deprived cells. This effect may be mediated through VDAC, as incorporation of Bcl-x_L with VDAC in planar phospholipid membranes inhibited voltage gating of VDAC channels at positive potentials. In this model, cytochrome c release would result indirectly from osmotic swelling, secondary to membrane hyperpolarization associated with the shift from state III (+ADP) to state IV (−ADP) respiration.

There have as yet been no attempts to reconcile experimentally these conflicting effects of Bcl-2/Bcl-x_L on VDAC pore function. Inhibition of VDAC was demonstrated with permeability assays using liposome reconstitution systems with recombinant Bcl-x_L and purified or recombinant VDAC (Shimizu et al., 1999). However, activation of VDAC was observed by electrophysiologic studies of planar phospholipid membranes incorporating purified VDAC and recombinant full-length Bcl-x_L (Vander Heiden et al., 2001). In the former case, permeability to uncharged substrates (sucrose and FITC-cytochrome *c*) was assessed, and in the latter, conductance of monovalent ions (KCl). It would be interesting to determine whether Bcl-2/Bcl-x$_L$ or their isolated BH4 domains alter the electrical properties of single VDAC channels in the former model, as might be predicted for a direct interaction. Conversely, the small (5%) decrease in single channel ionic conductance observed in the Vander Heiden study might be more significant for passage of uncharged organic molecules.

Why do Bcl-2 and Bcl-x$_L$ form channels in synthetic lipid membranes with similar electrical properties as Bax, but studies of outer mitochondrial membranes reveal only Bax-associated channels? There are some existing examples of membrane proteins with specific transport functions *in situ* that behave as nonspecific or unregulated channels in artificial lipid membranes. The adenine nucleotide transporter (ANT) has a strict requirement for cardiolipin for exchange activity. Calcium has been proposed to convert ANT to a large nonspecific channel by binding to the phosphate head groups of cardiolipin normally associated with ANT (Hoffmann *et al.*, 1994; Brustovetsky and Klingenberg, 1996). Uncoupling proteins 1, 2, and 3 exhibit Cl$^-$ transport activity, but not H$^+$ transport in the absence of the cofactor coenzyme Q (Echtay *et al.*, 2001). One may speculate that a specific transport function of Bcl-2/Bcl-x$_L$ could require a cofactor missing in artificial lipid membranes, with purified recombinant protein behaving as a nonspecific, and perhaps nonphysiologic, pore. Similar questions can be raised for some variability in the reports concerning the oligomerization state of mitochondria-localized Bax or Bid (Desagher *et al.*, 1999; Grinberg *et al.*, 2002).

2.9 Antiapoptotic functions: what color is your parachute?

If the antiapoptotic functions of Bcl-2/Bcl-x$_L$ are not attributable to Bcl-x$_L$ pores and, at least under some circumstances, are independent of heterodimerizations with proapoptotic BH members, what is left? An intriguing observation in several laboratories is the ability of mitochondrial Bcl-2 to inhibit translocation of cytoplasmic Bax to mitochondrial membranes. Nomura *et al.* (1999) reported that cytosol from apoptotic cells facilitated Bax targeting to isolated mitochondria. Mitochondria obtained from Bcl-2 transgenic mice were targeted as efficiently by recombinant Bax in the presence of apoptotic cytosol as those from nontransgenic littermates. Conversely, cytosol harvested from Bcl-2-transfected cells following an apoptotic stimulus did not support Bax translocation to mitochondrial membranes. As Bcl-2 protein was not detected in the cytosol samples, these results suggest that mitochondrial Bcl-2 modifies one or more cytosolic factors (which may originate in mitochondria) required for Bax translocation. The potential role of caspases was discounted by the failure of caspase inhibitors to block Bax translocation in whole-cell studies (Nomura *et al.*, 1999), although others have obtained opposite results (Goping *et al.*, 1998). Murphy *et al.* (2000) further localized the Bcl-2-dependent step as a conformational change in cytosolic Bax, detectable with antibodies to an N-terminal epitope, that preceded Bax translocation. The numerous demonstrations of a Bcl-2/Bcl-x$_L$ effect on cellular metabolism suggest some avenues for further investigation of an intrinsic biochemical function (Hockenbery *et al.*, 1993; Lam *et al.*, 1994; Shimizu *et al.*, 1998).

Mutational studies of Bcl-2 and Bcl-x_L have directed attention to the hydrophobic cleft bordered by α-helices 2–5 and 7–8 (including BH1-3 domains) (*Tables 3 and 4*). Single amino-acid substitutions in BH1 (Gly^{145}Ala in Bcl-2; Gly^{138}Ala and Arg^{139}Gln in Bcl-x_L) and BH2 (Trp^{188}Ala in Bcl-2) disabling antiapoptotic function occur at known contact sites for proapoptotic BH3 domain peptides and disrupt heterodimerization with proapoptotic BH proteins. However, not all of the residues with strong loss of function and heterodimerization mutant phenotypes are directly involved in BH3 peptide binding (e.g., Val^{15}Glu in Bcl-2), and not all mutations that block heterodimerization result in strong LOF phenotypes (e.g., Val^{126}Gly in Bcl-x_L). Most of the single residues necessary for antiapoptotic functions of these proteins are involved in interior contacts, both inter- and intrahelical. These interactions involve hydrogen bonds between a central hydrophobic helix and a peripheral helix, that is, α5 and α4 (Arg^{146}:Leu^{137} in Bcl-x_L), and van der

Table 3. Mutational studies of Bcl-2

	Bcl-2 wild-type	Bcl-2 mutant	Phenotype	Dimers	Reference
BH1	Gly 138	Ala 138	LOF	-Bax	(Yin *et al.*, 1994)
	Gly 145	Ala 145	LOF	-Bax/+homo	Ibid.
	Gly 145	Glu 145	-GOF	-Bax	Ibid.
BH2	Trp 188	Ala 188	LOF	-Bax	Ibid.
	Gln 192	Ala 192	None		Ibid.
	Glu 200	Ala 200	LOF	-Bax	Ibid.
BH3	Leu 97	Ala 97	LOF		(Cheng *et al.*, 1997)
	+Phe 104	+Ala 104			
BH4	Ile 14	Gly 14	LOF		(Lee *et al.*, 1996)
	Val 15	Gly 15	LOF		Ibid.
	Val 15	Glu 15	LOF	-Bax	(Hirotani *et al.*, 1999)
	Ile 19	Gly 19	LOF		(Lee *et al.*, 1996)
	Tyr 21	Asp 21		↓Bax	(Hirotani *et al.*, 1999)
	Leu 23	Gly 23	LOF		(Lee *et al.*, 1996)
	Tyr 28	Ala 28	↓ cell cycle		(Huang *et al.*, 1997)
		Ser 28	slowing		
		Phe 28			
	Trp 30	Ser 30		↓Bax	Ibid.
	Δ 1–34		GODF	-Bax	(Cheng *et al.*, 1997) (Hirotani *et al.*, 1999)
	Δala 31 or Δala 34 (caspase sites)		GOF		(Cheng *et al.*, 1997)
	Thr 56 (cdc2 phos site)	Ala 56	↓ cell cycle slowing		(Furukawa *et al.*, 2000)
	Ser 70 (PKC phos site)	Ala 70	LOF		(Ito *et al.*, 1997)
	Cys 155 +Cys 226	Ser 155 + Ser 226	LOF	- homo	(Maser et al., 2000)
	Δ30–79		None		(Hunter and Parslow, 1996)

LOF: loss of function; GOF: gain of function; GODF; gain of death function.

Table 4. Mutational studies of Bcl-xL

	Bcl-xL wildtype	Bcl-xL mutant	Phenotype	Dimers	Reference
BH1	Leu 130	Ala 130	wLOF	–Bax	(Sattler et al., 1997)
	Gly 138	Ala 138	LOF	–Bax,Bak/ +Bad/ +homo	(Yang et al., 1995); (Ottilie et al., 1997)
	Arg 139	Gln 139	LOF	–Bax	(Sattler et al., 1997)
	Phe 131	Val 131	None	–Bax,Bak	(Cheng et al., 1996)
	Asp 133	Ala 133	None	–Bax/+Bak	ibid
	Val 135 + Asn 136 + Trp 137	Ala 135 + Ile 136 + Leu 137	LOF	–Bax,Bak	ibid
	Phe 144 + Phe 146	Val 144 + Val 146	wLOF	–Bax,Bak	ibid
BH2	Gly 187	Ala 187	LOF	–Bax	ibid
	Glu 138 + Gly 187	Glu 138 + Ala 187	None		ibid
	Trp 188 + Asp 189	Gly 188 + Ala 189	wLOF	–Bax,Bak	ibid
BH3	Tyr 101	Lys 101	wLOF	–Bax,Bad	(Sattler et al., 1997)
	Glu 92 + Asp 95 + Glu 96	Gln 92 + Ala 95 + Asp 96	None		(Cheng et al., 1997b)
BH4	Δ3–23		LOF		(Shimizu et al., 2000)
	Val 10 + Asp 11	Gly 10 + Gly 11	LOF		(Shimizu et al., 2000)
	Tyr 22 + Gln 26 + Arg 165	Phe 22 + Asn 26 + Lys 165	GOF		(Asoh et al., 2000)
	Val 126	Ala 126	None	–Bax/+Bad	(Kelekar et al., 1997)
	Val 126	Glu 126	None	↓Bax/–Bad	ibid
	Val 126	Gly 126	None	–Bax,Bad	ibid
	Tyr 195	Gly 195	wLOF	–Bax	(Minn et al., 1999)

Abbrev.   wLOF weak Loss of Function.

Waals contacts between α1 and α5 (Val^{15}:Phe^{151} in Bcl-2), and α6 and α5 (Trp^{188}:Trp^{144} in Bcl-2). Additionally, these residues form hydrogen bonds at the NH$_2$-terminal end of α5, from the loop joining α4 and α5 (Arg^{136}:Asp^{133}, Gly^{134}, Asp^{136} in Bcl-x$_L$) or intrahelical bonds (Gly^{138}:Ala^{142} in Bcl-x$_L$; Gly^{145}:Ala^{149} in Bcl-2). Thus, an alternative explanation for the loss of survival function with these mutations may be destabilizing effects on the BH tertiary structure, especially in hydrophobic lipid environments. Of possible relevance to this interpretation, Asoh et al. observed a gain of function phenotype for three Bcl-x$_L$ mutations (Tyr^{22}Phe, Gln^{26}Asn, and Arg^{165}Lys) that were predicted to weaken hydrogen bonding to the distal portion of the α5-α6 hairpin (Asoh et al., 2000).

A recent development in this field is the identification of small molecule inhibitors of Bcl-x$_L$/Bcl-2 that act by binding to the hydrophobic cleft. There are interesting differences in the activities of these compounds, at the cellular level as well as with purified protein (*Table 5*). These properties correspond to the screen-

Table 5. Properties of Bcl-x$_L$ inhibitors

Compound	Binding Affinity	Disrupts peptide-protein interactions	Inhibits pore formation	Bcl-x$_L$ overexpression
HA14-1	9 µM (K$_i$)	Yes	ND	ND
BH3I's	3.3–15.6 µM	Yes	No	Resistant
2-methoxy antimycin A	0.8–2.4 µM	No	Yes	Sensitive
Compound 6	7–10.4 µM (K$_i$)	Yes	ND	ND

ing strategy used to identify each compound. Assays based on competitive binding with proapoptotic BH3 peptides (Degterev et al., 2001), or using docking simulations for Bcl-2 protein structures (modeled on the BH3 peptide:Bcl-x$_L$ complex) (Wang, J.L. et al., 2000; Enyedy et al., 2001), led to the identification of inhibitors of BH3 peptide:protein association, and in the case of the BH3Is, intracellular heterodimerization with Bax. The relative cytotoxicities of a panel of BH3Is are consistent with the above mechanism, and, as predicted, cell killing with these compounds is inhibited by increased expression of the Bcl-x$_L$ target protein (Degterev *et al.*, 2001). In contrast, the activity of 2-methoxy antimycin A on Bcl-x$_L$ was revealed by a cell screen for selective cytotoxicity based on Bcl-x$_L$ expression, demonstrating a direct relationship between AA cytotoxicity and Bcl-x$_L$ expression (Tzung *et al.*, 2001). Although this compound is shown to bind to Bcl-x$_L$ alone, with displacement by BH3 peptide, it appears to have little or no ability to displace the BH3 peptide from a peptide-Bcl-x$_L$ complex (D.H., unpublished results). Differences between the BH3Is and 2-methoxy antimycin A also appear in their effects on Bcl-x$_L$ pore formation, as 2-methoxy AA inhibits pore formation and the BH3Is are ineffective in these assays.

In summary, studies with these inhibitors suggest that occupancy of the Bcl-x$_L$ hydrophobic groove can inhibit structural rearrangement to form a membrane pore, as well as dimerization with proapoptotic partners. Detailed mapping of the ligand-binding sites may lead to new insights into the membrane topology and function of these proteins as well as refinement of strategies to exploit these proteins as drug targets.

2.10 Bcl-2 and Bax: identical twins?

It is surprising that proapoptotic and antiapoptotic BH proteins have highly similar three-dimensional structures. For the limited number of structures available (Bcl-x$_L$, Bcl-2, Bax, and Bid), it is not possible to predict function from the structure alone (Muchmore *et al.*, 1996; Chou *et al.*, 1999; McDonnell *et al.*, 1999; Suzuki *et al.*, 2000; Petros *et al.*, 2001). In one attempt to identify distinguishing features of pro- and antiapoptotic halves of this family, Korsmeyer proposed that packing of the amphipathic second α-helical domain from the NH$_2$-terminus constituted a critical determinant (McDonnell *et al.*, 1999). This domain contains the BH3 sequence that is conserved among all BH proteins. In this model, proapoptotic proteins have an exposed BH3 domain, while the BH3 domain is buried in antiapoptotic proteins. Exposure of the BH3 domain is believed to enable its hydrophobic face to interact with the groove formed by BH1, 2, and 3 domains in a heterodimerization partner. As the solved structures of Bid and Bax contain buried BH3 domains, they are classified as latent proapoptotic proteins. Activating steps, caspase-mediated cleavage in the former case and a conformational change dependent on tBid in the latter case, are predicted to expose the BH3 domain. Proteolytic removal of the NH_2-terminus of Bcl-2 and Bcl-x_L, by caspase-3, generates proapoptotic versions of these proteins (Cheng *et al.*, 1997a; Clem *et al.*, 1998; Kirsch *et al.*, 1999; Basañez *et al.*, 2001). The NH_2-terminal α-helix forms an undercarriage for the BH3 helix; thus, removing this portion of the protein may untether the BH3 domains. However, there are also examples of function switching not explained by proteolytic unraveling of a compact protein structure. Wang, N.S. *et al.* (2001) reported that transient transfection of Bcl-2 in human embryonic kidney 293 and breast cancer MDA-MB-468 cells triggers apoptosis despite mutation of Asp^{34} at the caspase cleavage site, preventing caspase cleavage. Even stranger was the report by Middleton *et al.* (1996) of a Bax survival function in growth factor-dependent neuronal cells. It is hard to avoid the conclusion that antiapoptotic and proapoptotic BH proteins have much in common, with intracellular targeting, susceptibility to proteolysis, and perhaps post-translational modifications, such as phosphorylation, providing their quite different cellular phenotypes, at least most of the time.

3. Location, location, location

The early debates over subcellular localization of Bcl-2 have quieted now with the recognition of Bcl-2 effects that can be attributed to specific sites within the cell. Bcl-2 is an integral membrane protein associated with outer mitochondrial membrane and endoplasmic reticulum (Chen-Levy and Cleary, 1990; Monaghan *et al.*, 1992). There is little evidence for a soluble form of Bcl-2. While there remain some questions concerning Bcl-2 distribution at inner mitochondrial or cristal membranes (Lombardi *et al.*, 1997; Motoyama *et al.*, 1998; Gotow *et al.*, 2000), much of the Bcl-2 fractionating with mitoplast preparations probably represents adherent contact sites between the inner and outer membranes (de Jong *et al.*, 1994). Nguyen *et al.* (1993) showed that the COOH-terminal hydrophobic domain of Bcl-2 functions as a mitochondrial targeting sequence, with much lower efficacy for ER protein import. Other antiapoptotic proteins have homologous COOH-terminal domains and appear to have similar intracellular distributions (Gonzalez-Garcia *et al.*, 1994; Akgul *et al.*, 2000; Duriez *et al.*, 2000; Lee *et al.*, 2001; O'Reilly *et al.*, 2001).

3.1 Endoplasmic reticulum targeting

Lee *et al.* (1999) selectively targeted Bcl-2 to mitochondria or endoplasmic reticulum (ER) with COOH-terminal fusions to organelle-targeting sequences from ActA and cytochrome *b*5. Certain apoptotic stimuli in Rat-1 fibroblasts were susceptible to both ER- and mitochondria-targeted Bcl-2 inhibition (serum-starvation plus c-myc overexpression, taxol, and ceramide), but others (etoposide) could be rescued only with mitochondria-targeted or wild-type Bcl-2. Annis *et al.* (2001) postulated that ER-targeted Bcl-2 rescued a subset of apoptotic deaths characterized by early mitochondrial depolarization prior to cytochrome *c* release.

Speculation about ER-specific functions of Bcl-2 has centered on calcium transport. Lam *et al.* (1994) initially reported in 1993 that Bcl-2 inhibited the efflux of ER calcium induced by thapsigargin (TG), an irreversible inhibitor of the ER Ca^{2+}-ATPase pump (SERCA). This result has not been consistently confirmed, with two groups in agreement (Foyouzi-Youssefi *et al.*, 2000; Nutt *et al.*, 2002), but other groups showing that Bcl-2 increases TG-induced ER Ca^{2+} efflux (Kuo *et al.*, 1998; Williams *et al.*, 2000) or has no effect (Ichimiya *et al.*, 1998). A detailed analysis of altered Ca^{2+} flux in Bcl-2-expressing cells has suggested increased ER Ca^{2+} uptake (SERCA-dependent or -independent) (He *et al.*, 1997; Kuo *et al.*, 1998; Kobrinsky and Kirchberger, 2001), increased mitochondrial Ca^{2+} buffering capacity (Ichimiya *et al.*, 1998; Zhu *et al.*, 1999), or increased permeability of ER membranes to Ca^{2+} (Foyouzi-Youssefi *et al.*, 2000; Pinton *et al.*, 2000) as mechanisms. Foyouzi-Youssefi *et al.* and Pinton *et al.* observed lower ER calcium concentrations in Bcl-2 compared to control cells by direct measurements with the Ca^{2+}-sensitive fluorescent probes, cameleons and aequorins, respectively.

One logical explanation for the antiapoptotic activity of ER-localized Bcl-2 is the binding of proapoptotic BH proteins and prevention of their translocation to mitochondria. Alternatively, Bcl-2 may interact with known ER proteins to carry out an organelle-specific antiapoptotic function. Ng *et al.* (1997) identified binding of Bcl-x_L/Bcl-2 to Bap31, an integral ER membrane protein. In the model of adenovirus E1A-induced apoptosis, Bap31 is cleaved by caspase-8 at two sites to yield a proapoptotic NH_2-terminal p20 fragment. The association of Bcl-x_L and Bap31, stabilized by procaspase-8 and the *C. elegans* Apaf-1 homolog Ced-4, prevented proteolysis of Bap31 (Ng and Shore, 1998).

Hacki *et al.* (2000) demonstrated a connection between ER stress pathways, induced by brefeldin A or tunicamycin, and mitochondrial cytochrome *c* release. This inter-organellar cross-talk was blocked by ER-localized Bcl-2. In a possibly related finding, Zuppini *et al.* (2000) related proteolytic cleavage of Bap31 cleavage to ER stress, and noted reduced processing in cells deficient in the ER lectin chaperone, calnexinS.

3.2 Sequestration at non-mitochondrial sites

Non-mitochondrial localization of proapoptotic BH proteins has become the rule rather than the exception. Bid, one of the first described family members possessing only a BH3 domain, lacks a signal/anchor sequence and is predominantly cytoplasmic (Wang, K. *et al.*, 1996). Caspase-8 – or several other proteases, including granzyme B (Barry *et al.*, 2000) and lysosomal proteases (Stoka *et al.*, 2001) – cleaves Bid to yield a COOH-terminal, 15-kDa, truncated Bid (tBid), which then is translocated to mitochondrial membranes (Li, H. *et al.*, 1998; Luo *et al.*, 1998). Phosphorylation of Bid by casein kinases I and II at serine/threonine residues flanking the caspase-8 cleavage site inhibits caspase-mediated proteolysis (Desagher *et al.*, 2001). McDonnell *et al.* (1999) proposed that alterations in surface charge and hydrophobicity of the truncated Bid are responsible for intracellular relocalization to mitochondrial membranes. Mutations in the tBid BH3 domain, which disrupt heterodimerization with Bcl-2 and Bcl-x_L, did not interfere with trafficking of tBid to mitochondria (Luo *et al.*, 1998). Two studies suggest that mitochondrial targeting of tBid does not require 'receptor'protein at the mitochondrial surface. Lutter *et al.* (2000) examined the affinity of tBid for liposomes with different lipid compositions. Maximal tBid binding was obtained only with liposomes containing >20% cardiolipin, a mitochondria-specific phospholipid normally restricted to the inner mitochondrial membrane. Intracellular targeting of tBid was also impaired in the cardiolipin-deficient CHO mutant cell line PGS. Zha *et al.* (2000) reported post-translational modification of tBid by myristoylation at an NH_2-terminal glycine generated after cleavage by caspase-8. Furthermore, the tight association of the two tBid fragments (p7 and p15) after caspase-8-mediated cleavage prevented efficient binding of non-myristoylated p15 tBid to mitochondrial membranes (Chou *et al.*, 1999; Zha *et al.*, 2000). Thus, post-translational myristoylation appears to be required for relocalization of the p7–p15 tBid complex to mitochondria. Since myristoylation is also employed for trafficking to other membrane compartments, the specificity of myristoylated tBid may reside in conformational effects on tBid (Zha *et al.*, 2000).

The proapoptotic Bad protein also lacks a COOH-terminal signal/anchor sequence but has two consensus 14-3-3 binding sites. Phosphorylation at serines 112 and 136 within the 14-3-3 binding site results in cytoplasmic sequestration of Bad bound to 14-3-3 proteins. Bad phosphorylation occurs downstream of growth factor signals and has been attributed to several kinases: protein kinase B/Akt (Ser^{136}) (Datta *et al.*, 1997; del Paso *et al.*, 1997), mitochondrial-anchored protein kinase A (Ser^{112}) (Zha,J. *et al.*, 1996; Harada *et al.*, 1999), p70S6 kinase (Ser^{136}) (Harada *et al.*, 2001), and PAK1 kinase (Schurmann *et al.*, 2000). Ser^{155} in the Bad BH3 domain is also phosphorylated by protein kinase A (Datta *et al.*, 2000; Virdee *et al.*, 2000; Zhou *et al.*, 2000). Phosphorylation at this site inhibits BH3-dependent interactions of Bad with antiapoptotic BH proteins. A fourth phosphorylation site, Ser^{170}, also regulates the proapoptotic activity of Bad through an unknown mechanism (Dramsi et al., 2002). Complementary phosphatase activity has been alternatively ascribed to PP1A (Ayllon *et al.*, 2000), PP2A (Chiang, C.W. *et al.*, 2001) and PP2B, or calcineurin (Wang, H.G. *et al.*, 1999).

Other BH3 proteins interact with distinct extra-mitochondrial targets, perhaps to sequester them from sites of action (mitochondria and ER) in healthy cells. Bim is localized to the microtubule dynein motor complex by an interaction with the dynein light chain, DLC 1 (Puthalakath *et al.*, 1999), and Bmf associates with dynein light chain 2 (DLC2) in the myosin V actin motor complex (Puthalakath *et al.*, 2001). Several apoptotic stimuli, including paclitaxel (Bim) and anoikis (Bmf), result in translocation of these BH3 proteins from their cytoskeletal sites to mitochondria. DLC1 and DLC2 appear to be released together with their BH binding partners from cytoskeletal attachments during apoptosis.

Despite its status as the first proapoptotic BH protein to be identified, we understand relatively little about mechanisms of Bax cytoplasmic retention in healthy cells and translocation during apoptosis (Pawlowski *et al.*, 2000). By analogy with Bcl-2, the COOH-terminal hydrophobic sequence in Bax was initially believed to function as a mitochondrial signal/anchor. Montessuit *et al.* (1999) demonstrated that, unlike Bcl-2, recombinant full-length Bax could be expressed as a soluble protein. Solution of the three-dimensional structure of full-length Bax by

NMR spectroscopy (Suzuki *et al.*, 2000) revealed that the COOH-terminal hydrophobic sequence is folded back, shielding the hydrophobic cleft in Bax formed by the BH1, BH2, and BH3 domains. Soon after stimulation of apoptosis, cytoplasmic Bax undergoes a conformational change affecting an NH_2-terminal epitope, forms homodimers/oligomers, and becomes an integral mitochondrial membrane protein (Goping *et al.*, 1998; Nechustan *et al.*, 1999). The structure of a homodimer of Bax has yet to be solved, and it is not clear whether proapoptotic homodimers utilize a BH3-hydrophobic groove interaction similar to the proposed mechanism of interaction for heterodimers. The exact sequence of these apoptotic changes in Bax is disputed (Eskes *et al.*, 2000; Murphy *et al.*, 2000). Mutations in the Bax COOH-terminal hydrophobic tail resulting in constitutive mitochondrial localization (ΔSer^{184} and $Ser^{184}Val$) require an additional apoptotic stimulus for the NH_2-terminal conformational change (Nechushtan *et al.*, 1999). Wolter *et al.* (1997) demonstrated the requirement of the 21-residue COOH-terminal tail for mitochondrial translocation. Substitution of the Bcl-2 COOH-terminal tail for the Bax COOH-terminal tail directs Bax constitutively to mitochondria (Goping *et al.*, 1998). However, the Bax COOH-terminal tail acts as a mitochondrial signal/anchor sequence when fused with dihydrofolate reductase, at least *in vitro* (Goping *et al.*, 1998; Nechushtan *et al.*, 1999). Goping *et al.* (1998) identified the NH_2-terminal region (ART) of Bax as an inhibitory domain for mitochondrial membrane targeting and insertion. Deletion of amino acids 1–19 from Bax resulted in mitochondrial localization of Bax in the absence of an apoptotic signal.

Bax translocation from cytosol to mitochondria appears to be a general phenomenon following diverse apoptotic stimuli. Two reports have identified general inducers of Bax mitochondrial targeting that may trigger this event more broadly in apoptosis. Saikumar *et al.* (1998) reported that Bax translocation was dependent on ATP depletion during hypoxia. Maintenance of glycolytic ATP production during hypoxia prevented Bax translocation, while uncoupling of mitochondrial oxidative phosphorylation resulted in Bax translocation even under normoxic conditions. These results suggest that an ATP-dependent chaperone may sequester cytoplasmic Bax. Khaled *et al.* (1999) linked Bax conformational changes to progressive cytosolic alkalinization during cytokine withdrawal. Incubation of cytosolic extracts containing Bax at pH > 7.8 for 15 min resulted in exposure of a cryptic NH_2-terminal Bax epitope and partitioning into Triton X-114 detergent, consistent with exposure of the COOH-terminal hydrophobic domain.

4. A finger in every pot

To further complicate hypotheses of BH protein function, there are clear and convincing effects of BH protein expression on nonlethal cell pathways and fates. Pro- and antiapoptotic family members alter cell-cycle parameters in opposing directions: killer proteins such as Bax and Bad accelerate G0 to S-phase transitions, and survival proteins (Bcl-2, Bcl-x_L) retard S-phase entry from a G0 state (Brady *et al.*, 1996; Mazel *et al.*, 1996; Vairo *et al.*, 1996). Bcl-2 also accelerates withdrawal of cycling cells from the cell cycle after growth factor deprivation (Vairo *et al.*, 1996; Lind *et al.*, 1999). These effects have been noted in cell-transfection studies of fibroblasts, solid tumor lines, and transgenic T and B cells (Pietenpol *et al.*, 1994; O'Reilly *et al.*, 1996). T cells from Bcl-2-knockout mice enter S phase more rapidly than lymphocytes from littermates, reinforcing the conclusion that modulation of cell-cycle pathways are bona fide functions of the BH family rather than errant findings in overexpression models. Huang *et al.* (1997) reported that a Tyr^{28} mutant Bcl-x$_L$ failed to produce the cell-cycle delay found with wild-type Bcl-x$_L$, yet retained wild-type survival function. This group proposed that the Tyr^{28} residue in the BH4 domain of Bcl-x$_L$ was a site of phosphorylation and/or protein interaction. Uhlmann *et al.* (1996) had previously reported a novel gain of function for Bcl-2 with deleted amino acids 51–85, resulting in increased cell proliferation. Results such as these have led some researchers to speculate that cell-cycle regulation is a primary function of BH proteins, with apoptosis a secondary effect. Similar currents of thought have taken hold in the broader field of apoptosis from time to time, despite the difficulty in linking cell-cycle control to many instances of apoptosis (quiescent cells, for example). It is more consistent, although perhaps less straightforward, to implicate mitochondria in a bidirectional flow of information between cytoplasmic and nuclear targets. This mechanism may be distinct from metabolic control of ATP levels, pH, or redox state. For example, retrograde transduction is a defined signal transduction pathway in yeast linking mitochondrial state to nuclear gene expression (Kirchman *et al.*, 1999; Sekito *et al.*, 2000).

Linette *et al.* (1996) reported an association between Bcl-2-induced delay of G0-S transition in T cells and the transcription factor NFAT (nuclear factor of activated T cells). NFAT is required for expression of the delayed early genes IL-2, IL-3, and GM-CSF following T-cell receptor-mediated mitogenic signaling. Increased levels of Bcl-2 inhibited nuclear translocation of NFAT, which requires dephosphorylation by calcineurin to unsheath a nuclear localization signal. Shibasaki *et al.* (1997) demonstrated a tight association between Bcl-2 and calcineurin, dependent on the N-terminal BH4 domain of Bcl-2. Bcl-2-bound calcineurin retained activity, but this group speculated that sequestration of calcineurin at mitochondrial sites prevented its acting on cytoplasmic substrates, including NFAT. Bax expression resulted in disruption of Bcl-2-calcineurin interactions. This model fits known examples of differential targeting of serine-threonine phosphatases via interactions with regulatory subunits (Cohen, 2002).

A complementary study in fibroblasts by Vairo *et al.* (2000) determined that Bcl-2-mediated delays in G1 progression require p27 and the pRB relative p130, and are associated with increased expression of these proteins. Bcl-2 overexpression increased formation of the p130-E2F4 complex, characteristic of quiescent cells, which may act as a transcriptional repressor in this circumstance (Lind *et al.*, 1999; Vairo *et al.*, 2000).

A possible regulatory role for Bcl-2 in cell-cycle events is indicated by the observation that Bcl-2 phosphorylation, initially described after treatment of cells with microtubule inhibitors, represents a cell-cycle-dependent phosphorylation during G2/M (Scatena *et al.*, 1998; Yamamoto *et al.*, 1999). Mutational studies suggest that phosphorylation of Bcl-2 at Ser^{70}, Ser^{87}, and Thr^{69} inactivates the prosurvival activity of Bcl-2 following taxol treatment (Yamamoto *et al.*, 1999). Both stress-activated kinases, ASK1 and JNK1, and the HHV8 viral cyclin-CDK6 complex are involved in G2/M phosphorylation of Bcl-2 (Yamamoto *et al.*, 1999; Ojala *et al.*, 2000).

5. Specialists or generalists?

Why is there such apparent redundancy among both pro- and antiapoptotic BH proteins? In the case of the proapoptotic proteins, it has been suggested that the large number of pro-death family members is indicative of specialization, rather than redundancy (Ferri and Kroemer, 2001). The emerging pattern for proapoptotic BH proteins is one of latent lethality requiring new targeting, post-translational modifications, or conformation changes for activation. The unique localization, protein associations, and mechanism of activation found for the individual proapoptotic members Bax, Bad, Bid, Bim, and Bmf support the hypothesis that each acts as a 'sentry' for distinct damage signals, thereby increasing the range of inputs for endogenous death pathways.

The antiapoptotic proteins Bcl-2 and Bcl-x_L may have nonhomologous functions in hematopoietic development. A reciprocal pattern of expression for these proteins has been described in developing T and B cells, and subsets of adult human and murine bone-marrow cells (Gratiot-Deans *et al.*, 1994; Merino *et al.*, 1994; Park *et al.*, 1995; Grillot *et al.*, 1996). BH proteins are linked to cell differentiation decisions independent of survival activity. Bcl-2 induces differentiation in human neural-crest derived tumor cells (Zhang *et al.*, 1996) and rat pheochromocytoma PC12 cells (Sato *et al.*, 1994). In a comparison of Bcl-x_L and Bcl-2 in staurosporine-treated neuroblastoma cells, Bcl-x_L, but not Bcl-2, was associated with robust neuronal differentiation of surviving cells (Yuste *et al.*, 2002).

The use of Bcl-2 or Bcl-x_L to maintain survival of FDCPMix(A4) cells (Spooncer *et al.*, 1986) also revealed remarkable differences between these proteins in a nonapoptotic phenotype (L. Haughn and D. Hockenbery, unpublished results). In this case, both Bcl-2 and Bcl-x_L enable hematopoietic differentiation, but along restricted and divergent pathways. High-level expression of Bcl-x_L in FDCPMix(A4) cells resulted in selective recovery of erythroid (E) cells in the absence of IL-3 or other growth factors, while Bcl-2 expression yielded granulocytic/monocytic (G/M) cells (*Figure 1*). This result was not due to differential survival since virtually complete viability was maintained in each instance.

Thus, specialization in antiapoptotic BH proteins may emerge from investigations of phenotypes associated with expression in nonlethal contexts. Specific interactions between Bcl-2 survival proteins and cytoplasmic signaling pathways may dictate multiple cell-fate decisions for the antiapoptotic proteins with redundant survival functions.

6. Summary

If there is a predominant theme to the history of investigations on the Bcl-2 family of proteins in the last 15 years, it is one of expansion into multiple disciplines, both unanticipated and enlightening. Bcl-2 research has incorporated approaches from mitochondrial bioenergetics, membrane electrophysiology, macromolecular structure, ion transport, metabolic control, cell structure and dynamics, protein

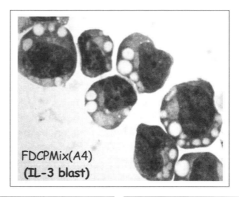

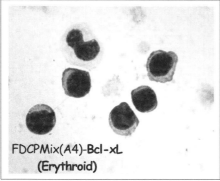

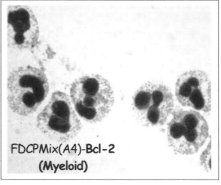

Figure 1. Bcl-2 and Bcl-x$_L$ support divergent cell lineage fates.
Bcl-2 and Bcl-x$_L$ in FCDPMix(A4) progenitor cells transduced with Bcl-2, Bcl-x$_L$, or vector control were deprived of IL-3 for 40 h. May-Grunwald-Giemsa stains; 1000× original magnification.

trafficking, and chemical biology. Among other tangible benefits, this has ensured a steady supply of new investigators to the Bcl-2 field bringing fresh perspectives.

There remains an underlying dichotomy in current efforts to understand how these proteins function in apoptosis (Vander Heiden and Thompson, 1999). On the one hand, there is the view that determining the critical event during apoptosis that is regulated by BH proteins (e.g., cytochrome *c* release via a defined mitochondrial channel) will reveal their essential functions in a straightforward manner. On the other hand, a different perspective on these proteins as 'integrators' of cell signaling and metabolic pathways, with less direct effects on apoptosis, informs attempts to look below the 'tip of the apoptosis iceberg' for more general functions. Undoubtedly, we can look forward to additional insights into the complex networks that govern cell survival in the next few years as these strategies bear fruit.

4. Functional domains in apoptosis proteins

Kay Hofmann

1. Introduction

The analysis of the modular structure of apoptotic signaling proteins has contributed on several levels to our current understanding of the signal transduction networks involved. The main purpose of this chapter is to discuss the functional domain types frequently occurring in proteins that are encoded by apoptosis genes. Besides the functional aspects of those domains, the structure of representative members will be discussed briefly, whenever available.

The first paragraph starts by introducing the domain concept, explaining the relationship between 'structural domains' and 'functional domains'. The methods being used to identify homology domains and to detect their occurrence in protein sequences are briefly discussed. While functional domains are present in all major signaling pathways, they are particularly abundant in apoptotic signaling, owing to the 'multiple-adapter' architecture of the pathways involved. The following paragraphs review the most commonly found domain types and discuss what is known about their *modus operandi*. The most upstream event of apoptosis signaling is the interaction between a death-inducing ligand and its receptor. This interaction is mediated by two particular domain types, one (TNH for T̲N̲F homology) found in the ligand, and the other one (TNFR-CRD for T̲N̲F receptor c̲ysteine-r̲ich d̲omain) in the respective receptors. The death receptors relay the apoptosis signal by their cytoplasmic region containing another domain type, the 'death domain' (DD). The next steps involve a number of adapter proteins harboring interaction domains of the six-helix bundle superfamily, namely, death domains (DD), death effector domains (DED), caspase recruitment domains (CARD), and pyrin domains (PYD). Several of these adapter proteins branch the signaling pathway towards nonapoptotic outcomes, including NF-κB activation. The presence of TRAF domains or Toll/IL-1R domain (TIR) is a hallmark of those proteins. Eventually, the recruitment of the death adapter proteins leads to the activation of caspases, a class of proteases specifically cleaving a number of substrates, including other caspases. A prominent class of physiologic caspase inhibitors containing the 'BIR' domain adds another level of regulation.

A point of convergence between receptor-induced apoptosis and that elicited by different stimuli is the triggering of the mitochondrial permeability pore. A central role in this process is filled by proteins of the Bcl-2 family, harboring several conserved motifs that are frequently referred to as the BH1 through BH4 regions. Among the factors liberated from the mitochondrion is cytochrome c, which binds to and activates another layer of adapter proteins that recruit downstream caspases. The domains involved in this process include CARD and the NB-ARC-type ATPases.

The apoptotic signaling pathway can be described by a multilayer model in which the different layers are populated by different domain types. A recurring mechanism in this pathway is the activation of downstream enzymes by 'induced proximity' (Salvesen and Dixit, 1999). Evolution has heavily exploited the modular nature of the signaling proteins by designing single-domain inhibitors. This phenomenon occurs not only in physiologic situations but also in pathologic processes. A deep understanding of the resulting signaling network is thus a prerequisite of any pharmacologic intervention targeting apoptosis.

1.1 The domain concept

The concept of protein domains originally comes from the analysis of three-dimensional protein structures. Most small proteins (<100 residues) have a 'monolithic' structure, consisting of various secondary structure elements, folded in a way that gives rise to a hydrophobic core and hydrophilic solvent-exposed regions. Larger proteins, by contrast, can follow two different architectural principles: in addition to forming a larger monolithic structure, they can also consist of several smaller folding units, the so-called domains. Each of these domains can fold independently from the rest of the protein and has its own hydrophobic core region. Most inter-residue contacts are satisfied within the domain, only a few interactions being found between domains. As a consequence of their autonomous folding capabilities, domains can typically be excised from their host protein and pasted into a different context, while maintaining both fold and function. In the course of evolution, this series of events has happened several times for many domain types. Evolutionary processes such as exon shuffling, together with the duplication, fusion, and fission of genes and gene regions, have helped to create the multidomain 'mosaic' structure found in many extant proteins.

In the absence of structural information, domains can frequently be detected by analyzing the protein sequences alone. When comparing two dissimilar sequences that both have acquired a domain of the same type by shuffling events, the domain typically appears as a region of localized sequence similarity. Such regions are frequently referred to as 'homology domains'. However, a region of local sequence homology is not necessarily a true homology domain: the possibility that the detected homology region is just the best-conserved part of two proteins with overall homology has to be excluded first. This question can be answered unambiguously only in those situations where the boundaries of the putative homology domain are well defined, as by the N- or C-terminus of the protein or by an adjacent well-characterized domain. Thus, not all local similarities claimed to be 'homology domains' really are domains in the structural sense, and even if they are, the position of the boundaries can deviate. Notwithstanding those caveats, most of the proposed homology domains that occur in diverse sets of proteins correspond nicely to domains in the structural sense. Since most domains, in their cut-and-paste process, preserve not only their structure but also their function, it is frequently possible to attach functional labels to particular domain types. In extreme cases, the property of a novel protein can be predicted from the functional domains contained in its sequence.

The term 'homology domain' or 'functional domain' should be used only for those protein regions either that are known to be domains in the structural sense, or that are at least predicted to fulfill that criterion. Conserved sequence regions that are too short to fold independently from the rest of the protein should rather be referred to as 'motifs'. Conserved motifs can have important roles, too: they are frequently recognized specifically by other proteins or domains.

1.2 Domain detection methods

As mentioned in the previous paragraph, homology domains can be detected by sequence analysis, where they appear as regions of detectable similarity embedded into sequences that are otherwise unrelated. Thus, any tool for local alignment, such as those using the Smith and Waterman algorithm (Smith and Waterman, 1981), is suitable for detecting domains if they are moderately well conserved. It is a widely held tenet that in distantly related proteins, the structure is much better conserved than the sequence. A similar observation can be made for a protein's function, some aspects of which are frequently maintained at evolutionary distances where the sequences no longer look similar. As a consequence, it can be expected that domains exist, both in the structural and functional sense, which cannot be spotted easily by sequence comparison alone. Although several 'structural genomics' projects have gained momentum, it is still not an option to wait for all relevant structures to be determined in order to identify a domain.

Recent years have seen not only a dramatic increase in the available sequence material, but also a constant improvement in the available sequence analysis techniques. In particular, the 'sequence profile' method (Gribskov et al., 1987) with its recent extension to 'generalized profiles' (Bucher et al., 1996), and various 'hidden Markov model' methods (HMM) (Baldi et al., 1994; Krogh et al., 1994; Eddy, 1998)have proved very useful for detecting very weak sequence similarities. Some of their properties make these two methods very well suited for domain detection purposes (Hofmann, 2000). A significant number of homology domains have been identified by either method, including most of the domains discussed in the following paragraphs.

The first-time identification of a new homology domain typically leads to its description by a generalized profile (GP) or an HMM. The domain descriptors are then deposited in one of the existing domain databases, which can be used to analyze newly identified sequences for their domain content (Hofmann, 1998; Attwood, 2000). Several competing domain and motif databases exist, including PROSITE, PFAM, and SMART, which contain descriptors for most if not all of the known domains involved in apoptosis signaling (Schultz *et al.*, 1998; Hofmann *et al.*, 1999; Bateman *et al.*, 2000). Recently, a new meta-database named INTERPRO has been established; it tries to combine the descriptors of several domain databases under a single user interface (Apweiler *et al.*, 2001). Pointers to the very useful search engines of the domain databases are provided in *Table 1*.

Table 1 Search engines of domain and motif databases

PROSITE	http://www.expasy.ch/prosite
PFAM	http://www.sanger.ac.uk/Pfam
SMART	http://smart.embl-heidelberg.de
INTERPRO	http://www.ebi.ac.uk/interpro

1.3 Functional domains in apoptosis

Many of the 'classical' signal transduction processes use a relatively straight path from receptor binding to the activation of corresponding transcription factors. Components found in many of those pathways include small GTPases, their regulator and effector proteins, kinases or even kinase cascades, and eventually transcription factors. By contrast, the signaling pathway leading to cell death follows a different paradigm. Here, most of the usual classes of signaling protein are avoided, and several layers of specific adapter proteins, together with a proteolytic caspase cascade, are used instead. The adapter proteins follow a specific domain-based interaction code, which is outlined in Section 6.

Interestingly, most of the domains occurring in the apoptosis adapters and active components are found exclusively in this pathway. Thus, a new protein found to contain one of the apoptosis-specific domains can be considered a new component in apoptotic signaling, or at least a very good candidate for this function. Consequently, the mining of sequence databases for new proteins containing apoptosis-specific domains has been a rich source of new apoptosis regulators.

2. Death ligands and receptors

The first step in the induction of receptor-mediated apoptosis is the binding of a trimeric 'death ligand' to a cognate 'death receptor'. The known death ligands are all members of a protein family named after its founding member, the tumor necrosis factor TNF. The receptors for proteins of the TNF family also belong to a single family that includes all known death receptors.

2.1 The TNF family

Two excellent reviews covering all aspects of the TNF and TNFR families have appeared recently (Locksley *et al.*, 2001; Bodmer *et al.*, 2002); they contain a far more detailed discussion of death-ligand and -receptor biology than can be provided here. Currently, the mammalian TNF family comprises 18 members, of which at least three are known to induce cell death upon binding to their cognate receptors (*Table 2*). The other ligands have roles in inflammation or cell differenti-

Table 2 The TNF and TNF-R families

Ligand	Receptor	Function
TNF	TNF-R1	Apoptosis, NF-κB activation
Lymphotoxin-α	TNF-R2	Pathogen response, NF-κB
Lymphotoxin-α1/β2	LTβ-R	Germinal center formation
LIGHT[a]	HVEM (LIGHT-R)	T-cell proliferation
FasL	Fas	Apoptosis
	Dcr3	Decoy
TRAIL[b]	TRAIL-R1 (Apo-2, DR4)	Apoptosis
	TRAIL-R2 (Killer, DR5)	Apoptosis
	TRAIL-R3 (LIT, DcR1)	Decoy
	TRAIL-R4 (TRUNDD, DcR2)	Decoy
?	TRAMP (Wsl-1, DR3)	Apoptosis
TWEAK[c]	Fn14 (TWEAK-R)	
CD40L (CD154)	CD40	Germinal center formation
CD30L	CD30	Lymphocyte differentiation
CD27L (CD70)	CD27	T-cell response
4-1BBL	4-1BB	T-cell response
OX40L	OX40	T-cell response
GITRL	GITR (AITR)	Inhibition of TCR Apoptosis
RANKL (TRANCE)	RANK	Osteoclast differentiation
	OPG	Decoy
APRIL	BCMA	B-cell activation
	TACI	
BAFF (Blys,TALL)	BAFF-R	B-cell activation
	TACI	
	BCMA	
EDA-A1	EDAR	Skin, hair, tooth formation
EDA-A2	XEDAR	Skin, hair, tooth formation
VEGI	?	Vascular endothelial cell growth
?	TROY (TAJ)	Hair, epithelium
?	RELT	
?	DR6	NF-κB activation, JNK
NGF	NGFR	Neuronal apoptosis

[a]LIGHT also binds to LTb-R and DcR3.
[b]TRAIL also binds to OPG.
[c]TWEAK might also bind to TRAMP.

ation; some of them are able to modulate or even inhibit cell death via induction of the NF-κB pathway. With one exception, all of the TNF ligands are type II transmembrane proteins; that is, their N-terminus is localized inside the cell while their C-terminus faces the outside. VEGI (vascular endothelial growth inhibitor) is a secreted protein. Several of the other ligands are also found in secreted forms, a fact which can be attributed either to alternative splicing or to proteolytic shedding of the ectodomain. In particular, the latter phenomenon is frequently observed, although the outcome depends on the ligand involved; while the soluble of FasL form is almost inactive (Tanaka *et al.*, 1998), ectodomain shedding appears to be required for activity of the EDA (ectodysplasin A) ligand (Chen *et al.*, 2001).

A defining feature of all members of the TNF family is the presence of a domain, sometimes referred to as the 'TNF homology domain' or THD. The THD is invariably localized at the extreme C-terminus, which in type II transmembrane proteins corresponds to the distal end of the ectodomain. In some of the smaller proteins, such as lymphotoxin-α or OX-40L, the ectodomain consists exclusively of the THD. Some of the larger proteins, in particular TRAIL and EDA, contain an extended 'stalk region' between the membrane and the THD. The active form of all ligands belonging to the TNF family is a trimer. The prevalence for trimerization is a property of the THD domain, as can be seen from the available three-dimensional structures of the ligands TNF (Jones *et al.*, 1989), lymphotoxin-α (Banner *et al.*, 1993), CD40L (Karpusas *et al.*, 1995), and TRAIL (Cha *et al.*, 2000).

The average THD region spans 150 residues and adopts a β-sandwich structure consisting of two layers of five β-strands each. Interestingly, the three-dimensional fold of the THD domain resembles that of the C1q domain found in proteins of the complement system (Shapiro and Scherer, 1998). There are reasons to assume that this relationship is biologically meaningful, since the C1q domain is able to form trimers like members of the TNF family. Moreover, the similarity between the two domain families is not restricted to the structure but can also be detected by sequence profile methods, a fact which normally indicates descent from a common ancestral domain. It is not unreasonable to speculate that some C1q-like domains might bind to receptors resembling members of the TNFR family. For another structural similarity of the THD, that to capsid proteins of small spherical plant viruses, the biologic relevance it not clear. The viral proteins form pentamers rather than timers, and a sequence relationship could not be established.

2.2 The TNFR family

All known receptors for members of the TNF family belong to a large and heterogeneous protein family with 29 members, the TNFR family. Conversely, all known members of the TNFR family act as receptors exclusively for members of the TNF family, with p75NGFR being the only exception (see below). Like the TNF family, the TNFR family is extensively discussed in two recent reviews (Locksley *et al.*, 2001; Bodmer *et al.*, 2002); see also *Table 2* for a list of known ligand/receptor combinations. While VEGI is the only known orphan ligand, there exist at least four orphan receptors (TRAMP/DR3, DR6, Troy/Taj, and RELT). Most members of the TNFR family are integral membrane proteins of type I, bearing an N-terminal signal sequence, followed by a more or less conserved ectodomain, a transmembrane region, and a highly variable C-terminal signaling domain exposed to the cytoplasm. Several TNFR proteins deviate from this architecture. Osteoprotegerin (OPG) and Dcr3 are not anchored to the membrane at all, and TRAIL-R3/Dcr1 uses a GPI anchor. These three proteins lack a signaling domain and thus are considered decoy receptors. XEDAR, BCMA, TACI, and BAFFR are type III membrane proteins with a topology equivalent to type I, but lacking a recognizable N-terminal signal sequence.

A defining feature of all members of the TNFR family is the presence of a particular type of cysteine-rich domain (CRD), which is involved in ligand binding. The CRDs are pseudo-repeats of 30–40 residues, which typically contain six Cys-residues forming three disulfide bonds. The known TNFR members contain between one and five copies of the CRD, which constitute the major part of the receptor's ectodomain. The CRDs can be classified into several subfamilies by virtue of their cysteine spacing (Naismith and Sprang, 1998). Some atypical CRDs also exist, but can be very hard to detect by sequence analysis methods due to their shortness and high sequence divergence. Several three-dimensional structures of members of the TNFR family have been solved, two them showing the receptor ectodomain in complex with its ligands: TNFR/TNF (Banner et al., 1993) and DR5/TRAIL (Hymowitz et al., 1999).

Members of the TNF ligand family form constitutive trimers, and there is little doubt that an active signaling complex involves at least a ligand trimer bound by a receptor trimer. It is less clear whether the receptors also have a tendency to trimerize in the absence of the ligand (Bodmer et al., 2002). If this is the case, the ligand binding would have to change the conformation of the intracellular region of the receptors in order to elicit a signal. If, by contrast, the receptor trimerization is induced by ligand binding, the induced proximity of the signaling domains would be the likely trigger. It is this intracellular part of the receptor that decides whether it signals apoptosis or another cellular event, such as NF-κB activation. In most receptors, the cytoplasmic C-terminal region is very dissimilar and no sequence motifs can be detected. However, there is one subset of receptors that harbors a well-defined homology domain in this region, the 'death domain' (DD). This subset is responsible for apoptosis induction and its members are frequently referred to as 'death receptors' (DRs). The death domain will be discussed in detail in the next paragraph. Currently, at least five functional death receptors are known: Fas, TNF-R1, TRAMP/DR3, TRAIL-R1/DR4, and TRAIL-R2/DR5. Two more death receptor candidates are DR6 and EDAR, whose role in apoptosis has not yet been established. A further protein, NGFR follows the typical death receptor architecture but is unusual in two respects. It does not bind to a member of the TNF family but rather to the unrelated nerve growth factor (NGF), and its role in neuronal apoptosis is still unclear. Even more unusual is osteoprotegerin (OPG), the putative decoy receptor for RANKL/TRANCE. This protein is secreted and contains the only known extracellular death domains. Since the two death domains of OPG have no access to components of the death-signaling pathway, OPG is clearly not a death receptor.

3. Adapter domains

The binding of the death receptors to their ligands induces the assembly of a number of adapter proteins at the cytoplasmic face of the receptor. The resulting multiprotein complex is sometimes referred to as 'DISC' (death-inducing signaling complex). The recruitment of the adapter proteins is mediated by a number of specialized interactions domains of the DD, DED, and CARD class, which share a common six-helix bundle fold (Fesik, 2000) and are also distantly related by sequence (Hofmann *et al.*, 1997).

3.1 The death domain (DD)

The death domain was originally defined as a region of sequence similarity in the cytoplasmic moieties of the two archetypal death receptors Fas and TNF-R1 (Itoh and Nagata, 1993; Tartaglia *et al.*, 1993). Clues to the function of this homology domain came from experiments showing that the death domain of both receptors is the site of interaction with their downstream targets. Mutagenesis experiments demonstrated the importance of an intact death domain for apoptosis induction, while the membrane-proximal part of the intracellular domain was less important (Itoh and Nagata, 1993; Tartaglia *et al.*, 1993). A missense mutation within the death domain of Fas is the cause of the lymphoproliferative disease phenotype observed in *lpr* mice (Watanabe-Fukunaga *et al.*, 1992).

Initially, three proteins (FADD, TRADD, and RIP) were found to interact with the receptor death domains (Boldin *et al.*, 1995; Chinnaiyan *et al.*, 1995; Hsu *et al.*, 1995; Stanger *et al.*, 1995). Interestingly, in all three interactors a region with homology to the death domain was identified, which also coincided with the region mediating the receptor interaction. This was a clear indication that death domains function by mediating a heterodimerization with other death domains.

The death domains of Fas, TNF-R1, FADD, TRADD, and RIP form a heterogeneous family with moderate but readily detectable sequence similarity in the order of 20–30% residue identity. Due to the importance of DD–DD heterodimerization as the first step of intracellular death receptor signaling, several groups have employed bioinformatic methods to screen sequence databases for new death-domain proteins, with the rationale that those proteins might constitute either additional death receptors or new adapter proteins. In two initial screens, several known proteins were found to contain regions related to the death domain, including the NGF-receptor, the death-associated protein kinase DAPK, and several proteins of the NF-κB activation pathway (Feinstein *et al.*, 1995; Hofmann and Tschopp, 1995). Nowadays, bioinformatic 'domain descriptors' for the death domain have been deposited in the databases listed in *Table 1*, and it has become a common practice to check experimentally found apoptosis regulators for death domains and also to use the descriptors for the screening of the EST and genomic databases. A current death domain census yields 32 human proteins, which are listed in *Table 3* together with their known interactors. A large body of data supports the idea that heterodimerization is a general *modus operandi* for death domains and is not restricted to the few cases listed above.

The typical death domain spans a region of approximately 90 residues with pairwise similarities between selected members being as low as 15%. An alignment of selected death domains is shown in *Figure 1*. Several three-dimensional structures of death domains have been reported, including those of Fas (Huang *et al.*, 1996), NGF-R (Liepinsh *et al.*, 1997), and FADD (Jeong *et al.*, 1999). All structures show a highly similar fold consisting of six antiparallel α-helices, forming a com-

Table 3 Death-domain proteins

Protein	Domain content	Function and DD-interaction
TNF Fas TRAMP (Wsl-1,DR3) TRAIL-R1 (Apo-2,DR4) TRAIL-R2 (killer, DR5)	CRD, TM, DD	Apoptosis, FADD-binding
EDAR	CRD, TM, DD	NF-κB activation, EDARAD-binding
NGF-R	CRD, TM, DD	Receptor for NGF
DR6	CRD, TM, DD, CARD	NF-κB activation
Osteoprotegerin (OPG)	CRD, DD	Decoy receptor for RANK-L
FADD	DED, DD	Adapter for death receptors
TRADD	DD	Adapter for TNF-R, NF-κB activation
RIP	Kinase, DD	Adapter for TNF-R, NF-κB activation
EDARADD	DD	Adapter for EDAR
RAIDD (CRADD)	CARD, DD	Adapter for RIP
MyD88	DD, TIR	NF-κB activation, IRAK recruitment
IRAK1 IRAK2 IRAK3 (IRAK-M) IRAK4	DD, kinase	NF-κB activation, MyD88 binding
NFKB1 (p105) NFKB2 (p100)	Rel, Ank, DD	NF-κB transcription factors
LRDD (PIDD)	LRR, ZO1, DD	p53 induced, proapoptotic
DAPK	Kinase, Ank, GTPase, DD	Apoptosis
p84	DD	Rb-binding
MLT (MALT1)	DD, Ig, paracaspase	
Ankyrin-1 Ankyrin-2 Ankyrin-3	Ank, DD	Cytoskeletal anchoring
Unc5H1 Unc5H2 Unc5H3 Unc5H4	Ig, Tsp, TM, ZO1,DD	Netrin receptor, neuronal guidance

Domain abbreviations: CRD, Cys-rich domain; TM, transmembrane domain; DD, death domain; DED, death effector domain; CARD, caspase recruitment domain; TIR, toll/IL-1R domain; Rel, DNA-binding domain of NFκB family; Ank, ankyrin repeat region; LRR, leucine-rich repeat region; Ig, immunoglobulin domain; Tsp, thrombospondin domain; ZO1, ZO1/Unc-5 domain.

pact bundle. The N- and C-termini of the domain are in close vicinity, a property that allows the DD to be inserted into a heterologous sequence with minimal structural disruption, a hallmark of truly modular domains. The domain boundaries derived from the DD structures are in good agreement with the boundaries of the observed sequence homology. Of particular interest is the recently solved structure of a complex between the death domains of the *Drosophila* proteins TUBE and PELLE (Xiao *et al.*, 1999). The interaction surface seen in this complex is probably

```
DD
FAS    Hs 204:SDVDLSKYITTIAGVMTLS.QVKGFVRKNGVNEAKIDEIKNDNVQDTA.EQKVQLRNWHQLHGKKEAYD.TLIKDUKKANLC.TUAEKI
FADD   Hs  93:GEEDLCAAFNVICDNVGKD..WRRUARQLKVSDTKIDSIEDRYPRNLT.ERVRESLRIWKNTEKENATVA.HLVGAURSCQMN.LVADLV
RIP    Hs 579:TTSLTDKHLDPIRENUGKH..WKNCARKLGFTQSQIDEIDHDYERDGLKEKVYQMLQKWVMREGIKGATVGKLAQAUHQCSRIDLUSSLI

DED
FADD   Hs   1:MDDPFLVLLHSVSSSUS....SSEUTELKFLCLGRVGKRKLERVQS.....GLDUFSMLLEQNDLEPGHT.ELLREULASLRRHDULRRV
Casp8  Hs   1:MD..FSRNLYDIGEQUD....SEDUASLKFLSLDYIPQRKQEPIKD.....ALMUFQRLQEKRMLEESNL.SFLKEULFRINRLDULITY
PEA-15 Hs   1:MVE.YGTLFQDLTNNIT....LEDUEQLKSACKEDIPSEKSEEITT.....GSAWFSFLESHNKLDKDNL.SIIEHIFEISRRPDULTMV

CARD
APAF-1 Hs   5:ARNCLLQHREALEKDIKTSYIMDHMISDGFLTISEEEKVRNEPTQQQ...RAAMUIKMILKKD..NDSYV.SFYNAULHEGYK.DUAALL
CED-3  Ce   5:RRSLLERNIMMFSSHUKVDEILEVUIAKQVLNSDNGDMINSCGTVRE...KRREIVKAVQRRG..DVAFD.AFYDAURSTGHE.GUAEVL
CED-4  Ce   6:ECRALSTAHTRLIHDFEPRDALTYUEGKNIFTEDHSELISKMSTRLE...RIANFLRIYRRQAS.ELG...PLIDFFNYNNQS.HUADFL

PYD
Pyrin  Hs  13:LEELVPYDFEKFKFKUQ....NTSVQK.EHSRIPRSQIQRARPVKM.....ATLUVTYYGEEYAVQLTL..QVLRAINQRLLAEEUHRAA
ASC    Hs  12:LENLPAEELKKFKLKUL....SVPURE.GYGRIPRGALLSMDALDL.....TDKUVSFYLETYGAELTA.NVLRDMGLQEMAGOUQAAT
Cryopyr Hs 12:LEDLEDVDLKKFKMHUE....DyPPQK.GCIPLPRGQTEKADHVDL.....ATLMIDFNGEEKAWAMAV.WIFAAINRRDLYEKAKRDE
```

Figure 1. Alignment of representative members of the DD, DED, CARD and PYD families.
Residues identical or conserved in more than 50% of all sequences are shown by black and gray shadings, respectively. The lines above the DD, DED, and CARD alignment indicate the α-helices derived from the respective domain structures. The lines above the PYD alignment indicate predicted α-helices.

not the only one being used for DD–DD interactions. There is evidence that, at least in apoptosis signaling, death domains are able to form heterotrimers, an architecture that requires two different DD–DD interaction surfaces. The problem of the interaction surface for death domains and the related domain families of the six-helix class is the topic of a recent review (Weber and Vincenz, 2001a).

Functionally, all studied death domains appear to mediate interactions of unclear stoichiometry with other death-domain proteins. DD–DD interactions have a strong preference for apoptosis signaling pathways but they also appear to play a role outside apoptosis. The NF-kB activation pathway, which also shares other features with apoptotic signaling, makes heavy use of death domains, too. This is true for the 'developmental' pathway in flies, where the TOLL receptor signals via the DD proteins TUBE and PELLE, as well as for the 'immunologic' pathway in mammals, where IL-1 receptor and TOLL-like receptors (TLRs) signal via the IRAK kinases and MyD88 (O'Neill and Greene, 1998). In addition, there are several orphan DD proteins with no recognizable link to either apoptosis or TOLL/IL1R signaling. In these cases, which include ankyrin and the unc-5 type receptors, the interaction partners and function of the death domains are not known.

3.2 The death effector domain (DED)

The 'death effector domain' (DED) was originally defined in a purely functional sense as the N-terminal region of the death adapter protein FADD, without any reference to sequence homology. Overexpression of FADD leads to the induction of apoptosis in cell lines. When the N- and C-terminal halves of FADD were overexpressed separately, only the former was capable of inducing apoptosis. As a consequence, a model of FADD function was derived in which FADD couples to the receptor by its C-terminal death domain and induces cellular death by its N-terminal half (Boldin et al., 1995; Chinnaiyan et al., 1995; 1996b). By contrast, the death-inducing property of the TNF-receptor associated protein TRADD is not caused by a death effector domain in that protein but rather by a secondary interaction between TRADD and the DED-containing FADD (Varfolomeev et al., 1996).

The important role of the death effector domain of FADD prompted a search for interaction partners, the rationale being that those proteins would most likely constitute the adjacent downstream component of death signaling. In parallel, bioinformatic screens were used to search for additional proteins harboring regions with similarity to the DED. Those proteins would be candidates for additional death effectors, maybe working in the context of other death receptors. The experimental approach was faster and yielded results that fitted both bills: By using a yeast two-hybrid screen, a DED-interacting protein with a very interesting domain structure was identified (Boldin et al., 1996; Muzio et al., 1996). The identified protein, originally called Mach or FLICE, is now known as caspase-8. It was immediately clear that caspase-8 was a component of apoptosis signaling, since related caspases, particularly caspase-3, were already known to have a central role in this process (Cohen, 1997; Salvesen and Dixit, 1997). Interestingly, the N-terminus of caspase-8 carried two copies of a tandem repeat, a strong resemblance to the death effector domain of FADD. This finding established the death effector domain also as a homology domain. The fact that this region was identified as the active FADD interactor suggested that the death effector domain is able to mediate the interaction with other death effector domains, in a manner similar to that of the death domain.

The bioinformatic searches yielded a small number of additional proteins containing the DED homology domain. Besides the astrocytic phosphoprotein PEA-15, a number of small viral proteins were identified, all of them containing two copies of the death effector domain in tandem arrangment (Thome et al., 1997). The viruses encoding these proteins belong to two different classes: most of them are γ-herpesviruses, while the poxvirus MCV (molluscum contagiosum virus) encodes two of the short double-DED proteins (Thome et al., 1997). The viral proteins were of particular interest in the light of the preceding discovery of caspase-8, with which they share the tandem DEDs and considerable sequence relationship. Several independent studies provided evidence that the viral proteins were able to inhibit FADD mediated apoptosis by binding to the death effector domains of FADD and/or caspase-8 (FLICE), and thus disrupting their interaction (Bertin et al., 1997; Hu et al., 1997; Thome et al., 1997). This class of viral proteins was termed v-FLIPs, for viral FLICE inhibitory proteins. Subsequently, a cellular homolog of the v-FLIPs was identified and named c-FLIP, or, alternatively, CASH, MRIT, CLARP, CASPER, I-FLICE, or usurpin (Goltsev et al., 1997; Han et al., 1997; Inohara et al., 1997; Irmler et al., 1997; Rasper et al., 1998). The c-FLIP gene produces several alternative splice forms. A short form corresponds directly to the viral FLIP proteins, while a long splice form generates a protein with high similarity to caspase-8, albeit with an inactive protease domain. This long form of FLIP is a particularly good inhibitor of caspase-8 since it interacts with the enzyme both via its death effector domains and via the caspase-like domain.

Despite several attempts, no additional death effector domain proteins have been identified since then, making the DED family relatively small by comparison (see *Table 4*). Like the death domain, the death effector domain spans approximately 90 residues. Some representative DEDs are aligned in *Figure 1*. The three-dimensional NMR structure of the FADD death effector domain shows a fold

Table 4 Death effector domain proteins

Protein	Domain content	Function and DED-interaction
FADD	DED, DD	Death adapter, Caspase-8/10 recruitment
PEA-15	DED	Apoptosis regulation ?
Caspase-8 (FLICE) Caspase-10	} DED, caspase	Apoptosis, FADD binding
FLIP (MRIT, CFLAR, CASH . . .)	DED, caspase (inactive)	Apoptosis inhibition, FADD binding
DEDD (DEDPRO1)	DED	Apoptosis control, nucleolus ?
DEDD2 (DEDPRO2)	DED	?

Domain abbreviations: DD, death domain; DED, death effector domain.

related to that of the death domain (Eberstadt *et al.*, 1998), confirming an earlier prediction derived from subtle sequence similarities (Hofmann *et al.*, 1997). The ordered part of the NMR structure corresponds well with the domain boundaries predicted from sequence analysis. This was to be expected, since the tandem arrangement of DEDs in the very short viral FLIP proteins had allowed for a precise delineation of the homology domain. At present, no structures for DED–DED complexes are known, and only speculations on the possible interaction surfaces exist (Eberstadt *et al.*, 1998; Weber and Vincenz, 2001a).

Functionally, all known death effector domains are involved in binding to other death effectors. Despite their analogous fold and similar properties, no dimerization between a death domain and a death effector domain has ever been described. The reason for this specificity is still unclear and will probably remain so until the problem of the interaction surfaces has been solved satisfactorily. Another mystery is the stoichiometry of the DED interactions. Analogy to the case of the death domains would suggest a trimerization of DEDs, an idea that is supported by the presence of three death effector domains in the all-important interaction between FADD and caspase-8.

3.3 The caspase recruitment domain (CARD)

While the activation of caspase-8 was now understood as a consequence of the DED-mediated recruitment to the FADD-containing receptor signaling complex, the activation of the other, non-DED caspases remained unclear. Some light was shed on this problem by the discovery of another adapter domain. The starting point was the identification of a death-domain-containing adapter protein named RAIDD or CRADD (Ahmad *et al.*, 1997; Duan and Dixit, 1997). Similar to FADD, overexpression of RAIDD is able to induce apoptosis, but, unlike FADD, the RAIDD protein does not bind directly to the death domain of a receptor but rather to the RIP-containing signaling complex associated with the TNF receptor. A likely *modus operandi* was immediately apparent from the properties of RAIDD's N-terminal region: it both has similarity to and interacts with the prodomain of caspase-2 and recruits this enzyme to the death-signaling complex (Ahmad *et al.*, 1997; Duan and Dixit, 1997). A further region of similarity was identified early on in the prodomain of the *C. elegans* caspase ced-3, suggesting that this caspase can be activated in a similar manner. The great importance of this homology domain became apparent later, again by bioinformatical analysis, when regions with significant sequence similarity were identified in many other caspases and several other proteins (Hofmann *et al.*, 1997). Since the general function of this new domain appeared to be the connection of caspases to various upstream signaling complexes, the name CARD for 'caspase recruitment domain' was proposed. Except for the downstream, or 'executor', caspases, such as caspase-3, all mammalian caspases appear to have a cleavable prodomain that comprises either two death effector domains or one CARD domain.

Following the paradigm of the structurally and functionally related death effector domains, it was to be expected that the caspase-recruiting proteins should also contain a CARD domain. This prediction has now been confirmed in several cases, APAF1 being the most important example (Zou *et al.*, 1997). APAF1 was identified as a component of the so-called apoptosome, a multiprotein complex working downstream of the mitochondrium in apoptosis signal transduction (Zou *et al.*, 1997; 1999). Nowadays, the function of this complex is fairly well understood: upon triggering of the mitochondrial permeability pore, cytochrome *c* is released from the mitochondrium and binds to the WD-40 repeat region of APAF1 (Zou *et al.*, 1997). As a consequence, the N-terminal CARD of APAF1 recruits caspase-9 by its CARD-containing prodomain and thus activates the enzyme by induced proximity (Srinivasula *et al.*, 1998). Several other proteins with established or likely roles in apoptosis signaling were found to contain CARD domains, including two IAP-types apoptosis inhibitors (discussed in Section 4.2) and RIP2/RICK, a protein kinase closely homologous to RIP, but with a CARD instead of the death domain.

The function of the CARD domain does not seem to be strictly specific for apoptosis; instead, several CARD-containing proteins appear to have a role in interleukin-1 signaling and/or NF-κB activation, as in the situation encountered with the death domains. Nonapoptotic caspases, such as the founding member caspase-1 (ICE), also contain a CARD in their prodomain and are likely to be activated by a mechanism similar to the other CARD caspases. Caspase-1 is required for the cleavage and secretion of interleukin-1. The CARD is frequently found in the same proteins as the pyrin domain discussed in the next section, which has its main role in inflammatory signal transduction.

In the current public database, 31 different mammalian proteins with CARD domains are listed (*Table 5*). Like the related death adapter domains, the CARD domain spans a region of approximately 90 residues and is located preferentially at the N- or C-termini of proteins. An alignment of selected CARD domains is shown in *Figure 1*. All characterized CARD domains appear to work by mediating the interaction with other CARD domains, in a way analogous to the death and death-effector domains. The evolutionary relationship between DDs, DEDs, and CARDs had originally been proposed through sequence analysis (Hofmann *et al.*, 1997) and was confirmed by the solution of several three-dimensional structures of CARD domains. The first CARD structure to become available was that of RAIDD; it revealed a six-helix bundle similar to that of the other death adapter domains (Chou *et al.*, 1998). Of particular interest is the structure of an interaction complex

Table 5 Mammalian caspase recruitment domain proteins

Protein	Domain content	Function and CARD-interaction
RAIDD (CRADD)	CARD, DD	Recruitment of caspase-2
Caspase-1	CARD, caspase	IL-1 secretion, binds CARDIAK, CARD8, CARD12, inhibited by psICE, ICEBERG
Caspase-2	CARD, caspase	Apoptosis, binds RAIDD
Caspase-9	CARD, caspase	Apoptosis, binds APAF-1
Caspase-4 Caspase-5 Caspase-11 Caspase-12	CARD, caspase	Recruitment mode not known
ICEBERG psICE	CARD	Inhibitors of caspase-1
APAF-1	CARD, NB-ARC, WD-40	Recruitment of caspase-9
Bcl-10	CARD	Binds CARD9, CARD10, CARMA, CARD14
cIAP1 (MIH-C) cIAP2 (MIH-B)	BIR, CARD, RF	Apoptosis inhibitors
RIP2 (RICK, CARDIAK)	Kinase, CARD	Binds caspase-1
Helicard (MGA5) RIG-I	CARD, helicase	Apoptotic helicases
Nod1 (CARD4) Nod2	CARD, NACHT, LRR	NF-κB activation, immunity
NOP30 (ARC)	CARD	Nucleolar protein
ASC (Pycard, CARD5)	PYD, CARD	Apoptosis, inflammation
CARD6	CARD	
NALP1 (DEFCAP, NAC, CARD7)	PYD, NACHT, LRR, CARD	Inflammation
CARDINAL (TUCAN, CARD8)	CARD	Inhibitor of NF-kB activation
CARMA1 (CARD11)	CARD, PDZ, SH3, GuK	NF-κB activations, binds Bcl-10
CARMA2 (CARD14, Bimp2)	CARD, PDZ, SH3, GuK	NF-κB activations, binds Bcl-10
CARMA3 (CARD10, Bimp1)	CARD, PDZ, SH3, GuK	NF-κB activations, binds Bcl-10
CARMA4 (CARD9)	CARD, PDY, SH3, GuK	NF-κB activations, binds Bcl-10
CARD12 (IPAF, CLAN)	CARD, NACHT, LRR	Apoptosis, caspase-1 activation
CIITA, isoform I	CARD, NACHT, LRR	
KIAA1271	CARD	

Domain abbreviations: CARD, caspase recruitment domain; DD, death domain; NB-ARC, ATPase domain; NACHT, ATPase domain; BIR, IAP-repeat; RF, RING-finger; LRR, leucine-rich repeat region; PYD, pyrin domain; PDZ, PDZ domain; SH3, Src-homology 3 domain; GuK, guanylate kinase domain.

involving the CARD domains of APAF1 and caspase-9 (Qin *et al.*, 1999). Like the structure of the DD–DD interaction complex, this structure reveals a dimeric interaction mode. There is, however, a major difference in the interaction surface used, which is completely different in the two complex structures. Since the two observed interactions modes are not mutually exclusive, it has been suggested that all classes of death adapter domains can form trimeric complexes in physiologic situations, using both types of interactions simultaneously (Weber and Vincenz, 2001a; 2001b).

3.4 The pyrin domain (PYD)

The latest addition to the list of death adapter domains is the pyrin domain (PYD). In contrast to the domains mentioned previously, the pyrin domain was identified purely by bioinformatic approaches in the complete absence of interaction data or other functional information. A region of localized sequence homology was detected in a number of proteins that also contain a CARD domain, including ASC/Pycard, Nod1/CARD4, and NALP1. The grown recognition for the importance of functional domains in apoptosis proteins is reflected by the fact that the pyrin domain was discovered independently by five different groups and published almost simultaneously (Bertin and DiStefano, 2000; Fairbrother *et al.*, 2001; Martinon *et al.*, 2001; Pawlowski *et al.*, 2001; Staub *et al.*, 2001). Three reports have suggested the name 'pyrin domain', which is also used here, while PAAD and DAPIN are alternative names. Up to now, no three-dimensional structure for the pyrin domain has been available. Since several methods of secondary structure prediction and threading analysis all suggest a six-helix topology similar to DD, DED, and CARD, the grouping of the PYD with those domains appears to be appropriate.

The ASC/Pycard protein has the architecture of a typical adapter protein, with an N-terminal pyrin domain and a C-terminal CARD. The pyrin domain of this protein was shown to mediate homodimerization and also interaction with the PYD of the NALP1 protein (Martinon *et al.*, 2001). The current understanding of the physiologic importance of these proteins and their interaction is still insufficient. Another protein suggesting a functional relationship to other death adapter proteins is the CASPY caspase of zebrafish. This protein has the architecture of a typical caspase, but uses a PYD as the prodomain, while mammalian caspases use DEDs or CARDs instead. NALP1 and NALP2 appear to be the founding members of a large protein family with an analogous architecture consisting of an N-terminal pyrin domain, followed by a NACHT-type ATPase (see Section 5.2) and a leucine-rich repeat (LRR) region. An interesting link between the pyrin domain and a family of human diseases is its occurrence in the proteins pyrin/marenostrin and cryopyrin. The gene for the former protein is mutated in Mediterranean periodic fever syndrome (Consortium, T.F.F., 1997; Consortium, T.I.F., 1997), while the gene for cryopyrin is mutated in Muckle-Wells syndrome, another periodic fever disorder (Hoffman *et al.*, 2001). Cryopyrin is a NALP-like protein, while pyrin has a different architecture with an N-terminal pyrin domain, followed by a B-box zinc finger and a C-terminal SPRY domain. Interestingly, a third condition associated with periodic fever is caused by a mutation in the TNF-receptor, providing a further link between the corresponding pathways.

A distinct subfamily of pyrin domains, the 'IFI-domain', is found in a family of interferon-induced proteins. The original reports of the pyrin domain disagree in the judgment of whether these domains should be considered genuine pyrin domains. There is little doubt that these domains are all related and are much

Table 6 Mammalian pyrin domain proteins

Protein	Domain content	Function and PYD-interaction
Pyrin (Marenostrin, MEFV)	PYD, BBOX, SPRY	Periodic fever syndrome, inflammation
ASC (Pycard, CARD5)	PYD, CARD	Apoptosis, inflammation
NALP1 (DEFACP, CARD7)	PYD, NACHT, LRR, CARD	Inflammation
NALP2 (NBS1, Pypaf2)	PYD, NACHT, LRR	Inflammation
Cryopyrin (NALP3, Pypaf1)	PYD, NACHT, LRR	Periodic fever syndrome, inflammation
NALP4	PYD, NACHT, LRR	Inflammation
AIM2	PYD, IFI-CT	γ-interferon inducible
IFI-16	PYD, IFI-CT	γ-interferon inducible
MNDA	PYD, IFI-CT	γ-interferon inducible
IFI203 (mouse only)	PYD, IFI-CT	γ-interferon inducible
IFI204 (mouse only)	PYD, IFI-CT	γ-interferon inducible
IFI205 (mouse only)	PYD, IFI-CT	γ-interferon inducible

Domain abbreviations: PYD, pyrin domain; BBOX, B-box type Zn-finger; NACHT, ATPase domain; IFI-CT, C-terminal domain in interferon-induced proteins.

closer to each other than they are to the other six-helix death adapter domains. Thus, a grouping with the pyrin domain seems to be warranted. It would be interesting to know whether these IFI-type domains are able to interact with the conventional pyrin domains. The current protein sequence databases hold nine human pyrin proteins, which are listed in *Table 6*. This number is likely to grow considerably, since the poorly analyzed part of the human genome appears to contain many more proteins belonging to this domain class.

3.5 The Toll/IL-1R domain (TIR)

The TIR domain can with some justification be grouped with the death adapter domains, although it is neither related to the six-helix domain family nor does it strictly signal apoptosis. Despite these differences, there are also a number of common features. The name TIR stems from the identification of the domain as a homology region in the cytoplasmic portions of TOLL and IL-1 receptors. The TIR domain plays a crucial role in the signaling by these two receptor families, in a manner analogous to that of the death domain in death receptor signaling. Members of the IL-1 receptor family respond to the binding of interleukin-1 (both α and β) as well as interleukin-18. Several members of this receptor family are orphans in the sense that their ligands have not yet been identified. Members of the TOLL family, often called TLR ('TOLL-like receptors'), recognize factors indicating a bacterial or fungal infection, among them lipopolysaccharides and related polymers. The different ligand preferences are reflected in fundamentally different ligand-binding domains. While the interleukin-1 receptor family has Ig-type modules in their ectodomain, TOLL and the TLRs have leucine-rich repeats (LRRs) instead. Both receptor classes elicit similar intracellular responses, with NF-κB activation being the most intensively studied effect. This common aspect of signal transduction is mediated by the TIR domain, which is common to both receptor classes (O'Neill, 2000; Schnare *et al.*, 2001).

Like the death domain, the TIR domain apparently works by recruiting a TIR-containing adapter protein to the receptor. Until the recent discovery of the Mal protein (Fitzgerald *et al.*, 2001), the MyD88 gene product was the only adapter protein known to interact with TIR receptors. MyD88 is still considered the major adapter in TIR signaling, as it seems to play a role in NF-κB signaling by both receptor classes. Interestingly, MyD88 adapts the TIR/TIR-based interaction to one based on death domains. While the C-terminal TIR domain binds to the receptor, the N-terminal death domain apparently binds to the death domain of the interleukin-1 receptor-associated kinases (IRAKs). The four mammalian IRAKs are homologous to the PELLE gene, which plays a role in TOLL-induced NF-κB activation in *Drosophila*. In this well-studied model pathway, an additional death-domain protein called TUBE binds to the death domain of PELLE and plays an important role in TOLL signaling. Mammals seem to lack a TUBE homolog, while a MyD88 homolog has recently been described for *Drosophila* (Horng and Medzhitov, 2001).

The structure of the TIR domains of the TOLL-like receptors TLR2 and TLR3 have been described (Xu *et al.*, 2000). The TIR domain contains a central five-stranded parallel beta sheet, surrounded by five helices. It is thus very different from the short, six-helix adapter proteins, although is serves a similar purpose and also appears to favor trimerization. Currently, 22 mammalian proteins harboring a TIR domain are known; they are listed in *Table 7*.

4. Caspases and inhibitors

The main purpose of the death adapter proteins seems to be the signal-induced oligomerization and subsequent activation of caspases by binding to their DED- or CARD-containing prodomains. The fully active caspases are then able to cleave their substrates, including other caspases. The death-inducing effect of an untimely activated caspase can be very dangerous for an organism and a number of endogenous inhibitors of caspase inhibitors exist for keeping these enzymes in check.

4.1 The caspase catalytic domain

The group of apoptotic proteases that was formerly called 'ICE-like proteases' is now referred to as 'caspases', alluding to their properties as cysteine-proteases with a cleavage specificity for an Asp-based motif (Earnshaw *et al.*, 1999; Nicholson, 1999; Grutter, 2000). The members of this protein family can be classified by at least two different criteria: their mode of activation and their cleavage specificity. While the former is a property encoded by the prodomain (DED, CARD, or neither), the specificity is encoded in the catalytic domain. A typical caspase consists of three different parts. The prodomain is removed after activation, while the remainder of the protein is separated by a further activation cleavage into two subunits, p20 and p10. The three-dimensional structures of several active caspases are

Table 7 Mammalian Toll/IL-1R domain proteins

Protein	Domain content	Function and TIR-interaction
IL-1R1	Ig, TM, TIR	IL-1 receptor, MyD88 binding
IL-1RAcP	Ig, TM, TIR	IL-1 receptor accessory protein
IL-18R1 (IL-1RRP)	Ig, TM, TIR	IL-18 receptor
IL-18RAcP	Ig, TM, TIR	IL-18 receptor accessory protein
Oligophrenin-4 (IL-1RAPL)	Ig, TM, TIR	Role in hippocampal memory
T1/ST2	Ig, TM, TIR	Role in TH2 cells
IL-1RAPL2	Ig, TM, TIR	
IL-1RRP2	Ig, TM, TIR	IL-1 delta and epsilon receptor
SIGIRR	Ig, TM, TIR	
TLR1	Lrr, TM, TIR	Part of Gram-positive response
TLR2	Lrr, TM, TIR	Response to fungi and Gram-positives
TLR3	Lrr, TM, TIR	Response to dsRNA
TLR4	Lrr, TM, TIR	LPS receptor
TLR5	Lrr, TM, TIR	
TLR6	Lrr, TM, TIR	Part of Gram-positive response
TLR7	Lrr, TM, TIR	
TLR8	Lrr, TM, TIR	
TLR9	Lrr, TM, TIR	Response to bacterial DNA
TLR10	Lrr, TM, TIR	
MyD88	DD, TIR	Major adapter for most IL1Rs and TLRs
Mal (TIRAP, Wyatt)	TIR	Adapter for TLR-4
ANIC-BP1 ligand	Arm, SAM, TIR	

Domain abbreviations: Ig, immunoglobulin domain; TM, transmembrane region; TIR, TOLL/IL-1R domain; DD, death domain; Arm, armadillo-repeats; SAM, sterila alpha module.

available, including those of caspase-1 (Walker *et al.*, 1994; Wilson *et al.*, 1994), caspase-3 (Rotonda *et al.*, 1996), caspase-7 (Wei, Y. *et al.*, 2000) and caspase-8 (Blanchard *et al.*, 1999; Watt *et al.*, 1999). All of them were crystallized in the presence of a small molecule inhibitor. According to the structure, an activated caspase consists of two p20/p10 heterodimers. In the structure, the C-terminus of the p20 subunit is far away from the N-terminus of the corresponding p10 subunit, but close to the N-terminus of the other p10 subunit. This arrangement either requires a large conformational change upon activation or, more likely, indicates an exchange of subunits (Fesik, 2000). Other information that can be obtained from the caspase structures includes a general idea of the reaction mechanism and an explanation for the preferred cleavage after aspartate residues. The caspases differ from most other proteases family in the recognition of a complex cleavage signal. While the receptor-proximal caspases, or 'initiator caspases', prefer the motif [LV]ExD, the downstream caspases, or 'executor caspases', prefer DExD instead. This difference in specificity can also be explained by properties of the respective x-ray structures.

Recently, the relationship between caspases and several other protease families was established by sequence analysis (Uren *et al.*, 2000). The members of the newly identified families are cysteine proteases as well and cleave after an aspartate residue. By definition, those proteins should also be considered caspases. A major difference between the classical caspases and the new 'paracaspases' and 'metacaspases' is the lack of a p10-like sequence in the latter families. The paracaspases include the death-domain-containing MLT gene, while the metacaspases are more ancient and even include a yeast protein. Another ancient family of cysteine proteases with relationship to caspases are the insulinase-like enzymes.

Analysis of the domain architecture of the *C. elegans* caspase ced-3 provides another interesting insight into the modularity of caspase evolution. The catalytic domain of ced-3 resembles the catalytic domain of the mammalian executor caspases and reflects the role of the ced-3 gene product as the major (if not sole) apoptotic caspase. By contrast, the prodomain of the ced-3 protein does not resemble the executor caspases at all, but is related to the caspase-9 CARD domain instead. Apparently, ced-3 combines the activation mode of mammalian initiator caspases with the cleavage properties of caspase-3 and other downstream caspases.

4.2 The BIR domain

The BIR domain family, which plays an important role in the inhibition of caspases, had first been identified as a repeat region in baculoviral apoptosis inhibitors, the IAP proteins (Crook *et al.*, 1993). While the viral IAP proteins carry a RING-finger at the C-terminus, the N-terminal region contains two copies of a novel domain, which is now known as BIR (for baculoviral IAP repeat). Soon, the first mammalian homologs, XIAP, cIAP-1, and cIAP-2, were identified and found to have a role in the regulation of apoptosis, too (Rothe *et al.*, 1995; Duckett *et al.*, 1996). The cellular IAPs contain three copies of the BIR repeat, as well as a C-terminal RING-finger; cIAP-1 and cIAP-2 additionally contain a copy of the CARD domain, for which no binding partner has been identified yet. NAIP and ML-IAP are additional mammalian IAPs, which have been identified more recently; ML-IAP contains only a single BIR and a RING-finger, while NAIP is a much longer protein with three BIRs and a NACHT-type ATPase domain (see Verhagen *et al.* [2001] for a survey).

By sequence analysis, the BIR family could be extended considerably, although most of the newly identified proteins appear to have a role in the control of mitosis, rather than in apoptosis. A well-known member of the latter class is survivin, a short, single-BIR protein that was initially considered an apoptosis regulator. The average BIR repeat spans approximately 70 residues and contains a conserved $CxxCx_{15-16}Hx_{6-8}C$ motif, which resembles a CCHC-type zinc finger. This idea was confirmed by the three-dimensional structures of several BIRs from XIAP and cIAP-1 (Hinds *et al.*, 1999; Sun *et al.*, 1999), which show a coordinated Zn-ion embedded in a typical Zn-finger fold. Of particular interest are the structures of caspase-3 and caspase-7 in complex with the BIR domains of XIAP (Chai *et al.*, 2001a; Riedl *et al.*, 2001b). Unexpectedly, the point of contact to the caspases does not lie within a conserved BIR repeat but rather in the linker region upstream of BIR2. Another interesting structure is that of XIAP's BIR3 with the N-terminus of SMAC/DIABLO, a negative regulator of the inhibitory activity of the IAP proteins (Liu *et al.*, 2000; Wu, G. *et al.*, 2000). SMAC binds to a groove in the BIR3 domain and prevents access of XIAP to the caspase. SMAC is liberated from the mitochondria, along with cytochrome *c*, and has a proapoptotic effect by inhibiting the caspase inhibitors.

5. Other important domains

The domains discussed in the previous sections serve to initiate an apoptosis signal and to relay it to the initiator caspases. The following paragraphs deal with some important domain families working further downstream in the apoptosis pathway.

5.1 The Bcl-2 motif family

It has been known for a long time that the product of the Bcl-2 oncogene plays an important role in negatively regulating apoptosis. Soon, a large family of Bcl-2 related proteins was identified, some members of which apparently act in a proapoptotic way, while others resemble bcl-2 in its antiapoptotic effect (reviewed in Chao and Korsmeyer, 1998; Gross *et al.*, 1999; Adams and Cory, 2001). A hallmark of the Bcl-2 relatives is the presence of three particularly well-conserved sequence motifs known as the BH1, BH2, and BH3 regions (Bcl-2 homology). Occasionally, these regions are also referred to as BH domains, but this name should be avoided since they are certainly not domains by any definition. The BH regions are just patches of relatively high conservation contained in an α-helical context, as can be seen from the three-dimensional structure of several Bcl-2 homologs (Muchmore *et al.*, 1996; Chou *et al.*, 1999; McDonnell *et al.*, 1999).

Structurally and functionally, the members of the Bcl-2 family (*Table 8*) can be subdivided into at least three different classes. The first class has an overall

Table 8 Mammalian members of the Bcl-2 family

Protein	Subregions	Function and comments
Bcl-2	BH4, BH3, BH1, BH2, TM	Antiapoptotic, binds BH3 proteins
Bcl-xL	BH4, BH3, BH1, BH2, TM	Antiapoptotic, binds BH3 proteins
Bcl-w	BH4, BH3, BH1, BH2, TM	Antiapoptotic, binds BH3 proteins
Bfl1 (A1)	BH3, BH1, BH2	Antiapoptotic
Mcl-1	BH3, BH1, BH2, TM	Antiapoptotic
Boo (Diva)	BH1, BH2, TM	Antiapoptotic, bnds Bax, Bak
Bax	BH3, BH1, BH2, TM	Proapoptotic, p53 activated
Bak	BH3, BH1, BH2, TM	Proapoptotic
Bak2	BH3, BH1, BH2, TM	Proapoptotic
Bcl-Rambo (Mil1)	BH3, BH1, BH2, TM	Proapoptotic by C-terminal extension
Bcl-G	BH3, BH1, BH2	Proapoptotic
Bok	BH4, BH3, BH1, BH2, TM	Proapoptotic
Bid	BH3	Proapoptotic
Bad	BH3	Proapoptotic
Bim	BH3, TM	Proapoptotic
Bik (NBK)	BH3, TM	Proapoptotic
Blk (BikLK)	BH3, TM	Proapoptotic
Hrk (harakiri)	BH3, TM	Proapoptotic
Puma (Jfy1, Bbc3)	BH3	Proapoptotic, p53 activated
Noxa (PMAIP1)	BH3	Proapoptotic, p53 activated
Bmf	BH3	Proapoptotic, role in anoikis
Bcl-2L12	BH2 only (?)	?

Domain abbreviations: BH1, BH2, BH3, and BH4, homology regions in the Bcl-2 family; TM, transmembrane region.

similarity to Bcl-2, typically carries a C-terminal membrane anchor, and has an antiapoptotic function. Many members of this class also contain an N-terminal extension of the Bcl-2 homology, the so-called BH4 region. The second class groups proapoptotic proteins with overall Bcl-2 similarity. Most of these proteins carry BH1-3 regions but lack the BH4 region. This lack of BH4 conservation has been proposed as a distinction between pro- and antiapoptotic family members (Adams and Cory, 2001), but the trend is not absolute since the proapoptotic Bok protein contains a recognizable BH4 motif (Hsu, S.Y. *et al.*, 1997). A third class of Bcl-2-related proteins is very heterogeneous and groups several proteins where only sequence of the BH3 region is conserved; all members of this class appear to be proapoptotic. As more structures of Bcl-2 relatives become available, it will most likely be appropriate to subdivide this latter class into proteins with Bcl-2 related folds, which have lost conservation in the BH1 and BH2 regions, and proteins unrelated to Bcl-2, which have adopted a sequence motif related to the BH3 region. An obvious example of the former subgroup is the BH3-only protein Bid, whose structure can clearly be superimposed on that of all-BH proteins (Chou *et al.*, 1999; McDonnell *et al.*, 1999). For the latter subgroup, no structures are available, but Hrk appears to be a good candidate, as the sequence is far too short to fit into the Bcl-2 fold.

As mentioned before, the structure of the Bcl-2 family is monolithic and does not contain domains. It consists of two central hydrophobic helices, which are surrounded by several amphipathic helices carrying the BH motifs. Of particular interest is a structure of the antiapoptotic Bcl-xL protein in complex with the BH3 motif of the proapoptotic Bax protein (Sattler *et al.*, 1997). The interactions seen in this structure can explain the propensity of Bcl-2-like proteins to heterodimerize.

The early observation that many proapoptotic Bcl-2 homologs bind to Bcl-2 and other negative regulators leads to two contradicting models: do the antiapoptotic Bcl-2-like members inhibit apoptosis by binding and sequestering the proapoptotic members, or do they inhibit apoptosis by a different mechanism that can be prevented by the proapoptotic Bcl-2 relatives? Initially, the latter idea seemed more likely, as the nematode *C. elegans* contains only one Bcl-2 homolog, the antiapoptotic ced-9 gene product. Since there are no proapoptotic binding partners, a different inhibitory pathway must be operative, at least in the nematode. However, there is accumulating evidence that several proapoptic family members, such as Bax, are able to trigger the mitochondrial permeability pore autonomously, although the mechanism is not entirely understood (Martinou and Green, 2001). An interesting clue came from the observation of a structural similarity between the Bcl-xL fold and that of bacterial pore-forming toxins (Muchmore *et al.*, 1996). A model wherein proapoptotic Bax-like family members can form membrane pores by themselves is supported by the observation of large structural changes in the Bcl-xL structure upon binding to detergent micelles (Losonczi *et al.*, 2000). Alternative explanations are also possible, including the regulation of pre-existing membrane pores of the mitochondrion by association with proapoptotic family members (Shimizu *et al.*, 1999; 2001).

Combining all available information, a three-step regulatory model appears likely, at least for vertebrates. In the most downstream step, the all-BH proapoptotic members, such as Bax, trigger the mitochondrial permeability by an as yet unknown mechanism. This apoptosis-inducing action can be prevented by association with the antiapoptotic all-BH proteins such as Bcl-2 and Bcl-xL. The inhibitory effect of these proteins can, in turn, be abrogated by interaction with proapoptotic BH3-only proteins. The variety of BH3-only proteins can be regulated in different fashions, reflecting the fact that the mitochondrium is really at the intersection of many apoptosis pathways (Kroemer, 1999). Bid, a BH3-only protein with an important role in receptor-induced apoptosis, is activated by caspase-8 cleavage (Li, H. *et al.*, 1998; Luo *et al.*, 1998). Other BH3-only proteins, such as Noxa and Puma, are transcriptionally induced and play a likely role in p53-induced apoptosis (Oda *et al.*, 2000; Nakano and Vousden, 2001; Yu *et al.*, 2001). In nematodes, the ced-9 protein seems to work by a totally different mechanism, although it is related to bcl-2. Cytochrome *c* release does not play a role in nematode apoptosis, where, instead, ced-9 keeps ced-4 inactive by a direct interaction. There is no evidence that vertebrate bcl-2 members can work in a similar fashion by directly regulating APAF1 function. The ced-9 protein itself is regulated by binding of the egl-1 gene product, which is transcriptionally regulated, and bears some resemblance to BH3 regions (Liu and Hengartner, 1999a). So it seems that at least this regulatory aspect is conserved between animal lineages.

5.2 The NB-ARC and NACHT NTPase domains

The NB-ARC domain, also known as AP-ATPase domain, is a homology domain found in a number of apoptosis regulators (van der Biezen and Jones, 1998; Aravind *et al.*, 1999). It was first observed in the central region of APAF1, where it is located between the N-terminal CARD and the C-terminal WD-40 repeat domain, and the nematode Ced-4 protein, which is related to APAF1 but lacks the WD-40 repeats. Interestingly, this domain is also found in a number of plant-derived 'R' gene products, which are involved in the hypersensitive response, an apoptosis-related, host defense mechanism of plants (Heath, 2000). As will be discussed in Section 6, these plant R-proteins have an overall architecture similar to animal apoptosis regulators. Additional copies of the NB-ARC domain are found in a number of bacterial proteins of unknown function, which frequently also contain putative DNA binding motifs. A typical NB-ARC domain spans 300–350 residues and contains all the hallmarks of an ATPase, including a conserved P-loop (Walker A motif) and a moderately well-conserved Walker B motif. Since these conserved elements are typical of both ATPases and GTPases, and a specific nucleotide hydrolysis preference for these proteins has not been described, the assignment as ATPases is speculative.

Shortly after the discovery of the NB-ARC ATPases, a related domain family was described, which tends to occur in a similar set of proteins and at a similar position. This domain family was called NACHT domain (Koonin and Aravind, 2000). NACHT domains are found in several CARD proteins, where they are embedded between the CARD and a C-terminal LRR region. A NACHT domain is also found in the central region of the IAP-type protein NAIP and in the telomerase-associated protein TP1. The 400-residue NACHT domain has overall similarity to the NB-ARC domain. It, too, encodes a putative NTPase and has the

Walker-A and -B motifs well conserved. At least one NACHT protein, the Het-E1 gene product from the fungus *Podospora*, has been shown to bind GTP instead of ATP (Espagne *et al.*, 1997). This finding might point towards a function of the NACHT domain as GTPases.

6. A domain-centric view of apoptosis

As has been demonstrated in the previous paragraphs, most proteins involved in apoptotic signaling share specific domains belonging to a relatively small number of domain families. While, at first glance, the domain arrangement might appear almost random, there are nevertheless a number of recurrent domain architectures observed in different branches of apoptotic and inflammatory pathways.

A first example of a common domain architecture with variable domain types are the caspases, which play a central role in various branches of the apoptosis pathway. They are translated as mostly inactive zymogens that need to be activated by proteolysis – a process that apparently can be stimulated efficiently by bringing two procaspases into close contact (Muzio *et al.*, 1998; Salvesen and Dixit, 1999). This 'induced proximity' can be caused by a number of different interaction domains: mammals have two caspases with an N-terminal DED (Casp-8 and -10), which are recruited by a DED adapter protein (FADD). Another type of caspase (at least seven mammalian enzymes) uses the CARD domain instead, which is recruited by a CARD adapter protein (RAIDD, APAF1). In the zebrafish, at least one caspase with an N-terminal PYD is found. While the recruitment mode of this enzyme is not yet known, it is safe to assume that the corresponding adapter protein will also contain a pyrin domain. Therefore, in the organisms studied to date, at least three different domain types with similar properties are combined with the caspase catalytic domain and serve a common purpose – the recruitment of two or more caspase molecules into an activation complex.

A second example of a conserved meta-architecture is the two-domain adapter proteins, which are being used to change from one 'interaction mode' to another one. FADD changes from a DD-mediated interaction to a DED type, RAIDD switches from DD to CARD, and MyD88 switches from TIR to a DD. The adapter protein ASC/Pycard is not understood well enough to judge the direction of signaling, but it most likely connects a PYD interaction with a CARD interaction.

A third type of multidomain architecture is particularly frequent, although detailed knowledge about the purpose is still lacking. Some examples of this three-domain architecture are shown in *Figure 2*. The N-terminus of the proteins is occupied by an adapter-type domain, which can be either CARD, PYD, TIR, or BIR. The central part of the proteins consists of a NTPase domain, either of the NB-ARC or the NACHT type, while the C-terminal part consists of a variable number of repeats, typically of the WD-40 repeat or leucine-rich repeat (LRR) type. An attempt to infer the function of these protein from the known properties of the constituting domains gives only a rough idea: WD-40 and LRR domains are generic protein interaction domains, which bind to a wide variety of other proteins. NB-ARC and NACHT domains are NTPases, possibly ATPases, of unknown function. Generally, NTPases are involved in reactions that require energy, including

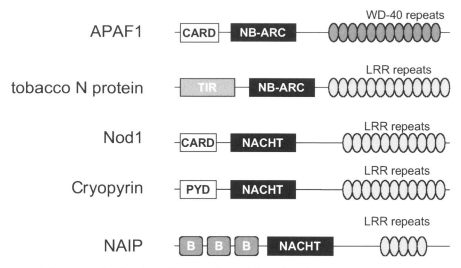

Figure 2. Representative domain architectures of several three-domain apoptosis regulators.
The domain abbreviations are CARD, caspase recruitment domain; NB-ARC, ATPase domain; TIR, Toll/IL-1R type domain; LRR, leucine-rich repeat domain; PYD, pyrin domain; and B, BIR domain.

folding/unfolding of proteins, DNA, or RNA. The best understood part is the N-terminal adapter domain, which most likely mediates the recruitment of another protein of the same domain type, frequently with the purpose of eventually inducing proximity of some catalytically active component. It can, however, not be predicted whether these three-domain proteins are recruiting something, or if they are being recruited to a complex.

APAF1 is the only protein of this architecture for which the mode of action is reasonably well understood. The WD-40 repeat portion is thought to bind cytochrome *c* released from mitochondrial stores. This binding must trigger a change in conformation that allows the formation of the apoptosome, recruiting and activating caspase-9 by its N-terminal CARD domain. The role of the NB-ARC domain in this process is not understood – it might be involved in facilitating the conformational change, maybe by extracting the CARD domain out of an inhibitory interaction adopted in the nonsignaling state. It is tempting to speculate that other proteins of this architecture act in a similar manner: the C-terminal LRR domains could be sensors for as yet unknown factors that trigger the recruitment of CARD-, PYD-, or TIR-containing proteins into a complex similar to the apoptosome. Future experiments are going to show whether there is some truth to these predictions made from the domain content.

The domain descriptions and architectural considerations presented in the previous sections can give only a very cursory overview of the present knowledge. For more focused analysis, the reader is referred to some topical reviews on domain structure (Liang and Fesik, 1997; Fesik, 2000), domain evolution (Aravind *et al.*, 1999; 2001), and domain interaction patterns (Darnay and Aggarwal, 1997;

Hofmann, 1999; Weber and Vincenz, 2001a). We cannot help but be fascinated by the intricate ways that nature reuses, shuffles, and reassembles a limited domain repertoire, resulting in a finely tuned signaling network that controls the most important aspect of cell biology: life or death.

5. The role of the endoplasmic reticulum in apoptosis

David G. Breckenridge and Gordon C. Shore

1. Introduction

When a metazoan cell encounters an external or internal stress that disrupts its physiologic processes, it is forced either to correct the problem or undergo programmed cell death to ensure the health of the organism as a whole. To achieve this, the cell has evolved sentinels that coordinate communication between subcellular compartments, thereby linking diverse stress signals to the core death machinery. Proper communication between organelles is also essential for the cell to synchronize different stages in the initiation and execution of apoptosis, allowing a rapid and efficient demise. Mitochondria are central regulators of apoptosis that store key regulators of caspase activation and nuclear DNA degradation within their intermembrane space (IMS) (see Chapter 7). In most cell-death pathways, the commitment to execute apoptosis occurs when prodeath signals converge on mitochondria, resulting in permeabilization of the outer membrane (OMM) and release of apoptotic IMS proteins such as cytochrome c (Cyt.c) (Green and Reed, 1998; Ferri and Kroemer, 2001). Bcl-2 family members reside in the OMM and control this step, with proapoptotic members facilitating the release of Cyt.c and antiapoptotic members blocking it. In the cytosol, Cyt.c complexes with procaspase-9 and APAF-1, forming the apoptosome complex, which triggers a cascade of caspase activation and subsequent proteolysis of caspase substrate molecules (Green and Reed, 1998).

Emerging evidence suggests that the endoplasmic reticulum (ER) also regulates apoptosis by communicating with mitochondria and by initiating cell-death signals of its own. The contribution of the ER to apoptosis is highlighted by the fact that Bcl-2 family members have been localized to ER membranes, in addition to mitochondria, and demonstrated to influence ER homeostasis directly. Calcium release from the ER has been implicated as a key signaling event in many apoptotic models, and, depending on the mode of Ca^{2+} release, it may directly activate death effectors or influence the mitochondria to undergo apoptotic transitions. A growing number of ER proteins have been shown to regulate apoptosis, some by inter-acting with Bcl-2 family members, and others being caspase substrates. Moreover, recent studies on how stress in the ER is coupled to apoptosis have demonstrated that the ER, like mitochondria, can directly initiate pathways to caspase activation and apoptosis.

In this review, we attempt to summarize and clarify what is known about the involvement of the ER in different apoptotic systems. The review is divided into three sections. The first section highlights recent data on the mechanism by which stress in the ER caused by conditions that compromise protein folding is converted into an apoptotic response. The second section discusses the involvement of ER calcium signals in apoptosis, with specific attention paid to Ca^{2+} signals between the ER and mitochondria. Finally, we overview some of the protein complexes located at the ER that have been proposed to play a role in apoptosis.

2. ER stress-induced apoptosis

2.1 The unfolded protein response

The lumen of the ER provides an optimal environment rich in chaperones for the proper folding and modification of proteins of the secretory pathway. These matu-ration steps are required for secretory proteins to reach their intended cellular des-tination and perform their designated function. An inability to fold efficiently secretory proteins places a cell in a hazardous situation; it would fail to produce properly targeted/functioning proteins for normal cellular processes, and poten-tially generate toxic, malfolded proteins harmful to the organism. Eukaryotic cells have, therefore, evolved an elaborate system, called the unfolded protein response (UPR), to monitor and regulate the folding environment within the ER. When unfolded proteins accumulate in the lumen, the UPR sends out survival signals to halt protein synthesis and specifically upregulate ER resident chaperones and other regulatory components of the secretory pathway (Travers *et al.*, 2000), giving the cell a chance to correct the environment within the ER (Patil and Walter, 2001). However, if the damage is too strong and homeostasis cannot be restored, the mammalian UPR can ultimately initiate apoptosis.

The UPR was first unraveled in yeast, where genetic screens identified a linear pathway involving three genes: Ire1, Hac1, and Rgl1 (see Sidrauski *et al.*, 1998; Kaufman, 1999; Patil and Walter, 2001 for excellent reviews). Ire1p is a trifunc-tional ER transmembrane protein with an N-terminal luminal sensing domain and a cytosolic serine/threonine kinase domain, followed by a C-terminal endoribonu-clease domain. In response to an accumulation of unfolded proteins within the ER, Ire1p oligomerizes, causing trans-autophosphorylation and activation of its kinase domain. This stimulates the endoribonuclease domain of Ire1p, which excises a short inhibitory intron from Hac1 mRNA though a nonconventional splicing reac-tion involving Rgl1p, a tRNA ligase. The modified Hac1 mRNA is now efficiently translated, generating Hac1p, a basic leucine zipper (bZIP) transcription factor that activates the transcription of ER chaperone genes by binding an unfolded pro-tein response element (UPRE) in their promoters.

The machinery of the mammalian UPR is somewhat more complex and involves several signal transduction pathways (Ma and Hendershot, 2001). Two mammalian homologs of Ire1p have been identified, Ire1α and Ire1β (Tirasophon *et al.*, 1998; Wang *et al.*, 1998a). Ire1α is ubiquitously expressed, while expression of Ire1β is restricted to the epithelium of the gut. Both proteins have conserved kinase and endoribonuclease domains and appear to function similarly to yeast Irep1. A wealth of recent studies suggests a model whereby under normal conditions the luminal domain of Ire1 is stably associated with the ER chaperone, BiP, but, following ER stress, BiP dissociates from Ire1 to bind unfolded proteins (Patil and Walter, 2001). Free Ire1 then oligomerizes and undergoes trans-autophosphorylation, and the activated endonuclease domain excises a short sequence (26 bp in man) from the mRNA of the X-Box binding protein (XBP-1), a bZIP transcription factor (Ma and Hendershot, 2001; Shen *et al.*, 2001; Yoshida *et al.*, 2001; Calfon *et al.*, 2002). This switches the reading frame of *XBP-1* mRNA and generates a new protein encoding the original N-terminal DNA binding domain and a new C-terminal transactivation domain that binds and promotes transcription at UPREs in target genes. The UPR in mammals also involves ATF6, an ER transmembrane protein with a cytosolic bZIP transcription factor domain that is cleaved off and released to the nucleus by S1P and S2P proteases following ER Stress (Ye *et al.*, 2000). Like XBP-1, cleaved ATF6 binds UPREs and upregulates ER chaperones (Yoshida *et al.*, 1998a). ATF6 has also been shown to upregulate *XBP-1* mRNA, presumably enhancing Ire1 directed cleavage (Yoshida *et al.*, 2001). ATF6 and Ire1 seem to represent redundant signaling pathways since $Ire1\alpha^{-/-}$ and $Ire1\alpha^{-/-}$, $Ire1\beta^{-/-}$ MEFs show no defect in the UPR induction of chaperone genes, although the $Ire1\alpha^{-/-}$ phenotype confers embryonic lethality, indicating the importance of this pathway in development (Urano *et al.*, 2000a).

The mammalian UPR has also evolved a third signaling branch that rapidly shuts down translation after an insult, preventing the continual accumulation of newly synthesized proteins into the ER when protein-folding conditions are compromised (Kaufman, 1999). This is achieved through a third transmembrane kinase, PERK (Harding *et al.*, 1999). The luminal domain of PERK is homologous to Ire1 and probably senses unfolded proteins in a similar manner. However, the cytosolic tail of PERK encodes a kinase domain that phosphorylates the translation initiation factor eIF2α, preventing the assembly of the 80s ribosomal initiation complex and halting general translation (Harding *et al.*, 1999), while selectively allowing the upregulation of ER chaperones and regulatory components of the secretory pathway (Harding *et al.*, 2000a; Niwa and Walter, 2000). PERK signaling also downregulates cyclin D levels, resulting in cell-cycle arrest in the G_1 phase (Brewer and Diehl, 2000), which may provide time for the cell to restore balance in the ER (Niwa and Walter, 2000).

2.2 ER stress-induced apoptotic signal transduction pathways

Through its three signaling arms, the UPR stalls synthesis of new proteins and arrests cells in G_1 while concomitantly upregulating chaperones, allowing the refolding or degradation of malfolded proteins within the ER. These survival signals may enable mammalian cells to survive physiologic ER stresses that place an excessive workload on the ER, such as the enormous secretion of immunoglobulins during differentiation of B cells into plasma cells (Ma and Hendershot, 2001; Calfon *et al.*, 2002). Experimentally, ER stress is induced by pharmacologic agents that inhibit N-linked glycosylation (tunicamycin), block ER-to-Golgi transport (brefeldin A), impair disulfide formation (DTT), or disrupt ER Ca^{2+} stores (thapsigargin, an inhibitor of the sacroplasmic/ER Ca^{2+} ATPase [SERCA] pumps, or A-23187, a Ca^{2+} ionophore). All of these agents eventually induce apoptosis within 24–48 h depending on the cell type, suggesting that if the damage to the ER is too great, or if balance is not restored within a certain window of time, an apoptotic response is ultimately elicited (Patil and Walter, 2001). This idea is supported by the observation that *PERK*-null MEFs undergo increased apoptosis following ER stress because of their inability to halt translation and the continuing buildup of unfolded proteins in the lumen (Harding *et al.*, 2000b). Numerous signaling molecules have been implicated in ER stress-induced apoptosis, as outlined below.

2.3 CHOP/GADD153

CHOP, also known as GADD153, is a nuclear transcription factor that forms a heterodimeric complex with members of the C/EBP family of transcription factors. CHOP is coordinately upregulated with ER chaperones following ER stress, suggesting that it is regulated by the Ire1/ATF6 pathways (Kaufman, 1999). CHOP involvement in apoptosis is underscored by the fact that tunicamycin-induced apoptosis is impaired in CHOP$^{-/-}$ MEFs, and CHOP$^{-/-}$ mice intraperitoneally injected with tunicamycin show decreased apoptosis in the renal tubular epithelium (Zinszner *et al.*, 1998). Of note, MEFs deficient in C/EBPβ, CHOP's main heterodimerizing partner, show a similar resistance to tunicamycin, consistent with CHOP's functioning as a CHOP/C/EBPβ transcriptional complex (Zinszner *et al.*, 1998). Overexpression of CHOP alone does not induce apoptosis, but it does arrest growth and sensitize cells to ER stress apoptosis. Recently, CHOP was reported transcriptionally to downregulate endogenous Bcl-2 in Rat1 cells and MEFs (McCullough *et al.*, 2001). CHOP could repress a Bcl-2-cat promoter reporter construct, and Bcl-2 protein expression was found to decrease within 16 h of CHOP expression. At present, the mechanism by which CHOP regulates the Bcl-2 promoter, and whether it depends on C/EBPβ, remain to be determined. In any case, since Bcl-2 potently inhibits ER stress-induced apoptosis (McCormick *et al.*, 1997; Srivastava *et al.*, 1999; Ferri and Kroemer, 2001; McCullough *et al.*, 2001), this finding provides an explanation of how CHOP sensitizes cells to ER stress and why CHOP$^{-/-}$ cells are more resistant to tunicamycin-induced apoptosis. Other potential transcriptional targets of CHOP have been identified, but at present it is unclear how they could affect cell growth and apoptosis (Wang *et al.*, 1998b).

2.4 Caspase-12

The mechanism by which stress in the ER is coupled to activation of caspases was for the most part a mystery until caspase-12 was characterized by Nakagawa and Yuan. Caspase-12 is ubiquitously expressed and, like all caspases, synthesized as an inactive proenzyme consisting of a regulatory prodomain, a large p20 subunit, and a small p10 subunit (Van de Craen *et al.*, 1997; Nakagawa and Yuan, 2000). However, unlike other caspases, caspase-12 is remarkably specific to ER stress-induced apoptosis since it is activated (as measured by proteolysis of the precursor into distinct cleavage fragments) by insults that elicit ER stress, but not by agents, such as staurosporin or TNFα/anti-Fas, that activate mitochondrial or death receptor pathways, respectively (Nakagawa *et al.*, 2000). Accordingly, caspase-12-null MEFs are partially resistant to apoptosis induced by Brefeldin A, tunicamycin, and thapsigargin, but not to staurosporin, anti-Fas, or TNFα. In addition, mice deficient in caspase-12 show increased survival and greater resistance to renal tubular epithelial cell death following intraperitoneal injection of tunicamycin (Nakagawa *et al.*, 2000). The partial protection observed in caspase-12-deficient mice clearly indicates the importance of this caspase in ER stress apoptosis, but also suggests that caspase-12 does not function alone to initiate apoptosis. This finding is not surprising given the redundancy of signaling in the UPR between Ire1 and ATF-6.

Caspase-12 is localized at the cytosolic face of the ER, placing it in a position to respond to ER stress as a proximal signaling molecule. Several groups have proposed different models of caspase-12 activation. Nakagawa and Yuan (2000) showed that when mouse glial cells undergo oxygen and glucose deprivation (OGD), a condition that induces ER stress, caspase-12 is cleaved from about 50-kDa size to two ~35-kDa fragments in a calpain-, but not caspase-, dependent manner. Calpains are calcium-activated cytosolic proteases that have been implicated in several forms of apoptosis (Wang, 2000). In addition to agents such as A-23187 and thapsigargin that directly mobilize ER calcium stores, other ER stress agents, such as tunicamycin, have also been found to increase immediately the levels of cytosolic Ca^{2+} (Carlberg *et al.*, 1996). Release of Ca^{2+} may be a conserved feature in the ER stress response that triggers calpain activation at the ER membrane and cleavage of caspase-12, or activation of other Ca^{2+}-dependent cell-death pathways (see below). m-Calpain can directly cleave caspase-12 *in vitro* at T132 and K158, releasing the prodomain from the p20/p10 subunits and generating the two ~35-kDa fragments observed *in vivo* with antibody against the p20 subunit. Calpain-cleaved caspase-12 may then autoactivate in *trans*, by self-cleavage at D316, separating the p20 and p10 subunits; therefore, they could form the active heterotetrameric p20/p10 complex characteristic of all caspases (Nakagawa and Yuan, 2000). Of note, however, the free p20 or p10 subunit of caspase-12 has not yet been observed *in vivo*; therefore, the second cleavage step remains theoretical. Consistent with this model, both calpain inhibitors (they block the first cleavage event) and the pan-caspase inhibitor zVAD-fmk (predicted to block the second cleavage step and caspase-12 activity) inhibited OGD-induced apoptosis. Furthermore, chelation of intracellular Ca^{2+} with BAPTA-AM has been reported to block cell death in response to thapsigargin, perhaps by inhibiting the calpain-dependent activation of caspase-12 (Srivastava *et al.*, 1999). Nakagawa and Yuan noted that Bcl-x$_L$ is also cleaved by m-calpain during OGD, an event that is predicted to convert Bcl-x$_L$ from antiapoptotic to proapoptotic. This raises the possibility that calpain-cleaved caspase-12 and Bcl-xL cooperate to induce apoptosis.

Rao *et al.* (2001) reported that in 293T cells exposed to thapsigargin or brefeldin A, caspase-7 translocates to the surface of the ER where it can bind and cleave caspase-12 at D94, releasing the prodomain. *In vitro*, the resulting C-terminal fragment appears to undergo self-cleavage at D341, separating the p20 and p10 subunits. Recombinant caspase-7 could activate an S-100 extract to cleave caspase-12 at D94 *in vitro*, but it was not reported whether caspase-7 could directly cleave caspase-12, and the effect of zVAD-fmk on caspase-12 cleavage was not tested *in vivo* or *in vitro*. One possible explanation that could reconcile the discrepancies between this study and that of Yuan's group is that cleavage of caspase-12 by calpain might be an early initiating event, whereas cleavage of caspase-12 by caspase-7 might be a late event involving feedback amplification, such has been described in other apoptotic systems. However, the observed contrasts might be related to the fact that the two studies were carried out in different species with different inducers of ER stress. Of note, Nakagawa and Yuan (2000) reported that in mouse cortical neurons, caspase-12 is expressed as a main isoform of ~50 kDa and a smaller isoform of ~42 kDa. Following OGD, a third larger isoform (~60 kDa) is upregulated. Murine caspase-12 expressed ectopically in 293T cells has also been reported to be differentially phosphorylated (Yoneda *et al.*, 2001). The human caspase-12 sequence has not been formally reported, but antibodies raised against murine caspase-12 detect a single 60-kDa protein in human 293T cells that is upregulated in response to ER stress prior to its cleavage (Rao *et al.*, 2001). Therefore, caspase-12 may be expressed as several isoforms, or as modified forms of a single species, that are differentially regulated.

Following stimulation, the cytosolic tail of Ire1 can recruit TRAF2 (Urano *et al.*, 2000b). When overexpressed in 293T cells, TRAF-2 can also interact with caspase-12 through its TRAFN domain and weakly induce caspase-12 oligomerization and cleavage (Yoneda *et al.*, 2001). Although the observed interactions need to be confirmed between endogenous components, it is possible that an activated Ire1/TRAF-2 complex recruits oligomers of casapase-12 that are cleaved by calpain or undergo autocatalytic activation. This scenario would allow the ER to couple the UPR survival response with an apoptotic cascade, and the fate of the cell could be determined by the strength of the two competing signals or by other secondary signals that modulate one of the responses. This theme is reminiscent of signals elicited by the TNF death receptor, which utilizes TRAF-2 to activate NF-κB and JNK survival signals, while simultaneously recruiting and activating procaspases-8, and the decision to execute cell death or proliferation is a complex interplay between these opposing signals (Baud and Karin, 2001). Consistent with this model of caspase-12 activation, when Ire1 is overexpressed, it autoactivates and potently induces cell death (Wang *et al.*, 1998a). However, *in vivo* Ire1 activation and caspase-12 cleavage occur with different kinetics; caspase 12 cleavage is a late event often not observed until 16 -24 h after treatment, whereas Ire1 is activated immediately and has been shown to return to an inactive, nonphosphorylated state within 5 h after application of ER stress (Harding *et al.*, 2000a). Clearly, the dependence of caspase-12 activation on Ire1 must be tested in *Ire1*-null MEFs.

2.5 JNK

Several groups have reported that ER stress activates the cJUN NH_2 terminal kinase (JNK/SAPK) pathway (Srivastava *et al.*, 1999; Urano *et al.*, 2000b). JNK signaling has been implicated in several forms of cell death, including UV irradiation, where *jnk-1,jnk-2*-null MEFs are resistant to Cyt.c release and apoptosis (Tournier *et al.*, 2000). JNK may exert its effect by phosphorylating mitochondrial proteins (Aoki *et al.*, 2002), including Bcl-x$_L$ (Ito *et al.*, 2001a). JNK activation can be detected within 1 h after treatment of cells with ER stress agents; therefore, it follows the kinetics of Ire1 stimulation (Urano *et al.*, 2000b). Ire1 overexpression induces JNK activation (Urano *et al.*, 2000b), which may be mediated through a TRAF2-dependent recruitment of c-JUN NH$_2$ inhibitory kinase (JIK) (Yoneda *et al.*, 2001). A TRAF-2 dominant-negative mutant, lacking its NH$_2$ activation domain, blocked Ire1's ability to activate JNK, and Ire1α$^{-/-}$ MEFs do not activate JNK in response to thapsigargin or tunicamycin, although the effect on subsequent apoptosis was not reported (Urano *et al.*, 2000b). However, interruption of JNK signaling prevented thapsigargin-induced apoptosis in human jurkat cells, supporting a role for this signaling pathway in ER stress-induced cell death (Srivastava *et al.*, 1999).

2.6 c-Abl

c-Abl is a tyrosine kinase implicated in several different forms of apoptosis depending on its subcellular localization. Nuclear c-Abl has been shown to exert an effect on DNA damage-induced apoptosis by activating the JNK pathway (Yuan *et al.*, 1996; 1999). However, cytoplasmic c-Abl participates in the release of Cyt.c from the mitochondria in response to reactive oxygen species (ROS) (Sun *et al.*, 2000). Recently, Ito *et al.* (2001b) convincingly demonstrated that in Rat-1 fibroblast approximately 20% of c-Abl resides at the surface of the ER and functions in the ER stress apoptosis network. After exposure to brefeldin A or A23187, ER localized c-Abl becomes activated and translocates to mitochondria, concomitant with the release of Cyt.c into the cytosol. c-Abl is an obligate requirement for ER stress-induced Cyt.c release because c-Abl$^{-/-}$ MEFs do not release Cyt.c and are resistant to apoptosis in response to A23187, brefeldin A, and tunicamycin (Ito *et al.*, 2001b). The mechanism by which c-Abl exerts its effect is unknown but may involve the inhibition of an antiapoptotic factor, such as Bcl-2, that prevents Cyt.c release. It also remains to be determined how c-Abl is activated in response to ER stress and whether it functions in conjunction (upstream or downstream) with Ire1 activated JNK signaling.

2.7 BAX, BAK, and BH3-only molecules

Generation of mice double deficient in the multidomain proapoptotic Bcl-2 members, BAX and BAK, has revealed what seems to be a striking dependence of ER stress-induced apoptosis on mitochondria (Wei *et al.*, 2001). Diverse death signals converge upon the activation of 'BH3-only' Bcl-2 family members, which translocate to mitochondria and induce homooligomerization of BAX and BAK, permeabilizing the OMM and allowing Cyt.c to pass into the cytosol (Korsmeyer *et al.*, 2000). Bcl-2 and Bcl-x$_L$ prevent the Cyt.c release either by directly binding and inactivating BH3-only molecules (Cheng *et al.*, 2001), or by blocking the actions of BAX and BAK (Ferri and Kroemer, 2001). Effector caspase activation and apoptosis are completely abrogated in *Bax*$^{-/-}$, *Bak*$^{-/-}$ MEFs in response to all intrinsic apoptotic stimuli tested, including tunicamycin, thapsigargin, and brefeldin A (Wei *et al.*, 2001), presumably because mitochondria fail to release Cyt.c. These findings are consistent with the ability of Bcl-2 and Bcl-xL to block Cyt.c release and cell death in response to ER stress agents (McCormick *et al.*, 1997; McCullough *et al.*, 2001). Caspase inhibitors do not affect the ability of ER stress to release Cyt.c, suggesting that caspase-12 may not be involved in this pathway (McCullough *et al.*, 2001). Assuming that BAX and BAK are not exerting a nonconventional function at the ER (see below), these findings predict that ER stress-induced apoptosis proceeds through the activation of one or more BH3-only molecules that, in turn, mediate cross-talk to the mitochondria. The BH3-only protein BAD is a good candidate for this role. In resting cells, BAD is kept in a phosphorylated state by several kinases, resulting in its sequestration to the cytosol by 14-3-3 (Zha *et al.*, 1996). Stimulation with thapsigargin or A23187, results in Ca^{2+}-dependent activation of calcineurin, which dephosphorylates BAD, resulting in its translocation to mitochondria (Wang *et al.*, 1999). A dominant negative mutant of calcineurin inhibited thapsigargin-induced dephosphorylation and translocation of BAD, and subsequent apoptosis. BAD may require the cooperation of other prodeath signals elicited by ER stress to release Cyt.c, since dephosphorylated BAD is found at the mitochondria within several hours of thapsigargin treatment, but cell death does not occur until a day after treatment. In addition, numerous BH3-only molecules are regulated at the transcriptional level (Vogelstein *et al.*, 2000); therefore, it remains possible that UPR transcriptional activates other BH3-only proteins that participate in Cyt.c release.

In summary, sustained ER stress can lead to an apoptotic reaction by unleashing a multitude of diverse signals (*Figure 1*). By activating protein kinases such as JNK and c-Abl, the ER may prepare the mitochondria to initiate cell death through the release of apoptotic IMS proteins. JNK and c-Abl may directly phosphorylate antiapoptotic targets at the surface of the mitochondria that normally inhibit BH3-only molecules, or, conversely, directly or indirectly activate proteins that encourage Cyt.c release. The increased cytosolic Ca^{2+} levels that follow ER stress result in dephosphorylation and translocation of BAD to the mitochondria. However, BAD-dependent activation of BAK and BAX, and Cyt.c release may not proceed until Bcl-2 is downregulated by CHOP. Meanwhile, following its activation at the ER membrane, mature caspase-12 is released into the cytosol, where it may directly cleave and activate downstream caspases in cooperation with the Cyt.c/Apaf-1/caspase-9 complex.

Alternatively, caspase-12 may cleave other regulatory molecules that work in conjunction with mitochondrial signals or that are relevant to the progression of apoptosis in some other way. Future identification of caspase-12 substrates should clarify this issue. It will also be important to understand how the ER coordinates prosurvival signals with prodeath signals, and how it is decided that the efforts to repair the secretory pathway are futile and that the death program should be triggered. One possibility is that UPR simultaneously primes the cell to undergo apoptosis via upregulation of CHOP, and activation of c-Abl, JNK, and BAD, while it upregulates chaperones to relieve the stress. If the stress is not adequately dealt with within a certain period of time, the ER might trigger a second signal that activates caspase-12, bringing on full-scale activation of downstream caspases and execution of apoptosis.

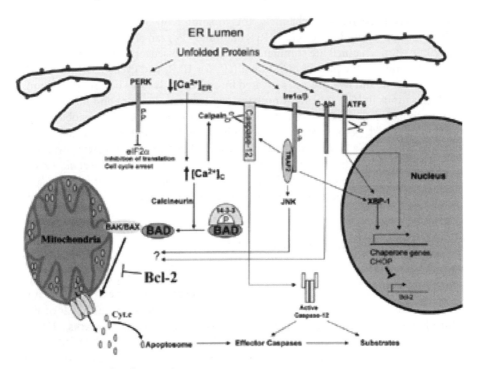

Figure 1. ER stress-induced apoptosis.
The accumulation of malfolded proteins in the ER triggers the activation of PERK, Ire1, and ATF6, which mount the unfolded protein response. PERK and Ire1 undergo autophosphorylation (P), resulting in the activation of their kinase domains. PERK shuts down translation by phosphorylating eIF2α, while the endonuclease domain of Ire1 cleaves *XBP-1* mRNA, generating a new XBP-1 protein competent in activating the promoters of chaperone genes. Cleavage of the cytosolic tail of ATF6 at the ER frees the transcription factor domain, which enters the nucleus and activates transcription of *XBP-1* and chaperone genes. Ire1 signaling also induces the recruitment and activation of JNK via TRAF2. Activated JNK and c-Abl may translocate to the mitochondria and phosphorylate unidentified regulators of Cyt.c release. Release of Ca^{2+} from the ER increases the $[Ca^{2+}]_c$, causing calcineurin-dependent dephosphorylation of BAD, which is released from 14-3-3 and translocates to mitochondria, inducing the homooligomerization of BAX and BAK. BAX/BAK pores in the OMM may induce the release of Cyt.c and other apoptotic IMS proteins into the cytosol, resulting in the activation of effector caspases. CHOP may assist this process by downregulation expression of the Bcl-2, which prevents the activation of BAX/BAK and Cyt.c release. Calpain is also activated by increased $[Ca^{2+}]_c$ and cleaves caspase-12 at the ER membrane, releasing a p20/p10 fragment into the cytosol, which may undergo self-processing, generating the p20/p10 heterotetrameric active enzyme. Active caspase-12 may cleave and activate downstream effector caspases or other unidentified substrates.

3. Regulation of mitochondrial apoptosis by ER Ca^{2+} signals

3.1 ER Ca^{2+} release

The ER, or its equivalent in muscle cells, the sacroplasmic reticulum (SR), represents the cell's largest calcium store. SERCA pumps located in the ER membrane maintain the $[Ca^{2+}]_{ER}$ up to three orders of magnitude higher than the $[Ca^{2+}]_c$ (Pozzan *et al.*, 1994) . The release of Ca^{2+} from the ER is primarily achieved by two well-characterized types of channels, the inositol 1, 4, 5-triphosphate receptor (IP$_3$R) and the ryanodine receptor (RyR) families (Pozzan et al., 1994; Berridge et al., 2000). Both channels are stimulated by second messengers generated by receptor stimulation at the plasma membrane (IP$_3$ for IP$_3$R, and cyclic ADP ribose for RyR). The IP$_3$R and RyR are also activated by cytosolic Ca^{2+} in a bell-shape fashion, with elevations in [Ca^{2+}]$_c$ being stimulatory up to a certain point and inhibitory at higher concentrations (Pozzan et al., 1994; Berridge et al., 2000). This feedback regulation by [Ca^{2+}]$_c$ permits communication between ER receptors generating coordinated bursts of Ca^{2+} release that can be organized into propagating waves varying in speed, amplitude, frequency, and spatiotemporal pattern (Berridge et al., 1998). By modulating these factors, the ER can generate a wide range of different Ca^{2+}-dependent cellular responses, including proliferation, fertilization, differentiation, and apoptosis (Berridge et al., 2000). For an in-depth understanding of ER Ca^{2+} signaling pathways, we refer the reader to recent reviews (Clapham, 1995; Berridge et al., 1998, 2000; Sorrentino and Rizzuto, 2001).

Release of calcium from intracellular stores has been observed in many forms of apoptosis and controlled oscillations of Ca^{2+} by IP$_3$R-mediated spikes may directly regulate the cell-death machinery. Jurkat cells deficient in type 1 IP$_3$R do not increase [Ca^{2+}]$_c$ in response to dexamethasone, ionizing radiation, and TCR or Fas stimulation, and resist subsequent apoptosis (Jayaraman and Marks, 1997). Moreover, targeted disruption of all three IP$_3$R isoforms in the chick DT40 B-cell line blocked Ca^{2+} mobilization and apoptosis by B-cell receptor cross-linking, and the degree of resistance increased with the number of IP$_3$Rs knocked out, suggesting that all three isoforms can contribute to apoptotic signaling (Sugawara et al., 1997).

The way in which Ca^{2+} signals engage cell-death pathways may largely be decided by the spatiotemporal pattern and intensity of ER Ca^{2+} release. IP$_3$R/RyR-dependent global increases in [Ca^{2+}]$_c$ can influence apoptosis by several mechanisms. For example, Ca^{2+}-dependent activation of calpains has been implicated in activating caspases in several forms of apoptosis (Wang, 2000), including ER stress- (Nakagawa and Yuan, 2000), B-cell receptor- (Ruiz-Vela et al., 1999; 2001), and radiation-induced apoptosis (Waterhouse et al., 1998). In addition, calpain cleavage of BAX has been reported to increase its proapoptotic activity (Wood et al., 1998), and in some cases calpain may cooperate with caspases in the execution phase of apoptosis (Wood and Newcomb, 1999). As discussed above, Ca^{2+} signals can lead to activation of BAD, and Ca^{2+}-dependent activation of the transcription factor MEF2 leads to upregulation of Nur77/TR3 (Youn et al., 1999), which can bind to mitochondria and induce Cyt.c release (Li et al., 2000). In contrast, privileged transport of Ca^{2+} between the ER and mitochondria may sensitize mitochondria to the effects of proapoptotic Bcl-2 family members (Hajnoczky et al., 2000).

3.2 Ca^{2+} signaling between the ER and mitochondria

Mitochondria have the capacity to accumulate high Ca^{2+} loads, and Ca^{2+} shuttling between the ER and mitochondria regulates normal physiologic processes, including oxidative respiration (Rizzuto et al., 2000). Ca^{2+} can freely diffuse across the OMM, and transport across the impermeable inner mitochondrial membrane

(IMM) is facilitated by a mitochondrial membrane potential ($\Delta\Psi_m$)-driven, low-affinity uniporter that is activated by elevated $[Ca^{2+}]_c$ (see Bernardi, 1999; Crompton, 1999; Duchen, 2000 for reviews). Calcium efflux from the mitochondria occurs through a Na^+–Ca^{2+} exchanger and also by a Na^+-independent mechanism (probably a Na^+–Ca^{2+} exchanger). Simultaneous measurement of $[Ca^{2+}]_m$ and $[Ca^{2+}]_c$ has revealed that after activation of the IP_3R or RyR, cytosolic Ca^{2+} spikes are simultaneously accompanied by parallel spikes in $[Ca^{2+}]_m$, highlighting the close coupling of ER Ca^{2+} release with mitochondrial uptake (Rizzuto et al., 1998). However, IP_3R and RyR channels only raise global $[Ca^{2+}]_c$ from ~100 nm to ~1 µM, an increase in $[Ca^{2+}]_c$ insufficient to activate the mitochondrial Ca^{2+} uniporter (Hajnoczky et al., 2000b). This issue has been resolved by a model accounting for privileged mitochondrial uptake sites that sense high local $[Ca^{2+}]_c$ in the proximity of ER Ca^{2+} release sites. Close contacts between the ER and mitochondria are clearly visible by thin-section electron microscopy, and in living cells mitochondria exist as a highly interconnected tubular network that is intimately associated with the ER network. Using live visualization of ER and mitochondria in HeLa cells transfected with appropriately targeted GFP constructs, Rizzuto et al. (1998) estimated that 5–20% of mitochondria are in close apposition with the ER. Furthermore, IP_3Rs and RyRs are not homogeneously distributed throughout the ER but concentrated within subdomains highly active in Ca^{2+} release, which are often in close apposition and facing mitochondrial surfaces (Mignery et al., 1989; Satoh et al., 1990; Takei et al., 1992; Ramesh et al., 1998; Sharma et al., 2000). It has been estimated that in some cell types, coordinated Ca^{2+} release by clusters of IP_3Rs or RyRs may elevate local $[Ca^{2+}]_c$ at ER-mitochondrial junctions as high as 20–50 µM (> 20-fold higher than global $[Ca^{2+}]_c$ rises), facilitating rapid uptake by the mitochondrial uniporter (Hajnoczky et al., 2000b).

3.3 Regulation of mitochondrial permeability transition by ER Ca²⁺ spikes during apoptosis

The high $[Ca^{2+}]_m$ that follows IP_3R/RyR spikes has recently been linked to mitochondrial release of apoptogenic factors through the opening of the permeability transition pore (PTP). The PTP is a high-conductance, nonselective ion channel that spans the IMM and OMM at points where the two membranes are in contact (Crompton, 1999). The voltage-dependent anion channel (VDAC), located in the OMM, and the adenine nucleotide translocator (ANT), located in the IMM, are the core constituents of the PTP. Other components include matrix-localized cyclosporine D (the target of the PTP suppressor, cyclosporine A [CsA]), hexokinase II, creatine kinase, and the benozodiazepine receptor. BAX and BAK have been reported to bind and stimulate opening of the PTP, while antiapoptotic Bcl-2 and Bcl-xL prevent its opening (reviewed in Zamzami and Kroemer, 2001). In vitro, BAX can form pores with either VDAC or ANT in planar lipid bilayers, and Bcl-2 can inhibit the formation of these pores (Brenner et al., 2000; Shimizu et al., 2000). Opening of the PTP is observed during most forms of apoptosis and has been proposed to encourage the release of apoptotic IMS proteins either indirectly by inducing matrix swelling and rupture of the OMM, or directly by forming pores in the OMM with proapoptotic Bcl-2 family members. Consistent with either model, inhibitors of the PTP, such as CsA, have been shown to prevent the release of Cyt.c during apoptosis, at least in some contexts (Zamzami and Kroemer, 2001). One area of contention, however, is the extent to which sustained PTP opening reflects an initiating as compared to an amplification event in different forms of apoptosis-signaling pathways.

It has been known for many years that mitochondrial Ca^{2+} overload is a strong inducer of PTP opening in isolated mitochondria (Crompton, 1999). Elegant studies by Hajnoczky and coworkers have recently demonstrated that ER Ca^{2+} spikes can facilitate PTP activation during apoptosis *in vivo* (Szalai *et al.*, 1999). In intact or permeabilized cells, high mitochondrial Ca^{2+} loads generated by IP_3R stimulation or exogenous Ca^{2+} spikes lead to weak PTP activation, characterized by small, transient mitochondrial depolarizations that are reversible and that do not release apoptotic proteins from the IMS. 'PTP flickering' may be a normal process in mitochondrial physiology (Crompton, 1999; Duchen, 2000). However, if cells are briefly challenged with apoptotic agents that act on mitochondria, such as C2 ceramide or staurosporin, mitochondria become sensitized to IP_3R-induced Ca^{2+} spikes, resulting in complete mitochondrial depolarization and release of Cyt.c. Inhibition of the mitochondrial Ca^{2+} uniporter with RuRed, or addition of the PTP inhibitor CsA, completely prevented both loss of $\Delta\Psi_m$ and Cyt.c release, suggesting that these events are dependent on mitochondrial Ca^{2+} uptake followed by PTP opening. Intriguingly, if the cells were subsequently washed to remove the apoptotic stimulus and the IP_3-inducing agonist (ATP), $\Delta\Psi_m$ fully recovered, but the cells eventually underwent apoptosis. These results suggest that during apoptosis IP_3R-driven Ca^{2+} spikes may cooperate with other mitochondrial apoptotic signals to induce transient opening of the PTP, facilitating release of IMS apoptotic agents into the cytosol, followed by resealing of the PTP and regeneration of $\Delta\Psi_m$, allowing continued production of ATP for the execution phase of apoptosis (Szalai *et al.*, 1999).

How IP_3R/RyR Ca^{2+}-release channels are activated during apoptosis, and whether these receptors actually induce Ca^{2+} spikes that are received by mitochondria during physiologic forms of apoptosis are questions that remain to be answered. Nonetheless, several other independent studies suggest that Ca^{2+} signals between the ER and mitochondria are important in diverse cell-death pathways. For instance, apoptosis induced by staurosporin in neural cells (Kruman and Mattson, 1999), C6 ceramide in U937 cells (Quillet-Mary *et al.*, 1997), glucocorticoid stimulation of lymphocytes, activation-induced death of T-cell hybridomas, and TNF-induced death of U937cells (Zamzami *et al.*, 1995) are all blocked by inhibition of mitochondrial Ca^{2+} uptake. In addition, changes in ER Ca^{2+} homeostasis have been shown to affect the sensitivity of cells to apoptosis. Stable overexpression of calreticulin, a major Ca^{2+}-binding ER chaperone, increases the ER Ca^{2+} storage and enhances agonist-induced IP_3R Ca^{2+} release, resulting in increased Cyt.c release, caspase activation, annexin V staining, and tunnel reactivity in response to thapsigargin, staurosporin, and etoposide (Nakamura *et al.*, 2000). In contrast, calreticulin-null MEFs are defective in IP_3R-dependent Ca^{2+} efflux (Mesaeli *et al.*, 1999), and resist etoposide-, staurosporin-, and UVB-induced apoptosis, showing decreased Cyt.c release and caspase activation (Nakamura *et* al., 2000). Moreover, overexpression of Bcl-2 alters ER Ca^{2+} homeostasis, and Bcl-2 targeted exclusively to the ER protects cells against many forms of apoptosis (see below). Therefore, IP_3R/RyR-mediated spikes in $[Ca^{2+}]_m$ may participate in the mitochondrial phase of apoptosis in many systems. It is important to emphasize, however, that at present the data suggest that ER Ca^{2+} signals sensitize mitochondria to respond to other apoptotic effectors in certain pathways rather than functioning as an obligate initiation event.

The involvement of the PTP in the permeabilization of the OMM is very controversial. Many groups have observed, both in isolated mitochondria and in intact cells, Cyt.c release in the absence of gross mitochondrial swelling or dissipation of $\Delta\Psi_m$, suggesting a PTP-independent mechanism for Cyt.c release (Zamzami and Kroemer, 2001). BAX can homooligomerize in synthetic liposomes, allowing passage of Cyt.c, and it has been proposed that BH3-only molecules induce the oligomerization of BAX and BAK in the OMM, forming Cyt.c-release channels that act independently of PTP opening (Korsmeyer et al., 2000). Thus, it is possible that both PTP-dependent and PTP-independent modes of OMM permeabilization exist. However, recent work by Scorrano et al. (2002) supports an alternative model for Cyt.c release and OMM permeabilization that involves both opening of the PTP and formation of specific BAX and BAK channels. Normally, Cyt.c is sequestered in regions of the IMS specialized for oxidative respiration and characterized by tight cristae junctions, raising the question of how Cyt.c becomes available for passage across the OMM in the absence of mitochondrial swelling and cristae unfolding during apoptosis. Using isolated mitochondria from wild-type or BAX/BAK-deficient mice, the authors found that the BH3-only molecule, tBID, overcomes this problem by activating two distinct steps; a BAX/BAK-independent structural rearrangement of the IMM that mobilizes Cyt.c stores out of cristae junctions, followed by a BH3-domain-dependent oligomerization of BAK and BAX into predicted pores that may allow transport across the OMM. Interestingly, CsA strongly inhibited tBid-induced mitochondrial reorganization of cristae, implicating the PTP in this step. Given the dependence of PTP opening on $[Ca^{2+}]_m$, IP_3R/RyR Ca^{2+} spikes from the ER may cooperate in physiologic settings with BH3-only molecules to allow transient openings of the PTP, facilitating cristae remodeling and mobilization of Cyt.c in the absence of large-scale mitochondrial swelling. *In vivo*, suboptimal activation of ER Ca^{2+}-release channels, causing transient openings of the PTP, may lead to small decreases $\Delta\Psi_m$ that are very difficult to detect (Szalai et al., 1999). Perhaps this explains why many groups do not observe mitochondrial depolarization prior to Cyt.c release. Large and sustained openings of the PTP, causing a complete loss of $\Delta\Psi_m$ and mitochondrial swelling, may follow Cyt.c-dependent activation of caspases and may represent a late stage in apoptosis that ensures the demise of the cell by crippling its energy production.

3.4 Coordination of Cyt.c release by ER-derived Ca^{2+} waves

After a population of cells is challenged with an apoptotic agent, the time at which any given cell will undergo Cyt.c release is highly variable and asynchronous. However, once initiated within a given cell, Cyt.c release is rapid and complete, with all mitochondria releasing Cyt.c stores within minutes (Goldstein et al., 2000).

Therefore, the decision to execute Cyt.c release is highly coordinated between mitochondria. The work of Pacher and Hajnoczky (2001) demonstrated that ER-borne Ca^{2+} spikes may organize this cross-talk within the mitochondrial population by eliciting Ca^{2+} uptake in proximal mitochondria, which then communicate to other mitochondria by propagating a secondary Ca^{2+} wave. The authors found that permeabilized or intact cardiac myotubes exposed to caffeine, a RyR agonist, initiate ER-derived, low-amplitude, high-propagation-rate $[Ca^{2+}]_c$ waves that are coupled to increases in $[Ca^{2+}]_m$ following the same spatiotemporal pattern as the $[Ca^{2+}]_c$ wave. If the cells are briefly pretreated with apoptotic stimuli, such as C2 ceramide, a second high-amplitude, low-velocity $[Ca^{2+}]_c$ wave spreads throughout the cell, and is accompanied by an 'apoptotic wave' of mitochondrial depolarization, Cyt.c release, and caspase activation, all following the same spatiotemporal pattern as the delayed $[Ca^{2+}]_c$ wave. Generation of the delayed $[Ca^{2+}]_c$ wave, and the accompanying apoptotic wave, was inhibited by CsA, Bcl-x_L, and an inhibitor of mitochondrial Ca^{2+} uptake sites, but not by caspase inhibitors. In addition, both of these events could be initiated by exogenous $CaCl_2$ pulses after the ER Ca^{2+} store was depleted in permeabilized cells pretreated with ceramide. Taken together, these results suggest that the delayed Ca^{2+} wave originates at mitochondria through Ca^{2+}-induced opening of the PTP. Interestingly, the authors found that mitochondria located at the initiation site of the apoptotic wave accumulate Ca^{2+} very efficiently, perhaps because they are in close apposition with ER Ca^{2+}-release sites and are intrinsically specialized for Ca^{2+} uptake, or because some mitochondria receive a greater degree of apoptotic stress, which switches them into an excitable state and marks them as initiation sites for wave propagation (Pacher and Hajnoczky, 2001). One might imagine that when a cell receives an apoptotic stimulus it may signal to BH3-only molecules to prime mitochondria to release Cyt.c and other IMS proteins. However, the coordinated release of these molecules is not initiated until privileged mitochondria receive an ER Ca^{2+} spike that causes a transient opening of the PTP and release of Cyt.c and matrix Ca^{2+}, initiating a second Ca^{2+} wave that is received and regenerated by neighboring mitochondria. These mitochondria, in turn, propagate the wave, and as this cycle is repeated over and over, a traveling wave of apoptotic mitochondrial transition spreads through the cell, allowing a coordinated and complete release of Cyt.c and cellular progression into the execution phase of apoptosis. This model reinforces the prediction that mitochondrial apoptosis in intact cells proceeds through a 'two-hit' mechanism (Pinton *et al.*, 2001) that requires a factor to operate directly on mitochondria as well as a costimulatory signal, in this case from the ER, that coordinates the release of Cyt.c between mitochondria (*Figure 2*).

3.5 Bcl-2 family members regulate ER Ca^{2+} homeostasis

Although the function of Bcl-2 and Bcl-x_L is best characterized at the mitochondrion, both proteins locate to other intracellular membranes such as the ER and nuclear envelope. Numerous reports suggest that Bcl-2 can alter ER Ca^{2+} homeostasis; however, there are discrepancies in how it does so. Several groups have reported that Bcl-2 increases the ER Ca^{2+} content and/or prevents ER Ca^{2+} release during apoptosis (Lam *et al.*, 1994; Distelhorst *et al.*, 1996; He *et al.*, 1997; Ichimiya *et al.*, 1998). In contrast, using probes that specifically target the ER, two groups

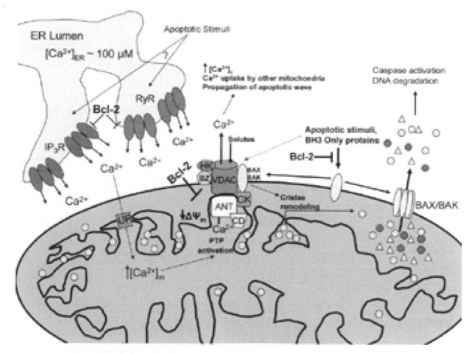

Figure 2. Regulation of the mitochondrial phase of apoptosis by ER Ca²⁺ spikes.
Mitochondria are sensitized to PTP opening by proapoptotic stimuli, possibly mediated by BH3-only proteins. Apoptotic stimuli may also activate clusters of IP₃R or RyR channels at the ER, generating Ca²⁺ spikes. The elevated $[Ca^{2+}]_c$ at ER–mitochondria junctions activates the IMM uniporter (UP), leading to accumulation of Ca²⁺ in the mitochondrial matrix. Bcl-2 prevents ER Ca²⁺ release either directly or by lowering the steady-state $[Ca^{2+}]_{ER}$. The high mitochondrial Ca²⁺ load activates PTP opening, facilitating BH3-only protein-dependent cristae remodeling and release of Cyt.c into unstructured regions of the IMS. Cyt.c (*circles*) and other IMS proteins, including SMAC, endonuclease G, and AIF, may traverse the OMM through BAX/BAK pores and activate cytosolic caspases and degradation of nuclear DNA. Alternatively, IMS proteins may be released through BAX/VDAC channels, or by mitochondrial swelling and OMM rupture caused by solute uptake during PTP opening. Opening of the PTP also releases mitochondrial Ca²⁺, which is taken up by neighboring mitochondria, propagating an apoptotic wave of mitochondrial transition throughout the cell.

found that overexpression of Bcl-2 reduces the steady-state level of $[Ca^{2+}]_{ER}$ by increasing the permeability of the ER membrane to Ca²⁺ in HeLa, mouse A20, and HEK-293 cells (Foyouzi-Youssefi *et al.*, 2000; Pinton *et al.*, 2000). Consistent with this latter finding, Pinton *et al.* (2001) found that manipulations that lead to decreased $[Ca^{2+}]_{ER}$ protect cells against apoptosis, whereas manipulations that increase $[Ca^{2+}]_{ER}$ sensitize cells to apoptosis. Thus, the primary role for Bcl-2 at the ER may be to lower the ER calcium reservoir, which could dismantle IP₃R/RyR-mediated signals during apoptosis. The studies described above were carried out with wild-type Bcl-2, which locates to nuclear, mitochondrial, and ER membranes; therefore, it cannot be ruled out that the observed effects on ER Ca²⁺ stores are caused by its actions at other cellular locations. Nonetheless, Bcl-2 can exert an antiapoptotic action at the ER, since Bcl-2 that is targeted exclusively to the ER by replacing its membrane insertion sequence with that of cytochrome $b5$ (Bcl-2cb5) can inhibit apoptosis induced by ER stress agents, cermide, Myc, ionizing radiation, or BAX overexpression, but not by etoposide or death receptor signals (Hacki et al., 2000; Annis et al., 2001; Rudner et al., 2001; Wang et al., 2001). Annis et al. (2002) observed that Bcl-2cb5 confers protection against apoptotic stimuli (Myc, C2 ceramide) that cause an obvious mitochondrial depolarization prior to the release of Cyt.c, but not against stimuli that induce Cyt.c in the absence of large mitochondrial depolarizations. Therefore, Ca^{2+} exchange between the two organelles may be required for Cyt.c release in some, but not all, pathways. Of note, Lee et al. (1999) could not detect ER Ca^{2+} release during Myc-induced apoptosis, but Bcl-2cb5 still blocked Cyt.c release and apoptosis, suggesting that in some cases ER–mitochondria cross-talk might be mediated by other events, such as lipid transfer or release of reactive oxygen species. It has also been reported that Bcl-2 can increase the capacity of mitochondria to store Ca^{2+} (Murphy et al., 1996; Ichimiya et al., 1998; Zhu et al., 1999), presumably by preventing the opening of the PTP, which releases matrix Ca^{2+}. Thus, by localizing to both ER and mitochondria, Bcl-2 might prevent apoptotic cross-talk between the two compartments by lowering the amount of free $[Ca^{2+}]_{ER}$ for IP$_3$R/RyR release and by increasing the tolerance of mitochondria to high Ca^{2+} loads. It cannot be excluded, however, that Bcl-2 also functions by sequestering proapoptotic members (Cheng et al., 2001), in which case the membrane location of Bcl-2 is irrelevant.

Proapoptotic, multidomain Bcl-2 family members may also exert their effects at the ER in some cases. When overexpressed in human PC-3 cells, BAX and BAK localize to both the mitochondria and ER, and induce caspase-independent emptying of ER Ca^{2+} pools concomitant with an increase in $[Ca^{2+}]_m$ (Nutt et al., 2002; Pan et al., 2001). Coexpression of Bcl-2/Bcl-x$_L$ inhibits this Ca^{2+} mobilization, and an inhibitor of mitochondrial Ca^{2+} uptake blocked BAX/BAK-induced increase in $[Ca^{2+}]_m$, Cyt.c release, and apoptosis (Nutt et al., 2002; Pan et al., 2001). Consistent with a role for these proteins at the ER, BAK and BAX interact with the cytosolic tail of the ER chaperone calnexin in yeast, and BAK-induced lethality in S. pombe is dependent upon this interaction (Torgler et al., 1997). Moreover, Bcl-2cb5 can inhibit BAX-induced apoptosis (Wang et al., 2001). BAX and BAK may have dual roles at the ER and mitochondria and help facilitate cross-talk between the two organelles. This could be achieved by stimulating IP$_3$R/RyR-release channels, or through formation of BAX/BAK pores in the ER membrane that facilitate Ca^{2+} transfer to the mitochondria. However, the ability of endogenous BAX and BAK to modify Ca^{2+} signals during physiologic apoptotic pathways needs to be confirmed; at present, it cannot be ruled out that the observed effects of BAX/BAK on ER Ca^{2+} homeostasis are either nonspecific or secondary to their effects on mitochondria.

4. Other ER regulators of apoptosis

4.1 BAP31

BAP31 is a ubiquitously expressed polytopic integral membrane protein of the ER that has been implicated as both a chaperone for protein export and a regulator of apoptosis. BAP31 was originally identified as an IgD interacting protein in B cells (Kim *et al.*, 1994), and was later proposed to influence cellubrevin and MHC I export from the ER (Annaert *et al.*, 1997; Spiliotis *et al.*, 2000), and control ER retention and degradation of the cystic fibrosis transmembrane conductance regulator (Lambert *et al.*, 2001). Our laboratory identified BAP31 as a Bcl-2 and Bcl-x_L interacting protein, a finding that suggested it could also function in apoptosis (Ng *et al.*, 1997). Further studies revealed that the C-terminal cytosolic tail of BAP31, which encodes a coiled coil domain with sequence similarity to the death effector domains (DEDs) of procaspase-8 and -10, could weakly associate with procaspase-8 in cotransfected cells (Ng and Shore, 1998). We have recently observed that in a physiologic setting BAP31 can recruit a novel procaspase-8 isoform, procaspase-8L (Breckenridge *et al.*, 2002). Procaspase-8L is characterized by the Nex domain, a 59-residue domain that extends procaspase-8/a at the N-terminus and facilitates a specific interaction with BAP31. Procaspase-8L is localized at the cytosolic face of the ER membrane in resting cells, and after apoptotic signaling by the E1A oncogene, endogenous procaspase-8L, but not procaspase-8, is recruited to BAP31. Concomitant with this recruitment, procaspase-8L undergoes processing by a mechanism inhibited by Bcl-2 but independent of FADD, an adapter molecule required for recruitment and activation of procaspase-8 at death receptors. Cells deficient in Bap31 and its cellular homolog, Bap29, do not process procaspase-8L in response to E1A and show reduced caspase-8 and -3 activity, and apoptosis. Oncogene-induced apoptosis is also known to proceed though the mitochondrial pathway (Soengas *et al.*, 1999); therefore, BAP31-induced caspase activation at the ER may cooperate with the apoptosome to bring about full activation of downstream executioner caspases. Activation of Procaspase-8L by BAP31 at the ER membrane may also place the caspase in a strategic location to cleave BAP31 itself, an event that can mediate cross-talk between the ER and mitochondria (below).

In addition to being a regulator of procaspase-8L, BAP31 is also a caspase-8 substrate (Ng *et al.*, 1997). The cytosolic tail of human BAP31 encodes two identical caspase recognition sequences that are rapidly cleaved by caspase-8 after activation of the Fas death receptor (Nguyen, 2000; our unpublished data). Fas initiates apoptosis by directly recruiting and activating procaspase-8 at the plasma membrane. Caspase-8, in turn, cleaves the BH3-only molecule, BID, generating tBID, which induces Cyt.c release from mitochondria and subsequent activation of downstream caspases through the formation of the apoptosome (Korsmeyer *et al.*, 2000). In some cell types, Fas signaling does not require the participation of the mitochondria, and caspase-8 can directly cleave and activate downstream caspases (Scaffidi *et al.*, 1998). We have found that stable expression of a caspase-resistant BAP31 (crBAP31) mutant strongly inhibits Fas-induced apoptosis in KB cells (Nguyen *et al.*, 2000), a cell line that requires amplification through the mitochondria. While crBAP31 has a very weak influence on Fas-induced caspase activation or cleavage of caspase substrates, it strongly inhibits Cyt.c release, mitochondrial depolarization, cytoskeletal reorganization, and membrane blebbing. The fact that crBAP31 inhibits these events, in the face of caspase activation and BID cleavage, suggests that in intact cells the ER exerts a restraint on apoptotic mitochondrial transition and cytoplasmic restructuring that is overcome by BAP31 cleavage in the

Fas pathway. Interestingly, caspase-cleaved BAP31 (p20BAP31) autonomously induces apoptosis by a mitochondria-dependent pathway (Breckenridge and Shore, unpublished), suggesting that caspase cleavage of BAP31 converts it from an inhibitor to an activator of mitochondrial dysfunction. This phenomenon has been observed with other antiapoptotic proteins such as Bcl-2 and Bcl-x_L (Cheng et al., 1997; Clem et al., 1998).

4.2 BIK

Our group has recently discovered that the BH3-only protein, BIK, is transcriptionally upregulated by p53 (Mathai et al., 2002). In healthy cells, BIK is undectable, but after oncogenic or genotoxic stress, BIK accumulates almost exclusively at the ER membrane. Adenoviral delivery of BIK potently triggers Cyt.c release and apoptosis independently of p53, but, unlike other BH3-only molecules, BIK carries out this function from the ER. Our recent studies suggest that BIK may activate mitochondrial release of Cyt.c and other death pathways by activating factors in both ER and cytosol (Germain et al., submitted). Thus, a post-mitochondrial supernatant (primarily containing cytosol and ER microsomes), but not an S-100 fraction, isolated from cells expressing BIK in the presence of caspase inhibitors, can release Cyt.c from mitochondria isolated from cells lacking BIK expression. In vivo, BIK induces Cyt.c release independently of caspases. The ER and cytosolic factors contributing to BIK-induced Cyt.c release are currently under investigation. Given the current model on the mechanism of action of BH3-only molecules at the mitochondria, it is surprising to find that BIK functions at the ER. As discussed above, this could reflect functions of BAX/BAK at the ER, or provide a mechanism to neutralize ER Bcl-2/Bcl-x_L, or represent hitherto unknown actions of certain BH3-only molecules at the ER required for communicating to the mitochondria. In this context, it is noteworthy that p53 upregulates numerous BH3-only proteins after genotoxic stress (Vogelstein et al., 2000); therefore, the actions of BIK on the ER, and of PUMA or NOXA at the mitochondria may be required in concert to trigger efficient Cyt.c release and caspase activation.

4.3 The reticulon family

The reticulon family members NSP-C and RTN-x_s were identified as Bcl-x_L- interacting proteins in a yeast two-hybrid screen (Tagami et al., 2000). Both proteins contain two transmembrane domains and localize to the ER. RTN-x_s can interact with both Bcl-x_L and Bcl-2 in cotransfected cells, and interactions between the endogenous proteins are weakly detected. NSP-C, however, interacts with Bcl-x_L, but not Bcl-2. Overexpression of RTN-x_s increases the amount of Bcl-2 and Bcl-x_L at the ER and inhibits the ability of both proteins to protect against tunicamycin- and staurosporin-induced apoptosis (Tagami et al., 2000). It remains to be determined whether reticulon family members inhibit the ability of Bcl-2/Bcl-x_L to alter ER Ca^{2+} homeostasis or some other function of these proteins at the ER.

4.4 Presenilins

Mutations in presenilin 1 (PS1) and presenilin 2 (PS2) have been linked to inherited forms of early onset Alzheimer's disease (AD). AD is caused by degeneration of cortical and limbic neurons of the brain, and is characterized by neurofibrillary tangles and extracellular aggregates or plaques of amyloid β peptide (Aβ), a processed fragment of the amyloid precursor protein (APP) (Hardy, 1997). Aβ induces synaptic dysfunction and neuronal apoptosis in experimental models by generating oxidative stress and disruption of ion homeostasis through lipid peroxidation (Mattson, 2000). The molecular mechanism by which presenilin mutations lead to AD seems to involve both increased production of the toxic Aβ peptide and sensitization of neurons to apoptosis. For example, PS1 has been shown to regulate the γ-secretase activity that is required to generate Aβ, and dominant mutations in PS1 and PS2, cause altered processing of APP and generate Aβ (Mattson *et al.*, 2001). In addition, neurons or transgenic mice expressing presenilin mutants are more sensitive to excitotoxicity and neuronal apoptosis induced by trophic factor withdrawal or exposure to Aβ, itself (Mattson *et al.*, 2001).

PS1 and PS2 are widely expressed and highly homologous integral membrane proteins of the ER containing eight predicted transmembrane domains with the N-terminus, C-terminus, and a hydrophilic loop between transmembrane domains 6 and 7 facing the cytoplasm. Presenilins have been reported to bind a number of ER Ca^{2+} regulatory proteins, including the RyR and its associated protein, sorcin, and the Ca^{2+}-binding proteins calsenilin and calmyin (Mattson *et al.*, 2001). Several lines of evidence suggest that the phenotype conferred by presenilin mutants may be the result of ER Ca^{2+} imbalances. Expression of presenilin mutants increases ER Ca^{2+} stores, resulting in greater elevations in $[Ca^{2+}]_c$ after IP_3R/RyR stimulation (Guo *et al.*, 1996; Wolozin *et al.*, 1996). Overexpression of Ca^{2+}-binding proteins, chelating cytosolic Ca^{2+}, or pharmacologically inhibiting IP_3R- or RyR Ca^{2+}-release channels inhibits the toxic effects of PS1 mutations (Mattson *et al.*, 2001). The increased release of ER Ca^{2+} caused by presenilin mutants may promote capacitative Ca^{2+} entry through plasma membrane voltage-dependent Ca^{2+} channels, leading to global perturbations in Ca^{2+} homeostasis. Importantly, sustained elevations in intracellular Ca^{2+} are strongly implicated in the pathogenesis of AD (Mattson *et al.*, 2001). In addition, several reports indicate that APP processing is regulated by intracellular Ca^{2+} levels, suggesting that presenilin mutants indirectly promote Aβ production by their effects on Ca^{2+} (Querfurth and Selkoe, 1994; Li *et al.*, 1995; Mattson *et al.*, 2001; Sennvik *et al.*, 2001). Of note, PS1 and PS2 are both cleaved by caspase-3 within the large cytosolic loop, an effect that might contribute to the apoptotic process through Ca^{2+} mobilization or by some other means (Wellington and Hayden, 2000).

Cells expressing PS1 mutants found in AD patients are defective in generating a UPR, apparently because they fail to activate Ire1, PERK, and ATF-6 (Imaizumi *et al.*, 2001). These observations could relate to the ability of mutant PS1 to perturb ER Ca^{2+} levels, and thus alter the protein-folding environment of the lumen, or to a direct role of wild-type PS1 in the UPR. An inability of neurons expressing AD presenilin mutants to mount a UPR would be expected to increase the vulnerability of these cells to undergo apoptosis in response to ER stress or other stress signals. Along this line, it is interesting that ER stress has been implicated in Aβ neurotoxicity, since cortical neurons from caspase-12-deficient mice are resistant to Aβ-induced cell death (Nakagawa *et al.*, 2000). Furthermore, PS1 and PS2 can bind m-calpain and inhibit its activation, and presenilin mutants lose this phenotype (Maruyama *et al.*, 2000), suggesting that, in some forms of AD, Aβ plaques could lead to increased calpain-dependent activation of caspase-12 and apoptosis. Therefore, presenilin mutations that lead to AD may not only increase the production of Aβ, but also sensitize neurons to the toxic effects of this peptide by deregulating Ca^{2+} homeostasis, preventing UPR survival signals, and by increasing calpain-mediated activation of caspase-12.

5. Concluding remarks

Intensive research into the role of mitochondria in apoptosis has revealed mitochondrial dysfunction as an essential component of almost all apoptotic pathways. Bcl-2 proteins are unquestionably the key guardians of mitochondria during apoptosis. Release of IMS proteins occurs when proapoptotic BH3-only and BAX/BAK proteins overcome the protective effects of antiapoptotic Bcl-2 proteins. Most models of Cyt.c release, however, stem from studies testing the effect of high concentrations of recombinant Bcl-2 family members on isolated mitochondria, where the dynamic nature of the mitochondrial network and its contacts with the ER and cytoskeleton are lost, or the effect of overexpression of Bcl-2 family members in intact cells, which represents a state of activation that these proteins are normally unlikely to achieve. Moreover, experiments in cultured cells that are overstimulated with high concentrations of apoptotic agents are probably poor representations of the delicate balance between prolife and prodeath switches that exist within the tissues of an organism. In these settings, proapoptotic Bcl-2 proteins are unlikely to function in isolation since the activation of proapoptotic BH3-only proteins may be suboptimal, and their ability to transform the mitochondria into a Cyt.c release-competent state may be largely dependent on the alliance with other costimulatory signals, which are also subject to positive or negative regulation. In this context, independent Ca^{2+} signals from the ER (and perhaps other modulators originating from the ER) may represent a critical determinant of how mitochondria respond to proapoptotic stimuli in physiologic circumstances.

Is the ER always working just to influence mitochondrial transition during apoptosis or can it function as a central integrator, like the mitochondria, in some pathways? In the case of ER stress, where the death signal originates within the ER itself, it is clear that the ER possesses its own set of apoptotic accessories to initiate several cell-death signals. Future studies should shed light on whether these ER signals, or other novel ER localized regulators, function in cell-death pathways induced by stress originating outside the ER. The massive surface area of ER membranes might provide a platform for the concentration and assembly of complexes that regulate activation of caspases, including caspase-12 and procaspase-8L. In addition, it is possible that Bcl-2 family members create pores in the ER membrane, releasing chaperones from the lumen that regulate caspase activation or other steps in the execution phase. Indeed, swelling and dilation of the ER is common during apoptosis (Weller *et al.*, 1995; Wu *et al.*, 1999; Chang *et al.*, 2000; Herrera *et al.*, 2000; Muriel *et al.*, 2000; Zeng and Xu, 2000), suggesting that ion homeostasis and volume control are deregulated. Unfortunately, elucidation of ER-controlled processes in apoptosis may be a slow process, since at present there is no assay (aside from calcium release) to measure changes in the organelle during cell death. In retrospect, this leads one to ask: what would our current knowledge of the involvement of mitochondria in apoptosis be had Cyt.c not been discovered as a cytosolic factor required for caspase activation *in vitro*?

6. Mitochondria in apoptosis induction

Bruno Antonsson

1. Introduction

Multicellular organisms are dependent on an ordered removal of unwanted and damaged cells. This is ensured by apoptosis or programmed cell death. Apoptosis is an energy (ATP)-requiring process that is conserved from low organisms such as the roundworm *C. elegans,* where apoptosis was initially identified, up to mammals, including man (Hengartner and Horvitz, 1994b). However, although apoptosis is essential for normal development and for maintenance of a balanced tissue home-ostasis in adult animals, it is also involved in a wide range of diseases. Increased apoptosis has been associated with stroke, myocardial infarction, reperfusion injury, arteriosclerosis, heart failure, infertility, diabetes, AIDS, hepatitis, renal fail-ure, and neurodegenerative diseases such as multiple sclerosis (MS), amyotropic lateral sclerosis (ALS), and Alzheimer's, Huntington's, and Parkinson's diseases (Kuhlmann *et al.*, 1999; Kockx and Herman, 2000; Yaoita *et al.*, 2000; Yuan and Yankner, 2000; Chandra *et al.*, 2001). However, impaired apoptosis is associated with various forms of cancer and autoimmune diseases (Krammer, 2000).

2. The mitochondria

The mitochondria are the power plants of the cells. Without the energy, in the form of ATP, produced by mitochondrial respiration, cells cannot survive under normal conditions. Mitochondria are complex, well-organized intracellular structures, comprising an outer membrane (MOM), an inner membrane (MIM), an inter-membrane space, and cristae structures formed by the folding of the inner mem-brane and the mitochondrial matrix on the inside of the inner membrane. The biochemical composition of the inner and outer membranes, in terms of lipid and protein composition, is different. The outer mitochondrial membrane is a fairly permeable membrane allowing free passage of molecules with a molecular mass of less than 1.5 kDa. The membrane contains two large channels or pores, the voltage-dependent anion channel (VDAC) and the protein import channel, the translocase of the outer membrane (TOM). These channels allow a controlled passage of large molecules, including proteins, across the membrane. In contrast, the inner membrane is a tightly sealed barrier. However, this membrane also contains proteins, such as adenine nucleotide translocator (ANT) and the protein import translocase of the inner membrane (TIM), which ensure that essential molecules, such as nucleotides and proteins, can enter the mitochondrial matrix.

Normal functioning mitochondria maintain an electrochemical gradient ($\Delta\Psi$m) across the inner membrane. The gradient is created through the efflux of H^+ ions from the matrix to the intermembrane space, resulting in a pH and voltage gradient. The main driving force for the efflux of H^+ ions is through the respiratory chain. However, several ion channels have been identified in the inner membrane, including the H^+/K^+ antiporter, the Cl^-/HCO_3^- antiporter, the Na^+/H^+ exchanger and the uncoupling proteins, which presumably also help to maintain the ion flux balance (Brierley et al., 1994; Garlid et al., 1998). An important protein complex involved in the regulation of the H^+ ion flux is the F_0F_1-ATPase/H^+ pump. This protein complex normally converts ADP into ATP, and the driving force is the flux of H^+ ions from the intermembrane space into the matrix. However, when the H^+ gradient is low, the protein complex can work in the reverse direction, pumping H^+ ions out from the matrix, whereby ATP is hydrolyzed into ADP.

Mitochondrial dysfunction is associated with both apoptotic and necrotic cell death. During necrosis, mitochondrial function is compromised through the loss of the mitochondrial membrane potential ($\Delta\Psi$m) as a result of the opening of large pores across the inner membrane, leading to mitochondrial permeability transition. This results in the incapacity of the organelle to synthesize ATP and provide the cells with energy. The mitochondrial permeability transition pore (PTP) is a multiprotein complex whose exact composition remains unclear. The complex is thought to contain VDAC in the outer membrane, ANT in the inner membrane, the matrix protein cyclophilin D, creatine kinase from the intermembrane space, and cytosolic hexokinase (Beutner et al., 1998; Crompton et al., 1999). However, VDAC, ANT, and cyclophilin D are often considered the core proteins of the pore. In addition, members of the Bcl-2 family, Bax, Bcl-2, and Bcl-X$_L$, have been found to co-purify with the PTP complex (Marzo et al., 1998b; Narita et al., 1998). In apoptosis, the mitochondrial dysfunction appears to be the result of a specific permeabilization of the outer mitochondrial membrane to large molecules, including cytochrome c. The loss of cytochrome c from the outer side of the inner membrane results in a dysfunction of mitochondrial respiration, effectively blocking the electron transport between complex III (cytochrome c reductase) and complex VI (cytochrome c oxidase). This not only leads to a disturbance of the membrane potential across the inner membrane but also increases the production of reactive oxygen species, resulting in increased lipid peroxidation (Hockenbery et al., 1993; Cai and Jones, 1999).

3. Apoptosis pathways

So far, two intracellular apoptosis-signaling pathways have been identified, the receptor pathway and the mitochondrial pathway (*Figure 1*).

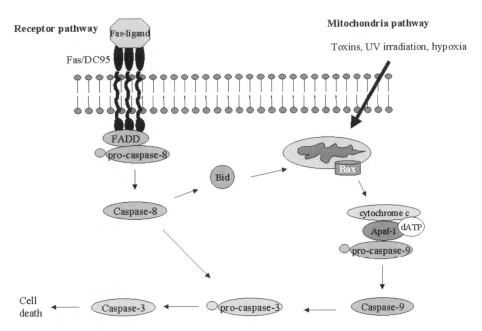

Figure 1. Apoptosis pathways.
Two intracellular apoptosis-signaling pathways have been identified, the receptor pathway and the mitochondrial pathway. (i) The receptor pathway is activated through ligand binding to plasma membrane receptors (Fas/CD95, TNFα, and TRAIL). Ligand binding induces oligomerization of the receptors and recruitment of adapter proteins and inactive procaspases, such as caspase-8. The procaspase is proteolytically activated at the receptor complex, and the active enzyme is released into the cytosol, where it activates downstream executioner caspases, such as caspase-3 or -7. (ii) The mitochondrial pathway is activated by a wide range of stimuli, including toxins, UV and gamma irradiation, hypoxia, staurosporine, and growth factor deprivation. All these stimuli result in the activation of pro-apoptotic multidomain Bcl-2 proteins, such as Bax or Bak. The proteins oligomerize and induce permeabilization of the outer mitochondrial membrane, through formation of large high-conductance channels in the membrane. Cytochrome *c* and other intermembrane proteins are released into the cytosol, where cytochrome *c* forms a complex with Apaf-1, procaspase-9, and dATP, referred to as the apoptosome. In the apoptosome, the inactive procaspase-9 is activated through proteolytic cleavage. The active enzyme subsequently activates downstream caspases, such as caspase-3. In some cell types, caspase-8 activated by the receptor pathway cleaves the 'BH3-domain-only' protein Bid. The C-terminal fragment of Bid (t'Bid) induces activation of the multidomain proteins and activates the mitochondrial pathway.

The receptor pathway (see Chapter 1) is activated through ligand binding to receptors in the plasma membrane (Schulze-Osthoff *et al.*, 1998; Schmitz *et al.*, 2000). Activation of the Fas/CD95, TNFα, and TRAIL receptors have been shown to trigger induction of apoptosis. Upon ligand binding, the receptors oligomerize, presumably forming trimers. This results in the recruitment of adapter molecules to the intracellular domains of the receptors. Inactive procaspases, such as caspase-8 or -10, form a complex with the adapter molecules, which is followed by the proteolytic activation of the caspases into their active forms. The active initiator caspases subsequently activate downstream executioner caspases, such as caspase-3 or -7. These, in turn, cleave cytosolic, structural, and nuclear proteins, triggering cellular, structural and morphologic changes ultimately leading to cell death.

The second pathway is the mitochondrial pathway. The Bcl-2 family of proteins controls this pathway. This group of proteins contains members with either pro- or antiapoptotic activity. The proapoptotic members can be further subdivided into multidomain and 'BH3-domain-only' proteins. The multidomain proteins, such as Bax and Bak, exert their function through permeabilization of the outer mitochondrial membrane, resulting in the release of proteins, including cytochrome *c*, apoptosis-inducing factor (AIF), adenylate kinase, endonuclease G, and Smac/Diablo from the intermembrane space of the mitochondria (Kluck *et al.*, 1997; Kohler *et al.*, 1999; Daugas *et al.*, 2000; Du *et al.*, 2000; Verhagen *et al.*, 2000; Li *et al.*, 2001). In the cytosol, cytochrome *c* forms a complex with Apaf1, dATP, and procaspase-9, referred to as the apoptosome (Li *et al.*, 1997; Zou *et al.*, 1999). The procaspase is processed into its active form, which subsequently activates downstream caspases such as caspase-3 and -7. Thus, at the lower level, the two apoptosis pathways converge into a common pathway. However, the upstream events regulating the activation of the proapoptotic Bcl-2 proteins are still largely unknown. Several stimuli, including, for example, hypoxia, production of reactive oxygen species, kinase inhibitors such as staurosporine, UV or gamma irradiation, growth factor deprivation, and several cytotoxic compounds, have all been shown to activate the mitochondrial pathway. However, how the individual stimuli, which are very different in nature, initiate and activate the signaling cascade remains unknown. In addition, the specific function of the individual multidomain members is elusive; it is possible that various family members are activated by specific stimuli.

The receptor and the mitochondrial pathways long appeared to be two independent pathways. However, it is now clear that cross-talk between the two pathways exists. This is mediated through the 'BH3-domain-only' protein, Bid. In some cell types, the so-called type 2 cells, only a small amount of caspase-8 is activated by the receptor pathway. The active caspase cleaves Bid into a C-terminal and an N-terminal fragment (Li, H. *et al.*, 1998; Schmitz *et al.*, 1999). The C-terminal fragment of Bid (t'Bid) activates multidomain proteins, such as Bax and Bak (Gross *et al.*, 1999b; Eskes *et al.*, 2000). In type 2 cells, the mitochondrial pathway functions as amplification for the receptor pathway. On the contrary, in type 1 cells, a large amount of caspase-8 is activated by the receptor pathway. The active caspase-8 then directly activates the downstream executioner caspases, without involvement of the mitochondrial pathway (Scaffidi *et al.*, 1998).

4. The Bcl-2 protein family

The founder protein of the Bcl-2 (*B-cell lymphoma gene 2*) family was identified as a proto-oncogen in follicular B-cell lymphoma. In the lymphoma cells, the Bcl-2 protein was overexpressed, rendering the cells resistant to apoptosis. The gene was found at the translocation site between chromosomes 18 and 14, resulting in it being under the expression control of the immunoglobulin heavy chain intron enhancer (Tsujimoto *et al.*, 1985). Subsequent studies identified Bcl-2 as a mammalian homolog of the apoptosis repressor ced-9 in *C. elegans* (Hengartner and Horvitz, 1994a). Bax (*Bcl-2 associated protein X*) was the first proapoptotic family member to be identified through coimmunoprecipitation with Bcl-2 (Oltvai *et al.*, 1993). The family now contains over 20 members (*Figure 2*). Although the overall amino-acid sequence homology is fairly low, the proteins contain up to four highly conserved domains, referred to as the *Bcl-2 homology* domains, or BH domains. The BH domains are essential for interactions between the proteins and for their activities (Yin *et al.*, 1994; Chittenden *et al.*, 1995; Hunter and Parslow, 1996; Hirotani *et al.*, 1999). In addition, some members have a hydrophobic C-terminal domain, thought to be involved in targeting the proteins to lipid membranes (*Figure 3*).

The family members have either anti- or proapoptotic activity. The antiapoptotic proteins, such as Bcl-2 and Bcl-X$_L$, contain all four conserved domains (BH1-4) as well as the hydrophobic C-terminal domain. Although both Bcl-2 and Bcl-X$_L$ contain the C-terminal hydrophobic domain, their intracellular localizations dif-

Anti-apoptotic	Pro-apoptotic	
	Multidomain	**"BH3 domain only"**
Bcl-2	Bax	Bid
Bcl-X$_L$	Bak	Bad
Bcl-w	Bok	Bim/Bod
A1	Bcl-X$_S$	Bik/Nbk
DIVA/BOO	Bcl-rambo	Hrk
NR-13		Blk
		Bmf
		Nix
		PUMA
		Noxa
		BNip3

Figure 2. The Bcl-2 protein family.
The Bcl-2 proteins can be divided into antiapoptotic and proapoptotic members. The proapoptotic proteins can be further subdivided into multidomain and 'BH3-domain-only' members.

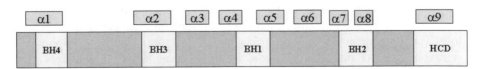

Figure 3. The protein domain structure of the Bcl-2 proteins.
The Bcl-2 proteins contain up to four highly conserved domains, referred to as Bcl-2 homology domains or BH domains (BH1–4). In addition, some of the proteins contain a hydrophobic C-terminal domain (HCD). The positions of the α-helices are indicated.

fer. Bcl-2 is exclusively found in intracellular membranes, including the outer mitochondrial membrane, the ER, and the nuclear envelope (Krajewski *et al.*, 1993). In contrast, in addition to membrane localization, Bcl-X$_L$ is also found as a soluble form in the cytosol (Hsu, Y.T. *et al.*, 1997). Translocation of the protein from the cytosol to the mitochondria was found during apoptosis. This indicates that the protein can exist in different conformations, favoring a soluble or a membrane-inserted protein form. In addition to Bcl-2 and Bcl-X$_L$, less well-characterized family members with antiapoptotic activity have been identified (*Figure 2*).

The proapoptotic subfamily can be further divided into two groups, the multidomain proteins, including Bax and Bak, and the 'BH3-domain-only' proteins, such as Bid and Bad. The multidomain proteins all contain the BH1-3 domains and the C-terminal hydrophobic domain, with the exception of Bcl-X$_S$, which, in addition to the C-terminal domain, contains only the BH3 and BH4 domains. Although Bax and Bak contain the same conserved domains, their subcellular localization is totally different. Whereas Bak is inserted into the outer mitochondrial membrane in normal cells, Bax is found predominantly in the cytosol as a soluble monomeric protein or, to a low extent, is loosely associated with the mitochondria (Hsu and Youle, 1998; Eskes *et al.*, 2000). Upon apoptotic stimulation, Bax undergoes conformational changes accompanied by a translocation of the protein form the cytosol to the mitochondria, where it is inserted into the outer mitochondrial membrane (Wolter *et al.*, 1997; Goping *et al.*, 1998; Antonsson *et al.*, 2001).

The 'BH3-domain-only' proteins are a more heterogeneous group. However, the common feature is that they possess only the BH3 domain. In addition, some of the members contain a C-terminal hydrophobic domain, whereas others, such as Bid and Bad, do not. The 'BH3-domain-only' proteins, at least partly, execute their proapoptotic function through activation of the multidomain proteins. The best-studied example is Bid, which will be discussed in more detail below. Bad promotes activation of the multidomain proteins in an indirect way, by sequestering the antiapoptotic proteins Bcl-2 or Bcl-X$_L$; this prevents them from binding to the multidomain proteins and inhibits their activation (Zha *et al.*, 1997).

5. Structure and channel-forming activity

The three-dimensional structures are available for four Bcl-2 family proteins, two antiapoptotic members (Bcl-X$_L$ and Bcl-2) and two proapoptotic proteins, one from the multidomain subgroup (Bax) and one from the 'BH3-domain-only' group (Bid) (Muchmore *et al.*, 1996; Chou *et al.*, 1999; McDonnell *et al.*, 1999; Suzuki *et al.*, 2000; Petros *et al.*, 2001). Surprisingly, the overall structures of all four proteins are very similar, despite their different activity in the regulation of apoptosis. The proteins show a fairly compact globular structure with two hydrophobic central helixes (α5 and α6) surrounded by amphipathic helixes. The helixes are connected by flexible loop structures, which show a larger variation among the different proteins. In Bid, the loop between helix 2 and helix 3 contains the caspase-cleavage site. In Bcl-X$_L$ and Bcl-2, the loop N-terminal of the BH3 domain contains several potential phosphorylation sites. The conserved BH domains, which are involved in protein interactions and are essential for the activity of the proteins, are all, as expected, located on the surface of the proteins. However, in Bax, the only protein structure to include the C-terminal hydrophobic domain, the BH3 cleft, is covered by helix $\alpha 9$ formed by the hydrophobic domain. Changes in the conformation of the protein might then regulate its susceptibility for interactions mediated through the BH3 domain and thus its activity.

The overall structure of the proteins resembles the structure of the pore-forming domains of some bacterial toxins, diphtheria toxin, and the colicins (Parker and Pattus, 1993). This suggested that the proteins might possess channel-forming activity. The surprising finding was that both the pro- and the antiapoptotic proteins show channel-forming activity in artificial membranes. However, at closer examination, both the channel-forming conditions and the channel characteristics were found to differ between the pro- and antiapoptotic members.

The antiapoptotic proteins have channel activity only at low pH, below pH 5.5 (Antonsson *et al.*, 1997; Minn *et al.*, 1997; Schlesinger *et al.*, 1997). This has raised the question of whether these proteins function as pore-forming proteins under physiologic conditions, since this is an extreme pH value in physiologic terms. Although decreases in pH have been reported during apoptosis, these are more moderate than those that appear to be required for channel formation by the antiapoptotic proteins, at least under *in vitro* conditions. In contrast, the proapoptotic proteins have channel-forming activity at neutral pH, although the channel activity is enhanced at lower pH.

Monomeric Bax shows no channel-forming activity. However, after exposure of the protein to certain detergents, Bax gains channel-forming activity. Channel-forming properties were shown to be associated with oligomerization of the protein. Recombinant Bax or cytosolic Bax exposed to Triton X-100 or octyl glucoside forms oligomers with molecular masses of 80 and 160 kDa respectively (Antonsson *et al.*, 2000). These would correspond to Bax tetramers and octamers. Oligomerization appears to take place only when the detergent is present over the critical micellar concentration. This is supported by the NMR structure study where a low octyl glucoside concentration did not induce any significant changes in the structure. However, at a concentration of 0.6 %, Bax structure changed dramatically, indicating aggregation or oligomer formation (Suzuki *et al.*, 2000). A study by Saito *et al.* (2000) showed that Bax oligomers were able to release cytochrome *c* from liposomes. The structure was estimated to be Bax tetramers. Channels formed by oligomeric Bax have multiconductance levels, ranging from a few pS up to 3–4 nS. They are pH-sensitive, slightly cation-selective, and Ca^{2+}-insensitive (Antonsson *et al.*, 1997; Schlesinger *et al.*, 1997). In a recent study, we tried to simulate Bax activation as it appears in Fas-induced apoptosis in type 2 cells. In this pathway, caspase-8 cleaves Bid into a C-terminal (t^cBid) fragment and a N-terminal (t^nBid) fragment. However, after cleavage, the two fragments stay together in solution (cut Bid) (Kudla *et al.*, 2000). Monomeric Bax was incubated with cut Bid in the presence of liposomes (Roucou *et al.*, 2002). As expected, monomeric Bax alone was unable to trigger carboxyfluorescein (CF) release from the liposomes; however, after co-addition of cut Bid, Bax permeabilized the liposomes to CF. Surprisingly, the quaternary structure of Bax activated by cut Bid was found to be monomeric;

no oligomeric Bax was detected, although the protein showed channel-forming activity. Further studies revealed that, although oligomeric Bax also permeabilized the liposomes to cytochrome c, the cut Bid-induced Bax channels were impermeable to cytochrome c. Electrophysiologic comparison of the channel activities showed that the cut Bid-induced Bax channel-forming activity was different from that of oligomeric Bax. Cut Bid-induced Bax channels showed low conductance levels only and were highly cation selectivity. Although the physiologic relevance of these new Bax channels remains unclear, these results indicate that Bax and possibly other multidomain proteins might form channels with different properties. It also shows that cut Bid alone in the presence of lipid membranes is not sufficient to induce formation of Bax oligomers. When monomeric Bax is incubated with cut Bid in the presence of mitochondria, Bax oligomers are formed; thus, it appears that additional factors, presumably from the mitochondria, are required to trigger the formation of large-conductance Bax channels (Eskes *et al.*, 2000). Furthermore, one study indicated that Bax can destabilize lipid membranes through reducing the linear tension without forming ion channels (Basanez *et al.*, 1999). A membrane destabilizing activity was also detected with the C-terminal fragment of the 'BH3-domain-only' protein Bid (t^{c}Bid) (Kudla *et al.*, 2000).

6. Activation of the multidomain Bcl-2 proteins

Regulation of the multidomain protein activity appears to occur mainly on the post-translational level, although Bax levels have been reported to change during apoptosis (Krajewski *et al.*, 1995b; Ekegren *et al.*, 1999). The Bax promotor has been shown to be transcriptionally activated by the tumor suppressor protein p53 (Miyashita and Reed, 1995). Interestingly, a large number of human cancers have mutations in the p53 protein. Changes in Bax expression have been reported in some pathologic conditions. For example, in one-third of breast adenocarcinomas, Bax expression levels have been found to be decreased (Krajewski *et al.*, 1995a). For some tumors, a low Bax expression has been correlated with a poor prognosis for the patients (Ito *et al.*, 1999). However, in most, if not all, cells, the multidomain proteins are present in inactive forms.

In normal cells or tissues, Bax is predominantly localized in the cytosol as a monomer (Hsu and Youle, 1998; Antonsson *et al.*, 2000). Bax translocation from the cytosol to the mitochondria after apoptotic stimulation has been demonstrated in several systems. Deprivation of interleukin-3 from FL5.12 hematopoietic cells induced Bax translocation (Goping *et al.*, 1998; Gross *et al.*, 1998). Translocation of Bax has also been demonstrated after withdrawal of the cytokine interleukin-7 from T cells (Khaled *et al.*, 1999). In this study, an increase of intracellular pH was found to coincide with Bax translocation. However, other studies have shown cytosolic acidification during apoptosis. Using a pH-sensitive, green fluorescent protein, Matsuyama *et al.* (2000) showed that mitochondrial dependent apoptosis resulted in cytosolic acidification and mitochondrial alkalinization. The pH changes were not detected in death receptor-triggered apoptosis. Whether the change in intracellular pH is a causal or resultant event in mitochondrial dependent apoptosis remains unclear.

Activation and translocation of Bax have been shown to be accompanied by conformational changes in the protein, resulting in changes in its quaternary structure. When Bax was activated through incubation with Bid, the conformation of the N-terminal domain changed, making this domain reactive to antibodies that do not react with the unactivated protein (Desagher *et al.*, 1999). Changes in the C-terminal domain have also been shown to occur after Bax activation (Nechushtan *et al.*, 1999). Upon activation, the C-terminal α-helix is removed from the BH3 cleft, making the BH3 domain accessible to interactions leading to complex formation (Suzuki *et al.*, 2000). After activation, Bax is found inserted into the outer mitochondrial membrane as large oligomers (Antonsson *et al.*, 2001). Bcl-2 inhibits Bax activation and oligomerization (Antonsson *et al.*, 2001; Mikhailov *et al.*, 2001). A study by Mahajan *et al.* (1998) demonstrated direct interactions between fluorescent-labeled Bax and Bcl-2 at the mitochondria by fluorescence resonance energy transfer (FRET). However, Mikhailov *et al.* (2001) could not show any direct interactions between Bax and Bcl-2 through immunoprecipitation or cross-linking in Bcl-2-overexpressing cells where Bax oligomerization was prevented. Thus, the molecular mechanism for Bcl-2 prevention of Bax activation remains unclear and might depend on cell type or the apoptotic stimulation. Tumor necrosis factor (TNF)-alpha-mediated cell death was recently shown to be associated with the formation of Bax protein complexes of 500 kDa. Conformational changes in both the C- and N-terminal domains were associated with complex formation. Furthermore, the adenovirus E1B 19K protein, a Bcl-2 homolog, inhibited Bax oligomerization and blocked TNF-alpha-induced cell death (Sundararajan and White, 2001).

In contrast to Bax, Bak is found inserted into the outer mitochondrial membrane of normal cells (Griffiths *et al.*, 1999). Similar to Bax, activation of Bak has been shown to be associated with changes in its tertiary and quaternary structure. When Jurkat or CEM-C7A cells were exposed to the apoptosis inducers staurosporine, etoposide, or dexamethasone, changes in the tertiary structure of Bak were detected (Griffiths *et al.*, 1999). Antibodies to the N-terminal domain of Bak were reactive only after apoptotic stimulation, suggesting that changes had occurred in this domain of the protein. Furthermore, although Bcl-X_L coimmunoprecipitated with Bak before stimulation, no coimmunoprecipitation was detected after stimulation. This result suggests that the proapoptotic activity of Bak is neutralized by binding to Bcl-X_L in the mitochondrial membrane, and that after apoptotic stimulation conformational changes in Bak result in dissociation from Bcl-X_L, leaving Bak free to exert its proapoptotic activity (Griffiths *et al.*, 1999). Another study showed that t^{c}Bid can activate Bak. t^{c}Bid translocated to the mitochondria and induced conformational changes in Bak, resulting in Bak oligomerization in the mitochondrial membrane and release of cytochrome *c* from the mitochondria (Wei *et al.*, 2000). Combined, these results show that conformational changes appear to be a major mechanism for activation of proapoptotic members of the multidomain group. So far, only one protein, Bid, has been shown to activate directly the multidomain proteins. However, it would be surprising if additional proteins were not also able to activate the multidomain proteins through direct interactions.

7. 'BH3-domain-only' proteins as multidomain protein activators

The 'BH3-domain-only proteins' appear to function mainly through activation of the multidomain proteins. Bid was shown a few years ago to provide a shunt between the two apoptotic pathways, the death receptor and the mitochondrial pathway (Schmitz *et al.*, 1999). In some cell types (so-called type 2 cells), Fas-induced apoptosis results in a low activation of caspase-8 at the membrane receptor. This appears to be insufficient for activation of the direct caspase pathway. In these cells, caspase-8 cleaves Bid, generating a C-terminal fragment (t'Bid) that interacts with Bax or Bak, and results in activation of these proteins and the mitochondrial pathway. This mechanism functions as an amplification loop to enhance the apoptotic intracellular signaling cascade. t'Bid has been shown to translocate to the mitochondria, where it is found in apoptotic cells (Gross *et al.*, 1999b). Whether the protein colocalizes only with Bax or Bak or whether it has an independent activity at the mitochondrial membrane still remains unclear. Fibroblasts from Bax and Bak double-deficient mice were completely resistant to t'Bid-induced cell death. The mice also showed a pronounced increased resistance to hepatocyte damage after treatment with anti-Fas antibodies (Wei *et al.*, 2001). At least in these two systems, Bid acts through activation of the multidomain proteins. However, in artificial membrane systems, t'Bid has a membrane-destabilizing effect (Kudla *et al.*, 2000). Whether such an activity is also present under physiologic conditions remains to be elucidated. It is conceivable that t'Bid, through destabilization of the mitochondrial membrane, could facilitate insertion of the multidomain proteins and thereby promote channel formation.

Other 'BH3-domain-only' proteins have been shown to activate the multidomain proteins in an indirect way. For example, unphosphorylated Bad binds to Bcl-2 and Bcl-X_L, neutralizing their antiapoptotic activity (Zha *et al.*, 1997). Survival factors such as interleukin-3 or NGF have been shown to induce phosphorylation of Bad. Phosphorylated Bad is released from the complexes with the antiapoptotic proteins Bcl-2 and Bcl-X_L, allowing the proteins to interact with the multidomain proteins, and prevent their proapoptotic activity. Another 'BH3-domain-only' protein, Bik, has also been found to be phosphorylated; in this case, phosphorylation reduced its proapoptotic activity (Verma *et al.*, 2001). The 'BH3-domain-only' protein Bim is bound to the microtubulin-associated dynein complex through interactions with the LC8 dynein light chain in normal cells. After induction of apoptosis by various stimuli, including UV irradiation, staurosporine, γ-irradiation, or growth factor deprivation, Bim was released from the microtubuli complex (Puthalakath *et al.*, 1999). In the cytosol, Bim is thought to bind to anti-apoptotic proteins, preventing their function.

8. How do the multidomain Bcl-2 proteins trigger the release of proteins from the intermembrane space of the mitochondria?

The first report indicating that mitochondria are essential for apoptosis was by Newmeyer *et al.* (1994), who used a cell-free model system. It is now clear that the proapoptotic Bcl-2 proteins execute their function at the mitochondria, although considerable controversy persists over the molecular mechanisms. Bax is primarily localized to the cytosol in normal tissues, but, after apoptotic stimulation, the protein specifically translocates to the mitochondria (Hsu, Y.T. *et al.*, 1997; Gross *et al.*, 1998; Zhang, H. *et al.*, 1998). In contrast, Bak is localized to the mitochondrial membrane at all times (Griffiths *et al.*, 1999). Endogenous Bcl-X$_S$ was recently shown to be localized in the cytosol in PC12 cells; however, when the protein was overexpressed, it was found associated with the mitochondria, suggesting that this multidomain protein also functions at the mitochondria (Lindenboim *et al.*, 2000). Antiapoptotic proteins such as Bcl-2 are localized in several intracellular membranes, including the mitochondria, ER, and nuclear envelope (Krajewski *et al.*, 1993). This poses the question of whether the antiapoptotic proteins might have additional functions, perhaps even unrelated to apoptosis. At the mitochondria, the multidomain proapoptotic proteins trigger the release of proteins from the intermitochondrial space, including cytochrome *c*, apoptosis-inducing factor (AIF), adenylate kinase, endonuclease G, and Smac/Diablo (Kluck *et al.*, 1997; Kohler *et al.*, 1999; Daugas *et al.*, 2000; Du *et al.*, 2000; Verhagen *et al.*, 2000; Li *et al.*, 2001).

8.1 Release mechanisms

The controversial question is how Bax and other multidomain proteins induce the release of cytochrome *c* and other proteins from the mitochondrial intermembrane space.

Two main mechanisms for permeabilization of the outer mitochondrial membrane have been proposed: formation of specific channels or pores in the outer membrane, or opening of the permeability transition pore (PTP) that results in mitochondria matrix swelling and rupture of the outer membrane. The following models have been proposed (*Figure 4*): (i) Bax forms channels itself. (ii) Bax destabilizes the mitochondrial membrane, inducing 'lipic holes' (Basanez *et al.*, 1999). (iii, iv) Bax forms chimeric channels with VDAC or ANT. (v) Bax triggers opening of the mitochondrial permeability transition pore (PTP). In the large number of publications addressing these questions during the last few years, arguments for and against each of these models can be found. Most studies are based on experiments performed on artificial membranes, isolated mitochondrial membranes, and extracted mitochondrial proteins or recombinant proteins. In intact mitochondria, and even more so in the cell, it is difficult to establish the molecular mode of action and distinguish between the various models.

8.2 Bax channels

As described above, recombinant Bax is able to form channels in artificial membranes without any additional proteins (Antonsson *et al.*, 1997). These 'Bax-alone' channels are able to permeabilize lipid membranes to cytochrome *c* (Saito *et al.*, 2000). However, when we closely examined the quaternary structure of artificially oligomerized recombinant Bax and Bax oligomers extracted from mitochondria of apoptotic HeLa or HEK cells, the oligomers differed in size (Antonsson *et al.*, 2001). The oligomers of recombinant Bax showed molecular masses of 80 and 160 kDa, respectively, whereas Bax oligomers extracted from mitochondria of apop-

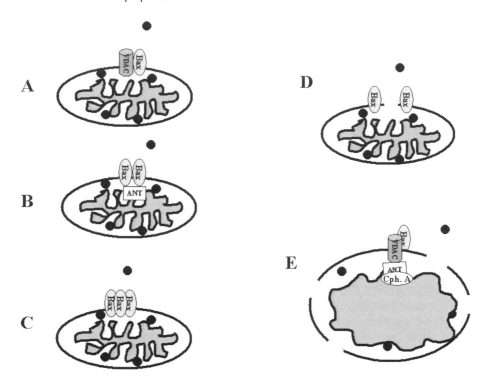

Figure 4. Models for permeabilization of the outer mitochondrial membrane during apoptosis. Several models have been proposed for how the proapoptotic multidomain proteins, such as Bax and Bak, induce the release of proteins, including cytochrome c, from the intermembrane space. (A) Bax forms a chimeric channel with VDAC. (B) Bax interacts with ANT in the inner membrane; either it forms a chimeric channel or the interactions induce or regulate a Bax channel. (C) A channel is formed by Bax oligomers. Whether these oligomers in the mitochondria are composed of Bax only, or contain other, not yet identified proteins, is unclear. However, those proteins are not VDAC or ANT. (D) Insertion of Bax destabilizes the membrane, inducing 'lipic holes' in the membrane. (E) The multidomain proteins trigger opening of the permeability transition pore, resulting in matrix swelling and rupture of the outer membrane. Models A–C propose formation of channels that could induce the release of specific proteins, whereas models D and E would permeabilize the membrane in an unspecific way, releasing proteins in an uncontrolled manner.

totic cells were found to be 96 and 250 kDa. This would suggest that either the number of monomeric Bax subunits is different or that Bax oligomers isolated from mitochondria contain additional proteins. Furthermore, cross-linking experiments suggested that the oligomers contain Bax reactive bands that do not correspond to multimers of Bax molecules, indicating the presence of other, still unidentified proteins. However, the Bax oligomers did not comigrate with either VDAC or ANT, showing that these proteins are not part of the stable Bax oligomers. In a study by Mikhailov *et al.* (2002), rat kidney cells were deprived of ATP to induce apoptosis and the cells subsequently treated with cross-linking reagents. In these apoptotic cells, Bax oligomers up to hexamers were identified on Western blots. All Bax-reactive bands corresponded to multimers of Bax molecules, suggesting that no other proteins had been cross-linked to the Bax oligomers under the conditions used in these experiments. However, a negative cross-linking result does not exclude interactions. Nevertheless, these data show that Bax forms large oligomers in the outer mitochondrial membrane in cells undergoing apoptosis. These oligomers are larger then those shown to be required to permeabilize liposomes to cytochrome c and could therefore be the cytochrome c-conducting channel in the outer mitochondrial membrane of apoptotic cells. Whether the Bax oligomers in the mitochondrial membrane are composed of Bax exclusively or contain additional proteins is still unclear. In a recent study using confocal and electron microscopy, Bax was found to form large clusters protruding from the mitochondrial outer membrane during apoptosis (Nechushtan *et al.*, 2001). The clusters were estimated to be composed of thousands of Bax molecules. In addition, Bak was found to redistribute from its dispersed membrane localization into the Bax clusters, effectively leaving the mitochondrial membrane. The cluster formation was prevented by coexpression of Bcl-X_L. The physiologic function of the clusters has not been determined.

In a recent study, a new high-conductance channel was identified in mitochondria from apoptotic cells (Pavlov *et al.*, 2001). The channel was localized in the outer mitochondrial membrane and its activity correlated with the onset of apoptosis. The mitochondrial apoptosis-induced channel (MAC) has a conductance of 2.5 nS and shows cation selectivity, and its pore size was estimated to be 4 nm. This size would allow the passage of cytochrome c and even larger proteins. The electrophysiologic properties of the channel resemble those of the high-conductance Bax channel but are clearly distinct from the main outer membrane channels, VDAC and TOM. These results provide the first proof of a specific apoptosis-related channel in the outer mitochondrial membrane.

Several studies have shown that the mitochondrial structures are not damaged during the release of proteins from the intermembrane space. When Bax is activated through NGF deprivation in SCG neurons, cytochrome c release is induced without mitochondria swelling. On the contrary, the mitochondria are smaller than in untreated cells. Furthermore, in the presence of caspase inhibitors, the cells are rescued, and after re-addition of growth factor, the mitochondria regain their normal cytochrome c content and size, indicating that no irreversible damage, such as rupture of the outer membrane, has been inflicted (Martinou *et al.*, 1999). Similar results were obtained by the addition of active recombinant Bax to isolated mouse-liver mitochondria. Finucane *et al.* (1999) showed that Bax induced cytochrome c release both when overexpressed in cells and when added to isolated mitochondria. No mitochondrial swelling was detected, demonstrating that opening of PTP was not involved. The release of cytochrome c was inhibited by Bcl-X_L and resulted in decreased cell death.

The permeabilizing activity of Bax is not inhibited by the PTP inhibitors cyclosporin A, EDTA, or Mg^{2+}. In fact, Mg^{2+} enhances the cytochrome c-releasing activity of Bax (Eskes *et al.*, 1998). However, some studies have suggested that Bax-induced cytochrome c release could be inhibited by cyclosporin A (Narita *et al.*, 1998). In these studies, Bax was active only in the presence of Ca^{2+}, a PTP opener. Furthermore, the quaternary structure of Bax was not defined. In isolated mitochondria, Ca^{2+} can induce cytochrome c release in a Bax-independent manner

(Eskes *et al.*, 1998). Thus, under the conditions used in this experiment, it might be difficult to distinguish the effect of Bax and of Ca^{2+}, which acts as a PTP opener.

In a study by von Ahsen *et al.* (2000), isolated mitochondria treated with Bax or Bid were depleted of cytochrome *c*; however, the mitochondria retained a fully intact protein-import machinery. Similar results were obtained when cells were treated with UV-irradiation or staurosporine. Mitochondrial protein import is dependent on $\Delta\Psi m$ and ATP. Thus, these results show that although the outer membrane has been permeabilized to allow the passage of cytochrome *c*, the inner membrane remains intact. It was subsequently shown by single-cell analysis that, in the presence of caspase inhibitors, the release of cytochrome *c* results in a drop in $\Delta\Psi m$, but, over the following 30–60 min, the potential recovered its original level (Waterhouse *et al.*, 2001). The results show that when the downstream caspase cascade is inhibited, the mitochondria remain functional long after cytochrome *c* has been released, indicating intact mitochondrial structures. In a study on mouse-liver mitochondria, Bid-induced cytochrome *c* release was shown to be Bak dependent. No cytochrome *c* release was detected in Bak-deficient mitochondria. Furthermore, Bid-induced cytochrome *c* release was not blocked by the PTP inhibitor cyclosporin A, whereas Ca^{2+} induced swelling, and subsequent cytochrome *c* release was blocked (Wei, M.C. *et al.*, 2000). Combined, these results strongly suggest that Bax, as well as Bak, can form ion channels in the outer mitochondrial membrane and trigger cytochrome *c* release independent of PTP or its components (*Figure 5*).

8.3 Bax-VDAC channels

The involvement of VDAC in Bax-induced apoptosis has been suggested by some studies. Shimizu *et al.* (1999) showed that Bax and Bak could induce cytochrome *c* release from liposomes in which VDAC had been incorporated, whereas neither of the proteins was active alone. They also showed that Bax was not able to induce cytochrome *c* release from mitochondria isolated from a VDAC1-deficient yeast strain, whereas cytochrome *c* was released from mitochondria isolated from wild-type yeast. In an additional study, Bax and VDAC reconstituted in liposomes were shown to form a new channel activity with a conductance 4–10 times larger then the individual proteins (Shimizu *et al.*, 2000a). Cytochrome *c* was able to pass through the chimeric channel, but it could not pass through channels formed by the individual proteins. Conversely, Sato *et al.* (2000) have shown that Bax oligomers are able to form cytochrome *c*-conducting channels in liposomes. The difference might be due to the quaternary structure of the Bax protein, since Bax monomers do not have a channel-forming activity. These results also contradict the results by Priault *et al.* (1999), who showed that Bax-induced cytochrome *c* release in yeast deficient of VDAC was as efficient as in wild-type yeast. The release was prevented by Bcl-X_L. They further showed that Bax did not induce permeabilization of the inner mitochondrial membrane. A study by Gross *et al.* (2000) also concluded that VDAC was not required for Bax killing activity in yeast. In a recent study, antibodies against VDAC were shown to prevent Bax-induced apoptosis but had no effect on Bid- or Bik-triggered apoptosis. It was further shown that binding of Bax and Bak to red blood cells was dependent on a plasma membrane VDAC protein (Shimizu *et al.*, 2001). Although these results suggest the involvement of

Apoptotic stress stimuli

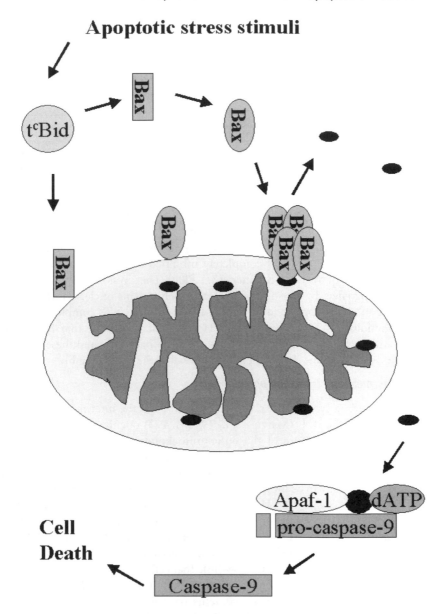

Cell Death

Figure 5. Model for the function of the multidomain proteins.
In normal cells, inactive monomeric Bax (rectangle) is present in the cytosol. When cells are exposed to apoptotic stress stimuli, upstream signaling cascades are activated, here exemplified by Bid. Bid is proteolytically processed into t^cBid, which interacts with Bax and induces conformational changes in Bax (oval). Whether this takes place in the cytosol or at the mitochondrial membrane is unclear. The conformational changes trigger Bax oligomerization and insertion into the outer mitochondrial membrane. Bax forms high-conductance channels through which cytochrome *c* (black oval) and other proteins from the intermembrane space are released into the cytosol. In the cytosol, cytochrome *c* forms a complex with Apaf-1, dATP, and procaspase-9. Upon complex formation, caspase-9 is activated. Caspase-9 subsequently activates executioner caspases, such as caspase-3, triggering a pathway ultimately leading to apoptotic cell death.

VDAC in multidomain protein activity, they do not show that VDAC is part of the channel-forming structure. VDAC might function as a receptor protein in the mitochondria membrane. This could explain the specific targeting of Bax to mitochondria during apoptosis.

8.4 Bax-ANT channels

Interactions between Bax and the adenine nucleotide translocator (ANT), a protein of the inner mitochondrial membrane and part of the PTP complex, have also been demonstrated. Bax coimmunoprecipitates with ANT, and the two proteins were shown to interact in a yeast two-hybrid system. Furthermore, Bax was shown to induce cell death when expressed in wild-type yeast, but not in an ANT-deficient yeast strain (Marzo et al., 1998a). Apart from its translocator activity, ANT can form channels in lipid membranes (Brustovetsky and Klingenberg, 1994; Ruck et al., 1998). Bax has been shown to enhance the ANT channel activity in artificial membranes. The channel activity was stabilized in the presence of the ANT inhibitor atractyloside and was completely inhibited by the natural physiologic ANT ligand, ATP, or by Bcl-2 (Brenner et al., 2000). The physiologic importance of the stimulating effect by Bax on the ANT channel remains unclear, since a non-physiologic inhibitor appears to be required, and the activity is completely inhibited at physiologic ATP concentrations. However, it is possible that Bax could interact with ANT in the inner membrane at the contact sites, and these interactions could modulate the activity of a Bax channel in the outer membrane.

8.5 The permeability transition pore

Proteins, including cytochrome c, can be released from the intermembrane space after opening of the PTP (Marzo et al., 1998b). PTP opening results in matrix swelling, leading to rupture of the outer mitochondrial membrane. The main arguments for the involvement of the PTP in Bax-induced apoptosis are early observations of a decrease in the transmembrane potential, and the finding that cytochrome c release can be inhibited by cyclosporin A, a PTP inhibitor (Halestrap et al., 1997; Narita et al., 1998). However, it now appears clear that, at least in most forms of apoptosis, the mitochondrial structures remain intact. These would suggest that activation of the PTP is not involved. However, the intracellular ATP concentration has been shown to influence whether a cell dies by apoptosis or necrosis (Nicotera et al., 1998). It is therefore possible that cells starting to die by the apoptotic pathway could switch to necrotic death, thus activating the PTP.

It appears that the multidomain proteins Bax and Bak form channels in the outer mitochondrial membrane independently of the PTP or its components. The multidomain proteins permeabilize the outer membrane, but the mitochondrial structural integrity remains intact. However, it cannot be excluded that under certain conditions the proteins might work in synergy with PTP or its components.

9. Activities of the mitochondrial proteins released

Cytochrome c is a component of the respiratory electron transfer chain, normally attached to the outer side of the inner mitochondrial membrane. During apoptosis involving the mitochondrial pathway, cytochrome c is released into the cytosol. The central role of cytochrome c in apoptosis has been demonstrated in cell-free systems *in vitro*, through injection of cytochrome c into cells, and in apoptotic cells (Liu, X. *et al.*, 1996; Bossy-Wetzel *et al.*, 1998; Zhivotovsky *et al.*, 1998). Using green fluorescent protein (GFP)-tagged cytochrome c, Goldstein *et al.* (2000) showed that cytochrome c release is a fast process. Once the release has been initiated, all mitochondria in the cell lose their cytochrome c within approximately 5 min. However, despite an apparently complete depletion of cytochrome c, the mitochondria retain or regain their membrane potential. In the cytosol, cytochrome c induces complex formation with Apaf-1, dATP, and procaspase-9. The procaspase is autoactivated, and the active enzyme subsequently cleaves and activates downstream caspases, such as caspase-3. The executioner caspases cleave various cellular substrates, ultimately leading to cell death (Li *et al.*, 1997).

Endonuclease G and AIF translocate to the nucleus after release from the mitochondria. The endonuclease induces DNA fragmentation in a caspase-independent manner (van Loo, 2001). Recent studies have identified a gene in *C. elegans*, cps-6, with strong similarities to the mammalian endonuclease G (Parrish *et al.*, 2001). AIF is an oxidoreductase and possesses no detectable DNAse activity; however, it has also been implicated in caspase-independent DNA fragmentation (Susin *et al.*, 1999; Miramar *et al.*, 2001). These results suggest the existence of a parallel apoptosis-signaling pathway independent of caspase activity. The endonuclease G pathway appears to be evolutionary conserved from the worm up to mammals.

Smac/DIABLO is a 25-kDa protein that, when released into the cytosol, binds to *i*nhibitor of *a*poptosis *p*roteins (IAPs). IAPs are a group of proteins that bind to activated caspases and in that way inhibit their activity. Smac/DIABLO competes for binding to the IAPs and thereby prevents the IAPs from inhibiting activated caspases (Srinivasula *et al.*, 2001). The first four N-terminal amino acids (Ala-Val-Pro-Ile-) of the mature Smac/DIABLO protein have been shown to interact with BIR domains in the IAP proteins (Chai *et al.*, 2000; Srinivasula *et al.*, 2001). The presence of IAPs in the cytosol ensures that the apoptotic signaling cascade is not accidentally induced, as by damaged mitochondria in a cell. However, when there is a massive release of cytochrome c and Smac/DIABLO from the mitochondria, Smac/DIABLO inhibition of the IAPs ensures a fast propagation of the caspase cascade and execution of the apoptosis death program.

10. Bax and mitochondrial respiration

Besides inducing permeabilization of the outer mitochondrial membrane, Bax is thought to have effects on components of the respiratory chain in the inner mitochondrial membrane. Activation of the Bax pathway induces changes in intracellular pH, matrix alkalinization, and cytosolic acidification. This could be the result of a disturbance of the mitochondrial respiration or of the ATPase. Oligomycin, an inhibitor of the F_0F_1-ATPase, binds to the F_0 subunit and inhibits the transport of H^+ ions, effectively shutting off the ATPase (Tzagoloff, 1970). Oligomycin was shown to block apoptosis induced by etoposide and dexamethasone, which involves the Bax pathway (Eguchi *et al.*, 1997; Leist *et al.*, 1997). Another study showed that oligomycin inhibits both acidification of the cytosol and the release of cytochrome *c* (Matsuyama *et al.*, 2000). Changes in intracellular pH are not detected in death-receptor apoptosis. Oligomycin has no effect on apoptosis triggered by anti-Fas antibodies, indicating that the effect is not a general inhibition of apoptosis, but specific to the Bax-dependent pathway. When Bax was expressed in wild-type yeast, cytosolic acidification, matrix alkalinization, and cell death were seen, but no effects were detected in F_0F_1-ATPase/H^+ pump-deficient yeast (Matsuyama *et al.*, 2000). In one study, calphostin *c* was shown to induce Bax translocation, homocomplex formation, cytochrome *c* release, and decrease of the mitochondrial membrane potential (Ikemoto *et al.*, 2000). Oligomycin inhibited Bax homodimerization and the decrease in membrane potential. In a study of Bax toxicity to yeast, it was shown that mutations of the mitochondrial proteins required for oxidative phosphorylation, which make the cells respiratory incompetent, decreased Bax toxicity (Harris *et al.*, 2000). However, mutations in mitochondrial proteins unrelated to the oxidative phosphorylation machinery did not affect Bax toxicity. Taken together, these results suggest that Bax and perhaps other multidomain proteins, in addition to their permeabilizing effect on the outer mitochondrial membrane, also have effects on components of the inner membrane, interfering with the respiration and/or ATP production. Whether these effects are reversible remains unclear. This is an important issue if we want to modulate apoptosis in diseases. Would inhibition of downstream caspases rescue the cells if severe irreversible damage has been inflicted on the mitochondrial respiration at an early point in the apoptotic signaling cascade? In a recent study, neurons and HeLa cells were exposed to apoptotic stimuli in the presence of BAF, a broad-spectrum caspase inhibitor (Xue *et al.*, 2001). Although the caspase inhibitor ensured short-term survival, mitochondria were selectively eliminated from the cells without any apparent effects on other intracellular structures. Cells deprived of mitochondria were irreversibly committed to death. In neurons, elimination of mitochondria was completely prevented by expression of Bcl-2, which works upstream of the mitochondria.

11. Conclusion

During the last few years, we have gained an enormous amount of information about apoptosis and in particular the mitochondrial apoptosis pathway. But much still remains unclear. The central role of the mitochondria and some mitochondrial proteins, particularly cytochrome *c*, is now fairly well understood. That proteins of the Bcl-2 family control the mitochondrial pathway, and that permeabilization of the outer mitochondrial membrane is triggered by the multidomain proteins are also well established. However, at the molecular level, how the mitochondrial proteins are released from the intermembrane space is still controversial. The events downstream of the mitochondria are now well characterized, although the direct effect of the various substrates degraded by the executioner caspases on the cell-death process still await to be elucidated. However, the upstream pathways between the initiating stimuli and activation of the multidomain proteins remain largely unknown. The mitochondrial pathway can be activated by a wide range of stimuli – how are these signals detected by the cells and how are they propagated to activate the multidomain proteins? Are the various multidomain proteins involved in specific apoptotic signaling pathways activated by specific stimuli, or are they only redundant proteins with identical function? It would be surprising if the answer to the latter question proved to be 'Yes'.

The antiapoptotic protein Bcl-2 is present not only in the mitochondria but also in other intracellular membranes, the ER, and the nuclear envelope. Could that indicate that this protein and maybe other Bcl-2 family proteins, in addition to their involvement in apoptosis, also have other, unrelated functions? The potential channel-forming activity of the antiapoptotic proteins has, so far, not been attributed any function in apoptosis, but could this activity be related to some other function?

7. Regulators and applications of yeast apoptosis

Kai-Uwe Fröhlich and Frank Madeo

1. Introduction

Apoptosis is a form of programmed cell death with a crucial role in the development and maintenance of metazoan animals. Intense research has revealed a complex network of regulators and effectors, which can be triggered by various toxins or external signals (e.g., ethanol, reactive oxygen species [ROS], and receptor ligands) and internal processes (e.g., mitotic catastrophe, replication failures, or developmentally programmed cell death). Regulatory pathways and inducers vary depending on tissue, developmental state, or host organism, resulting in diverse and sometimes contradictory models for the regulation of apoptosis. A simple model system, as in yeast, would be useful to clarify the dispute. However, apoptosis had been assumed to be confined to multicellular animals. For a unicellular organism such as yeast, a suicide mechanism seemed to be pointless, as it would result in the death of the whole organism. When the complete genome sequence of the yeast *Saccharomyces cerevisiae* became available in 1997, no relatives of the central players in apoptosis, such as the caspases, members of the Bcl-2/Bax family, or Apaf-1, were found, emphasizing the idea of a purely metazoan apoptosis.

It was a surprise, therefore, when cell death with the characteristics of apoptosis was first described in unicellular organisms. The extensive use of yeast for apoptosis research in very different areas of cell biology proves that the basic machinery of apoptosis is indeed present and functional in yeast. The genes functioning in the cell death of unicellular organisms have been confirmed as apoptotic regulators of metazoans. These results promise easier access to an understanding of apoptosis.

2. Expression of metazoan apoptotic genes induces cell death in yeast

The first application of yeast for apoptosis research was its use in two-hybrid studies to investigate the interactions between different apoptotic proteins. As an unexpected side result, the expression of several proapoptotic genes, including *bax* (Greenhalf *et al.*, 1996), caspases (Kang *et al.*, 1999), and Apaf-1/CED-4 (James *et al.*, 1997), has been found to be lethal for yeast.

Further research indicated that the resulting cell death is indeed of apoptotic nature. Ligr *et al.* (1998) have shown that *bax* (a Bcl-2 family member)-mediated cell death in *S. cerevisiae* is accompanied by typical features of apoptosis such as externalization of phosphatidylserine at the surface of the cytoplasmic membrane, membrane blebbing, chromatin condensation and margination, and DNA cleavage. Simultaneous expression of *bcl-x$_L$* prevents these effects and cell death.

In addition, human *bak* (another proapoptotic gene of the *bax*/*bcl*-2 family) induces cell death in *S. pombe* accompanied with an apoptotic phenotype: condensation and fragmentation of chromatin, vacuolization of cytoplasm, DNA cleavage, and dissolution of the nuclear envelope (Ink *et al.*, 1997). These alterations can also be suppressed by the expression of *bcl-x$_L$*.

Furthermore, mutant forms of Bcl-x$_L$, an antiapoptotic Bcl-2 family member lacking Bax-binding activity, can prevent *bax*-induced death in yeast (Tao *et al.*, 1997; Minn *et al.*, 1999) This indicates that Bcl-x$_L$ acts downstream of Bax, perhaps by competing for binding to a common target, which may be part of the conserved apoptotic machinery.

Recently, Pavlov *et al.* (2001) found a novel high-conductance channel of mitochondria. The activity of this channel correlates with the presence of Bax in the mitochondrial outer membrane and is absent in mitochondria from yeast cells overexpressing antiapoptotic *bcl*-2. The pore diameter of approximately 4 nm, inferred from the largest conductance state of this channel, is sufficient to allow diffusion of cytochrome *c* and even larger proteins. This channel, named the 'mitochondrial apoptosis-induced channel,' is a candidate for the outer-membrane pore through which cytochrome *c* and possibly other factors leave mitochondria during apoptosis (see Chapter 7).

Reed and coworkers used the cytotoxicity of Bax to *S. cerevisiae* to select antiapoptotic effectors from a human gene library and to identify yeast mutations preventing cell death. Both strategies were successful, and the results could be extended to mammalian apoptosis: *bax* expression was not lethal for a mutant in the mitochondrial F$_0$F$_1$-ATPase. Oligomycin, an inhibitor of this enzyme, prevented cell death in both *S. cerevisiae* and mammalian cells (Matsuyama *et al.*, 1998). Expression of a human gene, *BI*-1, rescued *bax*-expressing yeast. Its gene product is located predominately at the endoplasmic reticulum, with a small portion at the mitochondrial membrane. It interacts with Bcl-2, but not Bax or Bak. When overexpressed in mammalian cells, it suppresses apoptosis induced by Bax, growth factor withdrawal, or various drugs, but not by Fas (Xu and Reed, 1998). A *BI*-1 homolog has been described in plants. Its overexpression reduces cell death caused by heterologous expression of mammalian *bax* in *Arabidopsis* (Kawai-Yamada *et al.*, 2001).

Heterologous expression of a regulator caspase gene coding for human procaspase-8 or -10 in yeast results in production of the efficiently processed, active caspase (Kang *et al.*, 1999). Caspase-10 activity has little effect on yeast-cell viability, whereas expression of caspase-8 is cytotoxic. In contrast, autoactivation of executioner caspase-3 or -6 from proforms cannot be detected in yeast. However, coexpression of procaspase-8 results in procaspase-3 activation and produces a pattern of morphologic changes similar to mammalian apoptosis.

Expression of *CED*-4 from *C. elegans* in *S. pombe* also leads to an apoptotic phenotype, whereas coexpression of *CED*-9, a *bcl*-2 homolog, prevents chromatin condensation (James *et al.*, 1997). The possibility of rescuing yeast with antiapoptotic factors acting downstream in the mammalian apoptotic cascade indicates that Bax, Bak, CED-4, or caspases do not simply act as cytotoxic substances in yeast, but seem to activate the same or a similar mechanism as in metazoan organisms.

3. Mutations leading to apoptosis in yeast

All previously described examples of yeast apoptotic cell death involved heterologous expression of metazoan apoptotic inducers. In 1997, we found a yeast mutant dying with a typical apoptotic phenotype: exposure of phosphatidylserine, margination of chromatin, and formation of cell fragments (Madeo *et al.*, 1997). While the TUNEL test indicates massive DNA breakage, no DNA ladder is observed. Nucleosome linkers therefore appear not to be preferred targets of DNA cleavage in yeast, probably due to their short length (Lowary and Widom, 1989).

The mutation causing the apoptotic phenotype is a point-mutation of *CDC48* (*cdc48^{S565G}*), a *S. cerevisiae* protein belonging to the AAA family. AAA proteins are ATPases that extract proteins from protein complexes and are involved in protein degradation and in vesicle fusion. CDC48p is necessary for homotypic vesicle fusion in both ER and the Golgi apparatus (Latterich *et al.*, 1995), and participates in the extraction and degradation of misfolded proteins from the ER (ERAD) (Ye *et al.*, 2001; Rabinovich *et al.*, 2002).

Recently, Granot and colleagues showed that Bax-triggered apoptosis in yeast can be blocked by enhancement of vesicle trafficking. Moreover, a downregulation of vesicular transport enhances the susceptibility of yeast cells to apoptosis, providing an explanation for the *cdc48^{S565G}*-mediated apoptosis (Levine *et al.*, 2001). Future work will reveal whether vesicular trafficking also plays a role in metazoan apoptosis, and whether this process is closely linked to apoptosis, or whether the connection is rather indirect.

In the case of *CDC48*, yeast has demonstrated its potential to characterize new mammalian apoptotic regulators. In 1999, the antiapoptotic role of the human Cdc48p ortholog VCP/p97 and a related protein in *C. elegans* was described (Shirogane *et al.*, 1999; Wu, D. *et al.*, 1999). Intriguingly, VCP was found in a screen for new antiapoptotic proteins. A mutated form, similar to the *cdc48^{S565G}* mutation in yeast, induces apoptosis dominantly in B cells. This makes CDC48/VCP the first apoptotic regulator to be discovered by its function in yeast apoptosis. Recently, another AAA protein has been described that modifies Bax-induced apoptosis in yeast. According to Manon *et al.* (2001), expression of *bax* in yeast induces not only a release of cytochrome *c* but also a decrease in the amount of cytochrome *c* oxidase. They have shown that the decrease of cytochrome *c* oxidase is due to the activation of the mitochondrial AAA protease Yme1p, of which the cytochrome *c* oxidase subunit 2 (Cox2p) is a substrate. The absence of Yme1p slightly delays Bax-induced cell death, suggesting a role of this protease in yeast cell death and probably the orthologous function of its mammalian homolog.

Another mutation resulting in yeast apoptotic cell death is the deletion of *ASF1/CIA1*, coding for a histone chaperone. The disruptant arrests preferentially at the G_2/M transition, and the cell dies showing markers of apoptosis. Moreover, reduction of the mitochondrial membrane-potential, dysfunction of the mitochondrial proton pump, and release of cytochrome *c* to cytoplasm occur (Yamaki *et al.*, 2001). The human homolog of Asf1p/Cia1p, CIA, interacts with the largest subunit of TFIID, CCG1, which is involved in the regulation of apoptosis (Sekiguchi *et al.*, 1995).

Schizosaccharomyces pombe rad9 is a checkpoint gene preventing cells containing damaged or incompletely replicated DNA from entering mitosis. Rad9 protein (SpRad9) contains a stretch of amino acids with similarity to the Bcl-2 homology 3 death domain, which is required for SpRad9 interaction with human Bcl-2. Overexpression of *bcl-2* in *S. pombe* inhibits cell growth independently of *rad9*, but enhances the resistance of *rad9*-null cells to methyl methanesulfonate, and ultraviolet and ionizing radiation. The authors suggest that SpRad9 may be the first member of the Bcl-2 protein family to be identified in yeast (Komatsu *et al.*, 2000).

Ceramide is another potential bona fide apoptosis regulator shared by mammals and yeast. Some mammalian growth modulators, including tumor necrosis factor α, induce apoptosis or cell-cycle arrest via ceramide, which activates a specific phosphatase (Kishikawa *et al.*, 1999). Ceramide-induced G_1 arrest of *S. cerevisiae* is also mediated via activation of a protein phosphatase (Nickels and Broach, 1996).

4. ROS are central regulators of yeast apoptosis

ROS (reactive oxygen species) are well established as inducers of apoptosis. Treatment with low doses of H_2O_2 induces the apoptotic cascade in mammalian cell cultures. In addition, in neural cells deprived of nerve growth factor or potassium, ROS produced by the cell act as a late signal of the apoptotic pathway, downstream of the action of Bax and caspases (Schulz *et al.*, 1996). In *S. cerevisiae*, exposure to low doses of H_2O_2 or oxidative stress by glutathione depletion induces apoptosis. Inhibition of translation by cycloheximide prevents development of apoptotic markers in response to H_2O_2, indicating an active role of the cell in the death process.

Furthermore, increased formation of intracellular ROS occurs during yeast apoptosis, even in the absence of external oxidative stress. Yeast expressing *bax* and *cdc48*^{S565G} mutants is strongly stained after treatment with dihydrorhodamine 123, indicating an accumulation of ROS (*Figure 1*). The radicals appear to be necessary to induce the apoptotic phenotype, as anaerobic growth conditions and radical traps prevent both cell death and development of apoptotic markers (Madeo *et al.*, 1999). This puts ROS at a central position in yeast apoptosis. Yeast cells are also more sensitive to H_2O_2-induced apoptosis when vesicle fusion is blocked (Levine *et al.*, 2001). Hence *CDC48* mutation might trigger apoptosis in two distinct, synergistic ways: generation of radicals and arrest of vesicle fusion.

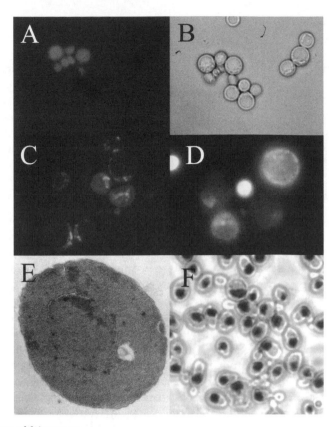

Figure 1. Facets of dying yeast.
Dihydrorhodamine 123 staining (A) reveals that some cells produce reactive oxygen to induce apoptosis. (B) shows the same region in phase contrast. Apoptotic chromatin condensation can be visualized by DAPI staining (C) or by electron microscopy (E). Staining with FITC-labeled annexin V lights up the cytoplasmic membrane of protoplasted yeast cells (cell wall removed enzymatically) early in apoptosis (D). The black nuclei in the TUNEL test indicate massive DNA fragmentation in apoptotic yeast (F).

Adenovirus E4orf4 protein has been reported to induce p53-independent apoptosis in transformed mammalian cells. By heterologous expression of human E4orf4, Kornitzer *et al.* (2001) could induce cell death in yeast by an irreversible growth arrest at the G_2/M transition accompanied by accumulation of ROS. E4orf4 expression in mammalian cells also leads to accumulation of ROS and induces G_2/M arrest prior to apoptosis, indicating that E4orf4-induced events in yeast and mammalian cells are highly conserved.

In mammalian cells, inhibition of proteasome-dependent proteolysis leads to either induction or repression of apoptosis, depending on the proliferative status of the cells. It has been suggested that in exponentially growing cells, proteasomes continuously degrade an activator of apoptosis (Drexler, 1997). To characterize which role the proteasome plays in yeast apoptosis, Ligr *et al.* (2001) screened for proteasomal substrates and found six genes which by overexpression in a proteasome-deficient background trigger apoptosis. One of the genes, *SAR1*, is required for vesicular transport from ER to Golgi; hence, it is involved in a similar process as *CDC48*. Another gene is *STM1*, a DNA-binding protein involved in DNA repair. *STM1* knockout strains show enhanced resistance to oxygen stress. Ligr suggests that STM1 protein is an activator of apoptosis triggered by exposure of cells to low concentrations of H_2O_2.

Another form of external stress resulting in apoptosis has been described by Ludovico *et al.* (2001). After treatment with low doses of acetic acid, yeast dies, showing markers of apoptosis. Again, this is accompanied by formation of ROS (our unpublished data).

5. DNA damage induces yeast apoptosis

Recently, the first evidence was reported that an apoptotic phenotype in yeast can be induced by DNA damage after UV irradiation, as monitored with the TUNEL test. Interestingly, a UV dose-dependent increase in the sub-G_1 population was found by flow cytometry. Sub-G_1 cells were isolated by flow sorting and analyzed by electron microscopy. This population showed condensed chromatin in the nucleus and cell shrinking (Carratore *et al.*, 2002).

Cytolethal distending toxins (CDTs) are multisubunit proteins produced by a variety of bacterial pathogens that cause cell-cycle arrest and apoptosis in mammalian cells. Recent studies suggest that they can act as intracellular DNases in mammalian cells. Expression of a CdtB subunit in yeast causes a G_2/M arrest, as seen in mammalian cells. CdtB toxicity is not circumvented in yeast genetically altered to lack DNA damage checkpoint control or that constitutively promotes cell cycle progression via mutant Cdk1, because CdtB causes a permanent type of damage that results in loss of viability (Hassane *et al.*, 2001).

6. Speculations about the origin of apoptosis

The importance of ROS in the regulation of apoptosis may indicate the origin and primary purpose of the suicide process. ROS are byproducts of respiration and occur in every aerobic organism. Because ROS are highly reactive and modify proteins, lipids, and nucleic acids, ROS-induced cell damage is a common and efficient event. Cells have developed mechanisms to detoxify ROS and to repair oxidative damage, but this can only reduce, not completely prevent, fatal cell damage. Most damaged cells will continue to metabolize for some time, even when they have lost the ability to proliferate. A rapid, active suicide of these cells would spare metabolic energy for neighboring cells, which, even in the case of unicellular organisms, are mostly clonal relatives, genetically identical to the damaged cell. That way, suicide of a unicellular organism could provide an evolutionary advantage for its genome.

Higher eukaryotes evaluate cell damage (via the p53 system) to decide whether suicide is advantageous. Before development of such a complex system, chemical reactivity of ROS themselves may have been used to trigger cellular suicide.

To induce cell death in situations without external oxygen stress, cells developed mechanisms to produce these signals autonomously. Such a suicide scenario has been described by Longo *et al.* (1997), who observed that stationary cells of *S. cerevisiae* survive for long periods in pure water but quickly lose viability in nutrient-depleted synthetic media. Bcl-2 delays the loss of viability. These cells accumulate ROS and die with an apoptotic phenotype (unpublished results). The source of the ROS is yet unidentified. As oxygen radicals are normal byproducts of respiration, a specific modulation of the respiratory chain may have been developed to increase the output of ROS as needed. Release of cytochrome *c* leads to an accumulation of reduced ubiquinone, increasing production of superoxide via the bc1 complex (Luetjens *et al.*, 2000). During a further refinement in the regulation of apoptosis, released cytochrome *c* itself became used as an apoptotic signal, perhaps in order to make the regulatory cascade less dependent on the redox state of the cell. With the development of multicellular organisms, a more flexible regulation of apoptosis became necessary, including responses to various external signals, resulting in additional regulatory steps upstream, downstream, or instead of ROS.

7. Apoptosis and aging in yeast

S. cerevisiae proliferates by budding. A 'mother cell' produces a bud that grows into a 'daughter cell', leaving a circular 'bud scar' on the cell wall of the mother cell. The number of scars therefore indicates the 'replicative age' of the respective mother cell. Yeast mother cell-specific aging has been intensively researched and reviewed (Jazwinski, 1999; Johnson, F.B. *et al.*, 1999; Costa and Moradas-Ferreira, 2001) in recent years. Similarities of morphologic and physiologic changes make yeast a simple model system for cellular and, perhaps, also organismic aging.

Old yeast cells (mother cells which have produced more than 30 daughter cells) are much larger than young 'virgin' cells (that have budded no daughter cells yet); their cell cycle as well as protein synthesis is slowed down, and the cell surface develops a loose and wrinkled appearance. The median life span of most laboratory strains of *S. cerevisiae* is about 25–35 generations or about three days. Nestelbacher *et al.* (2000) have shown that genetic and environmental changes that increase the burden of ROS on yeast cells result in a shortening of the life span of mother cells. Deletion of yeast genes coding for superoxide dismutase (Barker *et al.*, 1999) or catalase, changes in atmospheric oxygen partial pressure, and addition of the physiologic antioxidant glutathione all have the expected distinct effects on life span and indicate a role for oxygen in the yeast's aging process (Nestelbacher *et al.*, 2000). Recently, Laun *et al.* (2001) found that mitochondria are the source of ROS in senescent yeast cells. Mitochondria producing ROS are found in senescent cells, but not in virgin cells. This might suggest another physiologic role of apoptosis in yeast. It should be noted, however, that the portion of old cells in a yeast culture is extremely low. Suicide of old cells will not save much energy, but may prevent formation of genetically damaged daughters of very old mother cells. Currently, it is controversial whether human senescent cells die apoptotically or necrotically. The present results in yeast apoptosis favor an apoptotic age-related cell death. Consistently, primary human cells in culture undergo apoptosis, producing ROS when they become senescent (Jansen-Dürr, personal communication). The observation of radical-induced apoptosis in senescent yeast further strengthens the oxygen theory of aging and confirms the similarity of yeast and higher cell aging finally leading to apoptosis (for a review, see Fröhlich and Madeo, 2001).

8. Applications of yeast apoptosis to plant research

Higher plants also exhibit characteristic apoptotic features, such as DNA strand breaks and exposure of phosphatidylserine (O'Brien *et al.*, 1997), and can be protected from cell death by expression of the antiapoptotic genes, *bcl*-x$_L$ or *CED-9* (Mitsuhara *et al.*, 1999).

Osmotin, a protein involved in the plant's defense response, induces apoptosis in yeast (Narasimhan *et al.*, 2001). Induction of apoptosis is correlated with intracellular accumulation of ROS and can be linked to the RAS2/cAMP pathway. This is in accordance with the observation that hyperactivation of the RAS pathway in yeast results in cell death (Fedor-Chaiken *et al.*, 1990), as RAS signaling is induced by lack of nutrients.

An *Arabidopsis* cDNA library has been screened for functional suppressors of Bax-induced cell death in yeast cells. Pan *et al.* (2001) found proteins involved in the detoxification of oxygen radicals and identified the *A. thaliana* ethylene-responsive element-binding protein (AtEBP) as a dominant suppressor of Bax-induced cell death in yeast. Apoptotic phenotypes of *bax* expression in yeast could be abrogated by coexpression of AtEBP.

In a similar approach, Moon *et al.* (2002) identified soybean ascorbate peroxidase as a suppressor of Bax-induced cell death in yeast. Expression of the peroxidase prevents oxygen radical generation.

Δ1-pyrroline-5-carboxylate (P5C), an intermediate in biosynthesis and degradation of proline, is assumed to play a role in cell death in plants and animals. External supply of proline or P5C is toxic to *Arabidopsis,* suggesting a crucial role of P5C dehydrogenase in the process of preventing cell death by degrading P5C. Consistently, in a yeast mutant lacking P5C dehydrogenase, proline leads to growth inhibition and formation of ROS. P5C dehydrogenase is expressed at a basal level in all tissues analyzed, in agreement with a protective role against cell death (Deuschle *et al.*, 2001).

9. Applications of yeast apoptosis for medical research

The spread of HIV infection and the increasing number of patients treated with immunosuppressive drugs have caused infections by opportunistic fungi to flourish. The yeast *Candida albicans*, otherwise part of the normal human flora, has become a pathogen commonly causing systemic infections. Helmerhorst *et al.* (2001) found that the human salivary antifungal peptide histatin 5 kills *Candida* by induction of oxygen stress. The human salivary peptide histatin 5 exerts its antifungal activity through the formation of ROS. Indeed, *Candida* exhibits an apoptotic phenotype in response to classical metazoan apoptosis inducers (Mark Ramsdale, personal communication). The potential for apoptosis in *Candida* might lead to a new class of drugs against fungal infections.

In a recent review, Zhao and Elder (2000) summarize the contribution of research on yeast to the molecular understanding of HIV infection. Increasing evidence suggests that the HIV-1 viral protein R (Vpr) plays an important role in viral pathogenesis, as its functions are being linked to viral activation, suppression of human immune functions, and depletion of human CD4 lymphocytes, the major clinical manifestation of AIDS. *In vitro*, Vpr shows multiple activities in both mammalian and yeast cells, including nuclear transport, induction of cell-cycle G2 arrest, morphologic changes, and cell death. The occurrence of these activities in yeast indicates that Vpr interacts with highly conserved cellular processes to cause the effects, and allows Vpr activities to be studied in these genetically well-characterized organisms. Studies of Vpr in *S. pombe* and *S. cerevisiae* have helped to establish the following milestones: (i) Vpr induces G_2 arrest through inhibitory phosphorylation of the cyclin-dependent kinase by a pathway in which protein phosphatase 2A plays an important role. (ii) Vpr induces apoptosis by directly permeabilizing the mitochondrial membrane. (iii) Vpr also appears to kill cells by mitochondria-independent mechanisms. (iv) G_2 arrest and cell death induced by Vpr are two independent functions. Future studies of Vpr in yeast are expected to make additional contributions to understanding the mechanisms of Vpr activities and may also help address the importance of these activities during the course of HIV-1 infection (Zhao *et al.*, 1998; Elder *et al.*, 2001).

10. Perspectives

The finding of factors relevant to apoptosis in animals and yeast and even lower eukaryotes suggests that apoptosis developed in unicellular organisms long before the evolutionary separation between fungi, plants, and metazoan animals occurred.

In the near future, yeast studies promise the identification of more components of the basic, evolutionarily ancient stages of apoptosis. Yeast offers the opportunity to screen easily for substances acting directly on these basic components without diversion by a complex upstream network. This may result in the discovery of 'universal' activators or inhibitors of apoptosis that would be helpful for research, and potentially for medical applications as well.

8. Evolution of cell death: caspase-mediated mechanisms in early metazoans; noncaspase mechanisms in single-celled eukaryotes

Angelika Böttger and Charles N. David

1. Introduction

Apoptosis, or programmed cell death, has become a central topic in cell biologic research, and its biochemical mechanisms have been studied extensively in a number of model organisms including *Caenorhabditis*, *Drosophila*, and mouse, and in human cell lines. In these animals and cell lines, the main cell-death pathway is dependent on a family of cysteine proteases, the caspases (see Chapter 2). Recent work has now identified caspases and programmed cell death in Cnidaria, the oldest metazoan phylum, and thus pushed the origin of this process back to the threshold of multicellular evolution (Cikala *et al.*, 1999; Miller *et al.*, 2000; Seipp *et al.*, 2001).

Since caspases have not yet been identified in single-celled eukaryotes (Aravind *et al.*, 1999), it appears that the caspase-dependent cell-death pathway may have arisen coincident with the evolution of metazoan animals. The process of programmed cell death, by comparison, appears to have evolved much earlier and perhaps several times independently, since a number of single-cell eukaryotes, including the slime mold *Dictyostelium*, the ciliate *Tetrahymena,* and the protozoan *Trypanosoma brucei*, exhibit cell-death programs but appear to lack caspases (Davis *et al.*, 1992; Cornillon *et al.*, 1994; Ameisen *et al.*, 1995). These cell-death programs are accompanied by some, but not all, of the morphologic features characteristic of apoptosis. In the case of *Dictyostelium*, cell death has been shown to be associated with release of apoptosis-inducing factor (AIF) from mitochondria. This cell-death program appears to be an alternative to the caspase-dependent pathway. Since it is also present in higher metazoans (Susin *et al.*, 1999), it may represent an ancient form of programmed cell death which has been maintained in evolution.

The first part of this chapter will focus on apoptosis in cnidarians, specifically in the fresh-water hydrozoan polyp *Hydra* and in the colonial marine hydrozoan *Hydractinia*. We will begin with three examples of apoptosis in *Hydra* and *Hydractinia* that have been well analyzed in the past three years. In the second part of this chapter, we will consider examples of programmed cell death in three single-celled eukaryotes.

2. Cell death in Cnidaria and Porifera

Cnidaria are the phylogenetically oldest Eumetazoa. The phylum includes the first multicellular animals whose body plans are based on two important principles that define the body plans of all higher animals. These are first the formation of epithelial layers that later in phylogeny give rise to tissues and organs and second the presence of pluripotent stem cells that, in cnidarians, give rise to nerve cells, nematocytes, gland cells, and germ cells.

For developmental biologists, the most prominent cnidarian is *Hydra*. It is, in fact, the oldest model organism in developmental biology. Experiments carried out in the eighteenth century by the Swiss scientist Abraham Trembley revealed the almost unlimited capacity of *Hydra* to regenerate. This property led to more refined experiments to study pattern formation and stem-cell differentiation in *Hydra* (Gierer, 1974) and, more recently, to analyze the molecular parameters of development and regeneration (Broun *et al.*, 1999; Technau and Bode, 1999; Hobmayer *et al.*, 2000; Smith *et al.*, 2000). Recent work indicates that the cnidarians, including *Hydra* and *Hydractinia*, can also teach us a few lessons about the evolution of apoptosis and its role in growth regulation in multicellular animals, egg and sperm development, and metamorphosis.

2.1 Apoptosis in Hydra regulates cell numbers

Hydra reproduces asexually by budding, or sexually by fusion of male and female gametes. In laboratory culture, however, the preferred mode of reproduction is budding. Under these conditions, the number of buds an animal produces depends strictly on feeding. If animals are well fed, their numbers double every two to three days. The same is true for total cell numbers in such animals. If *Hydra* is starved, however, no buds are formed, and cell numbers remain roughly constant for a few days before beginning a slow decline.

Comparison of cell-proliferation rates in fed and starved animals revealed that they did not differ significantly (Bosch and David, 1984). This implied production of 'excess' cells under starvation conditions where the cell number did not increase in size; indeed, it could be shown that cells were eliminated from tissue by phagocytosis under these conditions. Phagocytic vacuoles filled with condensed cells containing apoptotic nuclei were found in epithelial cells (Bosch and David, 1984). The number of epithelial cells with phagocytic vacuoles depended directly on the feeding regimen and was about sevenfold higher in starved animals than in fed animals.

Closer examination of the apoptotic process in *Hydra* has been achieved by inducing cell death with colchicine or wortmannin. Rapidly cycling interstitial cells are particularly sensitive to colchicine treatment (Campbell, 1976) and were eliminated from treated tissue by apoptosis. Phagocytized corpses stained brightly with acridine orange and exhibited pyknotic nuclear morphology with DAPI staining, as shown in *Figure 1*. Chromatin degradation to nucleosome-sized fragments (laddering) was also observed (Cikala *et al.*, 1999). All these features of the cell-death

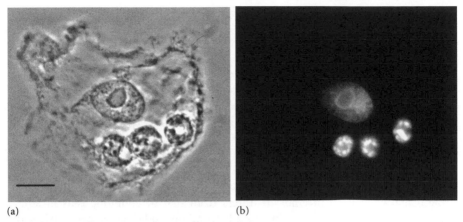

(a)   (b)

Figure 1. Single epithelial cell of *Hydra* containing three vacuoles, each with an apoptotic cell.
Hydra specimens were treated with wortmannin to induce apoptosis. Epithelial cells acting as
nonprofessional phagocytes rapidly engulfed the apoptotic cells. (a) Phase-contrast micrograph of an
epithelial cell from a treated polyp. (b) DAPI fluorescence of the same cell showing normal nuclear
morphology of the epithelial cell and pyknotic condensation in nuclei of the three apoptotic cells. Scale
bar: 10 μm.

phenotype in *Hydra* are indistinguishable from the apoptotic phenotype in higher
animals and suggest that the mechanism of apoptosis is the same in *Hydra* and
higher animals.

The use of fluorogenic substrates specific for caspase-3 (DEVD-AMC) has
demonstrated a dramatic increase in caspase activity in colchicine-treated animals.
Moreover, two cDNAs have been cloned from *Hydra* with homology to caspase-3
from higher animals. They encode typical procaspase sequences with an amino terminal prodomain followed by large and small subunits (Cikala *et al.*, 1999). More
recent experiments have shown that *Hydra* caspase-3A, when expressed in bacteria,
undergoes autocatalytic cleavage between the large and small subunit, leaving the
prodomain attached to the large subunit. Cleavage occurred at the C-terminal to
the sequence IRKD. This is clearly a caspase cleavage site, although the sequence is
somewhat atypical. In accordance with this unusual substrate specificity, active site
labeling could be achieved only by using the non-specific caspase inhibitors, z-
EK(biotin)D-AOMC and z-VAD-AMC. None of the commercially available specific caspase inhibitors, such as the caspase-3 inhibitor DEVD-FMK or the
caspase-1 inhibitor YVAD-FMK, could compete with the nonspecific inhibitors
(Böttger and David, unpublished observations).

2.2 Apoptosis during gamete development in Hydra

Oogenesis provides a further example of apoptosis in *Hydra*. During oocyte development, a large number of interstitial cells, the egg-restricted stem cells, become
committed to oogenesis. Several thousand of these interstitial cells give rise to a
macroscopically visible 'egg patch' on the side of the animal. One cell in the center
of the egg patch differentiates into an oocyte. The remaining cells differentiate into
nurse cells and are subsequently phagocytized by the oocyte (Honegger, 1989).

Because of their appearance, these phagocytized cells have been referred to as *Schrumpfzellen* ('shrunken cells'). After they are taken up by the oocyte, they remain enclosed in vacuoles in the cytoplasm of the oocyte (Zihler, 1972). The nuclei become pyknotic and chromatin is condensed at the edges of the nuclei. Moreover, the *Schrumpfzellen* within the oocyte cytoplasm stain strongly with acridine orange, suggesting that they are in a state of arrested apoptosis (Miller *et al.*, 2000). During cleavage, the *Schrumpfzellen* are distributed to blastomeres, ultimately becoming part of the endoderm and disappearing from it only 8–10 days after hatching.

Schrumpfzellen appear to be a source of nutrition for the developing embryo, which has to survive without food until it can catch prey for itself. How closely related nurse cell phagocytosis is to apoptosis is not yet fully understood. The DNA fragmentation pattern in embryos differs from that of apoptotic cells in adult *Hydra*. This DNA is only partially degraded, and no laddering has been observed. DNA strand breaks were not found by the TUNEL assay. Finally, it is still unclear whether or not caspases are involved (Technau, personal communication).

Apoptosis also occurs during spermatogenesis in *Hydra* (Kuznetsov *et al.*, 2001). In the developing testis, interstitial cells committed to sperm-cell differentiation accumulate in the ectoderm between epithelial cells. Spermatogonia multiply and differentiate to form cell layers representing progressive stages of spermatogenesis in the testis. Kuznetsov *et al.* have shown by acridine orange staining that sperm-cell differentiation is accompanied by extensive cell death. Epithelial cells that form part of the developing testis phagocytize the apoptotic corpses. Although the function of apoptosis during *Hydra* spermatogenesis is not understood at the moment, it should be noted that the same process occurs during mammalian sperm cell development, where it is regulated by specialized Sertoli's cells (reviewed in Print and Loveland, 2000).

2.3 Apoptosis during metamorphosis in Hydractinia

The elimination of larval structures during metamorphosis represents a classic example of the use of apoptosis during development. Metamorphosis in the tadpole, for instance, involves resorption of the larval tail concomitant with the building of new adult structures. Experiments have now shown that metamorphosis in *Hydractinia* is also accompanied by massive apoptosis (Seipp *et al.*, 2001).

Hydractinia is a colonial hydrozoan that forms a stolon network covered with polyps. The colonies consist primarily of sexual polyps and feeding polyps. Colonies arise from a single polyp formed by metamorphosis of a planula larva. In response to appropriate environmental signals, neuropeptides of the GLWamide family are secreted into larval tissue to induce metamorphosis (Schmich *et al.*, 1998). Recent experiments by Seipp *et al.* (2001) have demonstrated that metamorphosis is accompanied by degradation of large parts of the planula larva by apoptosis. *Figure 2* shows large numbers of TUNEL-positive cells in a planula larva undergoing metamorphosis. Apoptotic cells were not only removed by phagocytosis but also shed into the environment. Metamorphosis was accompanied by a dramatic increase in caspase activity measured with the fluorogenic caspase substrate DEVD-AMC (Seipp, Böttger, unpublished observations).

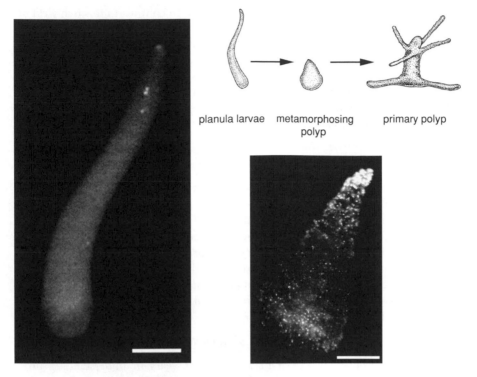

planula larvae   metamorphosing   primary polyp
                      polyp

Figure 2. Apoptosis during metamorphosis of *Hydractinia* planula larva to primary polyp.
The schematic diagram illustrates the process of metamorphosis, which takes about 24 h to complete.
The micrographs show, respectively, a TUNEL-stained larva before metamorphosis and a 3-h
metamorphosing larva containing large numbers of apoptotic cells. A planula larva contains
approximately 10^4 cells. Scale bar: 20 μm.

2.4 Programmed cell death in Porifera

Although species of the Porifera (sponges) are clearly multicellular, they lack the
structural features of the Eumetazoa. Here our knowledge regarding apoptosis is
still rather fragmentary. Caspases have not yet been discovered in sponges.
However, there are reports of apoptotic cells in sponges (Wiens *et al.*, 1999), and
two genes encoding bcl-2 and death domain-motifs have been reported in the
sponges *Geodia cydonium* and *Suberites domuncula* (Wiens *et al.*, 2000).
Unfortunately, no functional data regarding these genes in sponges are yet available.

The above results demonstrate clearly that the three basic functions for which
apoptosis is used in higher eukaryotes – cell number regulation, tissue sculpting
during embryogenesis (metamorphosis), and ooplasm formation – are also found
in the cnidarians *Hydra* and *Hydractinia*. There is no evidence yet for a role of
apoptosis in defense against infections. This may, however, simply be due to the
lack of an appropriate model system for infections in cnidarians. The results cited
above also demonstrate that caspases are present in *Hydra* and *Hydractinia*, and

that they are activated when apoptosis in initiated. In contrast to this rich yield of evidence for apoptosis and caspases in cnidarians, there is little evidence in sponges and single-celled eukaryotes (see below) for either. The lack of evidence of caspases is particularly telling since complete or almost complete genome sequences are available for yeast, *Dictyostelium*, and *Trypanosoma*. In none of these genome sequences has evidence been found of caspases. We conclude – tentatively – from these facts that caspase-dependent apoptosis arose coincident with the evolution of true multicellular animals – the Eumetazoa.

3. Cell death in single-celled eukaryotes

The idea that cellular suicide makes sense only if the dying cell sacrifices itself for the sake of an organism overlooks the fact that single-celled animals also perform differentiation programs and thus create more than one form of themselves. The survival of such organisms could depend on a sacrifice of cells in one developmental stage to allow survival of another stage that propagates the genome. It is therefore not surprising to find programmed cell death in single-celled eukaryotes. We will review here the data available about programmed cell death in the slime mold *Dictyostelium discoideum*, the ciliate *Tetrahymena thermophila*, and the kinetoplastid parasites *Trypanosoma* and *Leishmania*.

3.1 Differentiation and cell death in D. discoideum

Considerable progress has been made in recent years in understanding programmed cell death in the slime mold *D. discoideum*. *Dictyostelium* cells grow as single cells on bacterial substrates until food becomes limited. This situation triggers a differentiation program that involves aggregation of starving cells (mediated by cAMP signaling and chemotaxis) to form a large aggregate that differentiates into a fruiting body containing spore cells and a stalk. Cells forming the stalk are dead and exhibit some features of apoptosis, such as nuclear condensation (Cornillon *et al.*, 1994). However, DNA is not degraded to nucleosome-sized fragments, and the dead stalk cells are not phagocytized. Instead, they acquire a rigid cellulose cell wall reminiscent of plant cell walls.

Caspase inhibitors have been shown to block certain stages of this differentiation program, and active site labeling showed 'caspase-like' activity, a finding which, however, could not be confirmed with the standard fluorogenic caspase substrates, DEVD-AMV or YVAD-AMC (Cornillon *et al.*, 1994). Moreover, cell death was not inhibited by caspase inhibitors and therefore appeared to be caspase independent (Olie *et al.*, 1998).

Stalk-cell differentiation can be induced *in vitro* in suspensions of *Dictyostelium* cells treated with DIF (differentiation-inducing factor). Such cells are more amenable to experimental analysis, and Arnoult *et al.* (2001) have used them to demonstrate a decrease in cell size, degradation of DNA to oligonucleosome-size fragments, and loss of mitochondrial membrane potential in dying cells. They showed further that the *Dictyostelium* homolog of AIF (apoptosis-inducing factor [Susin *et al.*, 1999]) was released from mitochondria during the differentiation process, and that it could induce DNA fragmentation when added to nuclei *in vitro*.

These experiments strongly support the notion that a cell-death program accompanies stalk-cell differentiation in *Dictyostelium*. This program has distinct characteristics that distinguish it from caspase-dependent apoptosis in higher eukaryotes. It is caspase independent and appears to be mediated by release of AIF from mitochondria.

AIF is a highly conserved mitochondrial protein with NADH oxidase activity (Miramar *et al.*, 2001). Despite its normal localization in mitochondria, AIF contains a nuclear localization signal that appears to be responsible for its translocation to the nucleus after release from mitochondria in apoptotic cells. AIF transport to the nucleus leads directly to DNA degradation by a process different from that caused by caspase-dependent apoptosis because it results in larger DNA fragments. AIF has been shown to be essential for caspase-independent apoptosis in cleavage-stage mouse embryos (Joza *et al.*, 2001). Its presence in *Dictyostelium* cells suggests that it may represent an ancient cell-death pathway that is evolutionarily conserved. In higher animals, this pathway seems to coexist with the caspase-dependent apoptotic pathway. In *Dictyostelium*, however, which apparently lacks a classical caspase gene, the AIF pathway is used to organize the developmental cell-death program.

Caspase-mediated cell death and AIF-activated cell death have in common the fact that they both rely on mitochondrial disintegration (see Chapter 7). Caspase activation via the caspase 9/APAF1 pathway requires the release of cytochrome *c* from mitochondria. AIF-activated cell death requires release of AIF from mitochondria. In both cases, mitochondria are causally involved in apoptosis. This fact has led to an interesting proposal by Ameisen (1996; 1998) for the evolution of cell death. He has suggested that programmed cell death evolved during establishment of endosymbiosis between bacteria and the ancestors of the present-day eukaryote cell. Possibly some of the proteins involved in controlling apoptosis, such as bcl2 and AIF, are related to molecules involved in establishing the original endosymbiosis.

3.2 Programmed cell death in Tetrahymena

Conjugating *Tetrahymena* cells have been shown to carry out an unusual form of nuclear death that bears some resemblance to nuclear degradation during apoptosis. Conjugation in ciliates is unusually dynamic because ciliates possess two nuclei, a germ-line micronucleus and a somatic macronucleus. The micronucleus is transcriptionally inactive and serves to propagate the genome from one generation to the next. The macronucleus, by comparison, is transcriptionally active and contains amplified copies of chromosome fragments. During sexual reproduction (conjugation), the micronucleus undergoes meiotic cell divisions to yield four haploid micronuclei, three of which degenerate. The remaining micronucleus replicates, and one daughter nucleus is exchanged between conjugation partners. A new zygotic nucleus is then formed in each conjugating cell, and from this zygotic nucleus new macro- and micronuclei are generated. At this point, the old macronucleus is degraded.

This process of macronuclear degradation is especially intriguing because one nucleus in a cell survives while the other is degraded. Careful histologic observations have shown that the degradation process is similar to nuclear degradation in apoptotic cells. Nuclei become highly condensed and TUNEL positive. Nucleosome-sized DNA fragments are formed (Davis et al., 1992; Mpoke and Wolfe, 1996). The dying nucleus is surrounded by lysosomes, and nuclear acidification can be observed with the metachromatic dye acridine orange (Mpoke and Wolfe, 1997). The lysosomal marker acid phosphatase is enriched in the dying macronucleus, suggesting that final elimination of the old macronucleus is achieved by autophagy (Lu and Wolfe, 2001).

Recent work has shown that caspase inhibitors prevent degradation of the macronucleus in *Tetrahymena*, and caspase enzyme activity has been demonstrated in extracts of *Tetrahymena* cells (Ejercito and Wolfe, personal communication). Despite these similarities to apoptosis, caution is needed in drawing conclusions, since the caspase activity appears to be located in vacuoles, not in the cytoplasm, and to occur before nuclear degradation. Indeed, the observations are more similar to those for stalk-cell differentiation in *Dictyostelium* (see above), in which AIF, and not caspase, appears to the central regulator of the cell-death phenotype.

In addition to the nuclear death described above, there is evidence for cell density-controlled cell death in cultures of *Tetrahymena*. Cell death can be induced by cultivating cells at low cell density (Christensen et al., 1998) or by treatment with the protein kinase-inhibitor staurosporine, which presumably interferes with signal transduction (Straarup et al., 1997; Christensen et al., 2001). These results suggest that extracellular signals regulate cell survival in *Tetrahymena*. One such signal (*Tetrahymena* proliferation-activating factor, TPAF) has been isolated and shown to enhance survival at low cell density (Schousboe et al., 1998). Insulin-like signals have also been shown to prevent cell death in *Tetrahymena* cultures (Christensen et al., 1996). Despite the apparent similarity in survival signaling between *Tetrahymena* and higher eukaryotes, the phenotype of dying cells is quite different in *Tetrahymena*.

3.3 Programmed cell death in Trypanosoma

The last example of programmed cell death in single-celled eukaryotes that we consider concerns members of the Kinetoplastids, the parasites *Trypanosoma* and *Leishmania*, which cause disease in animals and man (Chagas' disease, African sleeping sickness, and kala azar). Kinetoplastids take their name from a giant mitochondrium, the kinetoplast, which is present in all members of this group. The life cycle of *Trypanosoma* involves alternation of hosts from insects to mammals. Epimastigotes are the proliferating stage of trypanosomes in the insect host. They differentiate into trypomastigotes that are arrested at G0/G1 and are transmitted to the vertebrate host. Here they differentiate into amastigotes that resume proliferation.

When the differentiation of epimastigotes into trypomastigotes was investigated in cell culture, it was observed that only a minority of epimastigotes were arrested at G0/G1 and differentiated into trypomastigotes. The majority underwent cell death that showed typical features of mammalian apoptosis, including TUNEL staining, DNA fragmentation into nucleosome-sized pieces, membrane blebbing, and cytoplasmic vacuolization (Ameisen et al., 1995). This form of programmed cell death is dependent on extracellular (autocrine) signals and has been interpreted as a means for epimastigotes to adjust their numbers in the insect host to levels that permit differentiation while ensuring the survival of the host. Similar observations were made with promastigotes of *Leishmania* (Moreira *et al.*, 1996) and with epimastigotes of *Trypanosoma brucei rhodesiense* (Welburn *et al.*, 1999).

Inhibitors of RNA and protein synthesis block cell death in trypanosomes, indicating that cell death requires *de novo* gene expression. Using a differential display method, Welburn *et al.* identified genes that were expressed in cells induced to die. Among those genes was a homolog of TRACK, a receptor for PKC, and a serine/threonine kinase, suggesting that cell–cell signaling could be involved in regulation of programmed cell death in *Trypanosoma*. A number of mitochondrial genes were also induced in cells during programmed cell death, including a mitochondrial RNA splicing protein, a mitochondrial transporter, cytochrome *c*, and prohibitin, a mitochondrial protein involved in cell-cycle regulation and tumor suppression (Welburn and Murphy, 1998).

These are all very interesting findings that may help to elucidate apoptotic mechanisms that arose very early in evolution. However, a complete picture is lacking at the moment. The difficulty in interpreting data relating to apoptosis in parasites in terms of evolution arises because these organisms are the result of a long process of adaptation to their multicellular hosts. For example, it has been shown that parasites can both induce and inhibit apoptosis in host cells, indicating that parts of the parasites' death machinery interacts with the death machinery of their host organisms (reviewed in (Barcinski and DosReis, 1999; Heussler *et al.*, 2001). It is hard to decide whether this is evidence for conservation of ancient pathways, or whether those pathways were acquired later as a result of the development of parasite–host adaptation.

4. Conclusions

In conclusion, we want to emphasize three points. First, it is now clear that forms of programmed cell death involving degradation of nuclear DNA and an organized elimination of dead cells were present very early in evolution before the development of multicellular organisms. Although our knowledge of these cell-death events is not complete, it appears that several apoptotic strategies were developed, some of which still operate in higher animals. Caspase-dependent apoptosis, however, seems to have evolved coincident with the development of the Eumetazoa, of which Cnidaria is the basic phylum.

Second, mitochondria appear to be causally involved in all forms of cell death, not only in the caspase-dependent pathway but also in caspase-independent pathways, one of which involves AIF. In the absence of caspases in higher animals, or in simpler life forms where caspases are not present, cell death can be induced by mitochondrial release of AIF, as, for example, in *Dictyostelium*.

Third, programmed cell death is a powerful instrument of 'social control' in cell populations, whether they are part of a colony or an organism. Thus, it remains a great challenge to understand the evolution of these mechanisms from their very first appearance to their remarkable refinement in higher animals, including man.

Acknowledgments

We thank our colleagues for permission to cite unpublished results. Dr. Stephanie Seipp provided the illustrations in Figure 2. Work from our laboratory has been supported by the DFG.

9. Programmed cell death in *C. elegans*

Anton Gartner, Arno Alpi and Björn Schumacher

1. Introduction

Apoptosis – the term will be used as a synonym for 'programmed cell death (PCD)' throughout this chapter – plays a crucial role in animal development and in maintaining tissue homeostasis. In this chapter, we will review studies on programmed cell death in *C. elegans*. We will start by summarizing classical studies that led to the definition of programmed cell death occurring during the development of *C. elegans*. We will emphasize the methods that are used to study programmed cell death in this model organism. In addition, we will describe biochemical interactions that occur during the execution of programmed cell death in *C. elegans*, and we will compare the molecular cell-death pathways in *C. elegans* to those in mammalian cells. Furthermore, we will discuss recent developments in the field of *C. elegans* cell death that mainly concern the engulfment of apoptotic corpses and the regulation of programmed cell death in the *C. elegans* germ line. Finally, we will try to elaborate on future research directions in the field of *C. elegans* cell death.

2. Identity and origin of cells undergoing programmed cell death during the somatic development of *C. elegans*

Genetic studies of programmed cell death in *C. elegans* take advantage of its highly reproducible and invariant somatic development. This reproducibility allowed the precise elucidation of the cell lineage that leads, within a time frame of approximately 3 days, to the generation of all of the 959 somatic cells that make up the adult hermaphrodite worm (*Figure 1*). Lineage analysis of embryonic and postembryonic cell-division patterns revealed that during somatic development of the hermaphrodite worm, 131 out of the total of 1090 cells born undergo programmed cell death (Sulston and Horvitz, 1977; Kimble and Hirsh, 1979; Sulston *et al.*, 1983). As is true of most other aspects of *C. elegans* development, these deaths show a high degree of uniformity with respect to the identity of dying cells as well as with

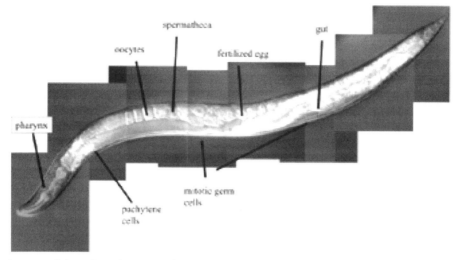

Figure 1. Adult *C. elegans* hermaphrodite as observed under DIC optics.
An adult worm measures about 1 mm in length. Various organs are indicated.

respect to the timing of each of these cell deaths during development. Although the apoptotic fate is restricted neither to particular cell types nor to any particular cell lineage, the bulk of apoptotic events affect neuronal cells and, to some extent, also hypodermal cells (Sulston and Horvitz, 1977; Kimble and Hirsh, 1979; Sulston *et al.*, 1983). The reproducibility of cell death provides unique advantages for genetic analysis: cell death can be studied on a single-cell level, and even mutations which cause only a very weak defect in programmed cell death, or which affect only a small number of cell types can be identified. Programmed cell death of somatic cells mostly occurs during embryonic development (113/131 deaths), which takes place within approximately 14 h under standard growth conditions. During embryogenesis, apoptosis occurs mostly between 220 and 440 min after fertilization (Sulston *et al.*, 1983). Programmed cell death also occurs, albeit to a lesser extent, during the transition through the four larval stages (Sulston and Horvitz, 1977; Kimble and Hirsh, 1979; Sulston *et al.*, 1983). Within the somatic tissues of the adult worm, neither cell divisions nor programmed cell deaths can be observed. However, recent studies suggest that apoptosis is very prominent in adult hermaphrodites during female germ-cell development (Gumienny *et al.*, 1999).

3. Identity and origin of cells undergoing programmed cell death in the *C. elegans* germ line

Only recently have the early observations, suggesting that programmed cell death can occur in the *C. elegans* germ line, been followed up (Gumienny *et al.*, 1999) (*Figure 1*). The *C. elegans* germ line proliferates both during larval development and in adult worms. Indeed, the germ line is the only proliferative tissue in the adult worm. In contrast to the almost invariably somatic *C. elegans* development, germ-line development is much more variable. As in mammalian tissues, populations of germ cells respond to growth-factor regulation and are subject to stochastic events, including the elimination of many cells by programmed cell death, to maintain tissue homeostasis. Furthermore, as in mammalian tissues, *C. elegans* germ cells, but not somatic cells, can respond to environmental stress, such as genotoxic stress, or to the stress caused by bacterial infections, with programmed cell death (see below). The only cell type within the germ line capable of undergoing programmed cell death is the meiotic pachytene stage (Gumienny *et al.*, 1999; Gartner *et al.*, 2000). Within the adult *C. elegans* hermaphrodite ovotestis, germ cells are only partially surrounded by a plasma membrane and are therefore part of a syncytium (Hall *et al.*, 1999; Seydoux and Schedl, 2001). By convention, these partially enclosed nuclei are referred to as cells. Within the germ line, germ cells progress through various stages of differentiation (*Figure 1*). At the distal end of the U-shaped ovotestis, mitotic precursor cells are generated throughout the worm's life. Upon passage through the transition zone, germ cells cease to divide and begin meiosis. The most abundant population of meiotic cells is the pachytene stage of meiosis I, residing between the transition zone and the bend of the gonad. Upon leaving the pachytene, germ cells progress into the meiotic diplotene, cellularize, undergo the final stages of oogenesis, and complete meiosis after fertilization upon passage through the spermatheca (Hall *et al.*, 1999; Seydoux and Schedl, 2001) (*Figure 1*). Under normal growth conditions, approximately 50% of female germ cells are doomed to programmed cell death (Gumienny *et al.*, 1999). These cells are eliminated rapidly by the corpse-engulfment machinery, resulting in a steady-state level of 0–4 morphologically apoptotic cells (Gumienny *et al.*, 1999).

4. Identification and quantification of programmed cell death in *C. elegans*

Cell death in *C. elegans* can be readily observed in living animals by standard Normasky optics (Sulston and Horvitz, 1977). The first sign of impending death is a decrease in the refractivity of the cytoplasm that occurs concomitant with a slight increase in the refractivity of the nucleus. Soon afterwards, both nucleus and cytoplasm become increasingly refractile until they resemble a flat disk. After about 10–30 min, this disk starts to disappear, and the nucleus of the dying cell decreases in refractility, begins to appear crumpled, and finally vanishes within less than 1 h (*Figure 2*) (Sulston and Horvitz, 1977). The morphology of the corpses as well as the kinetics of their disappearance is similar in somatic and germ cell apoptosis (*Figure 2*). However, as germ cells are only partially surrounded by a plasma membrane, the first step in germ cell death is the full cellularization of the apoptotic cell (Gumienny *et al.*, 1999).

Three distinct but related approaches have been used to quantitate programmed cell death in *C. elegans*. The first and most direct approach takes advantage of the highly reproducible anatomy of the worm, which allows the unambiguous identification of each cell in the body. The absence of programmed cell death can thus be scored indirectly by looking for the presence of extra 'undead' cells (cells that should have died but instead survived). Undead cells

(a)

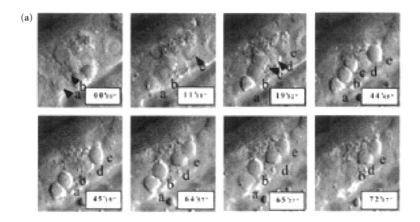

(b) **Genetic pathway for programmed cell death in *C. elegans***

egl-1 ➤ *ced-9* ····┊ *ced-4* ➤ *ced-3*   die (131 cells)

ced-9 ⊣ *ced-4* ···➤ *ced-3*   survive (959 cells)

Figure 2. Morphology of apoptotic cells and genetic pathway of programmed cell death
(a) Morphology of programmed cell deaths occurring in the germ line (adapted from Gartner *et al.*, 2000). Various morphologic phases of apoptosis under DIC (differential interference contrast) optics are depicted. Note, 'a' indicates a normal cell at 0 min, which becomes refractile at 11 min (the nucleus is still discernible at this stage). At 12 min, the nuclear structure merges with the surrounding cytoplasm, and this corpse-like structure persists until about 65 min and disappears within less than 1 min. Corpse morphology during somatic development resembles germ-cell corpse morphology (Gumienny *et al.*, 1999). (b) Genetic pathway for programmed cell death. The upper lane shows the status of cell-death signaling as a consequence of *egl-1* activation. The lower panel indicates the status of signaling in surviving cells. Solid arrows and solid T-bars indicate activation and repression, respectively. Dotted arrows and T-bars indicate that activation and repression do not occur. Reprinted from *Molecular Cell*, Vol. 5, A. Gartner, S. Milstein, S. Ahmed, J. Hodgkin and M. Hengartner. A conserved checkpoint pathway mediates DNA damage-induced apoptosis and cell cycle arrest in *C. elegans*, pp. 435–443 (2000), with permission from Elsevier Science.

cannot be distinguished from normal cells by their appearance under the microscope, but may sometimes be distinguished by their location (such as at a position where no cell is normally found). To increase the reliability of the assay, scoring of undead cells is usually performed in the pharynx, the animal's feeding organ, which is separated from the rest of the body by a clearly visible basement membrane (*Figure 1*). Within the pharynx, many programmed cell deaths occur; thus, a large number of deaths can be scored in a single animal, allowing the detection of even very weak effects on cell death (such as less than 2% extra cell survival) (Hengartner *et al.*, 1992; Hengartner and Horvitz, 1994). The drawback of this approach is that it scores the presence of cells that should not be there, rather than deaths per se. Consequently, care must be taken to confirm that extra cells are indeed the result of inhibition of death, rather than of extra cell divisions, or of aberrant cell migrations.

The second approach takes advantage of mutations in genes required for the efficient engulfment of apoptotic cells (see below). In these mutants, cells still die, but many dying cells fail to be engulfed and removed from the animal. These persistent, undegraded cell corpses are very obvious, even to the worm neophyte, and thus can be used as a simple assay for the extent of programmed cell death in the animal (Ellis *et al.*, 1991; Vaux *et al.*, 1992). Elimination of programmed cell death results in the absence of persistent cell corpses in these mutants. The main advantage of this assay is its ease of scoring. However, the number of persistent cell corpses is more variable than the number of surviving cells, and weak effects on cell death cannot be detected with this method.

The third method to determine programmed cell death is to identify dying cells by their distinct morphology with DIC (differential interference contrast) optics (*Figure 2*). For studying programmed cell death during somatic development, this approach is tedious, as only a few animals can be followed at a given time, and only animals at the proper stage of development yield useful information. In contrast, cell death occurring in the hermaphrodite germ line can be readily followed under DIC optics (Gumienny *et al.*, 1999; Gartner *et al.*, 2000) (*Figure 2*). Furthermore, apoptotic corpses can be visualized in living animals by staining with dyes such as acridine-orange (Gumienny *et al.*, 1999). Indeed, this staining method allows the detection of apoptotic corpses with standard GFP stereomicroscopes, enabling a pair of trained eyes to screen for the presence or absence of apoptotic corpses in a population of hundreds of worms within a few minutes.

5. Mutants define four distinct steps in the core apoptotic pathway

Genetic analysis has led to the identification of over 100 different mutations that affect programmed cell death. These mutations define more than 15 genes that affect all programmed cell deaths and a smaller number of genes that are needed to commit specific cells to the apoptotic fate (*Figures 2 and 3*). Since mutants which are defective in all programmed cell deaths are viable and now show obvious defect in their development or adult behavior, it was easy to combine various double-mutant combinations to build a genetic pathway for programmed cell death (Hedgecock *et al.*, 1983; Ellis and Horvitz, 1986; Desai *et al.*, 1988; Ellis *et al.*, 1991; Ellis and Horvitz, 1991; Hengartner *et al.*, 1992). Indeed, one of the future challenges in the *C. elegans* cell-death field will be the identification of essential genes that also affect programmed cell death.

Apoptosis can be separated into four distinct steps, each of which is defined by the analysis of various mutants (*Figures 2 and 3*). (i) Initially, specific cell types are committed to the apoptotic fate. (ii) Subsequently, the general apoptotic machinery, which is used in all dying cells, is activated. (iii) Later, the recognition and engulfment of dying cells by a neighboring cell proceeds. (iv) Finally, the remnants of engulfed cells are degraded. Genes of the first type generally affect very few cells or cell types, whereas the genes falling into the subsequent classes affect all cell deaths. These genes will be described below (*Figures 2 and 3*).

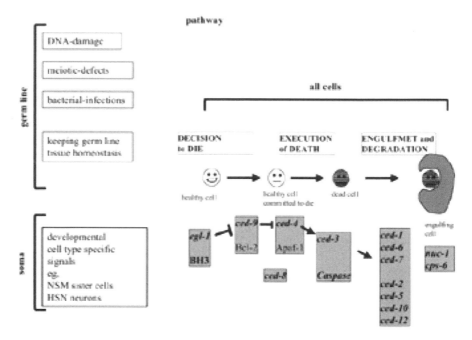

Figure 3. The genetic pathway of programmed cell death in C. elegans.
Various developmental and environmental clues that can activate programmed cell death are indicated in the left panels. Arrows indicate activation; T-bars indicate repression.

6. Four genes, *egl-1*, *ced-3*, *ced-4*, and *ced-9*, define a genetic pathway needed for (almost) all programmed cell deaths

The original isolation by Ellis and Horvitz (1986) of mutants defective in all cell deaths during nematode development was successful because all cells that are fated to die use a common pathway to execute programmed cell death (Ellis and Horvitz, 1986) (*Figure 3*). Searching for the desired mutants was simplified by screening in a genetic background where dead cells are easily recognized because they cannot be engulfed by neighboring cells (see section on mutants defective in the engulfment process). Later screens also focused on the identification of cell-death mutants with extra undead cells (Ellis and Horvitz, 1991; Hengartner *et al.*, 1992). As a result, these screens, initially numerous loss-of-function (lf) alleles of both *ced-3* (cell death abnormal) and *ced-4*, and a single gain-of-function (gf) allele of *ced-9* were identified (*Figure 3*) (Ellis and Horvitz, 1991; Hengartner *et al.*, 1992). The loss of function alleles of *ced-3* and *ced-4* can be classified into allelic series, the strongest alleles of which presumably result in a complete loss of gene function that leads to the complete inhibition of all programmed cell death. 'Undead cells' that fail to die tend to acquire a fate similar to their ancestors, but, interestingly, they lose the capacity for further division.

   The isolation of *ced-9* loss-of-function (lf) mutants was accomplished by looking for intragenic revertants of a rare *ced-9* gain-of-function allele (Hengartner *et al.*, 1992). The analysis of these revertants revealed extensive apoptosis of cells that normally do not die within homozygote *ced-9* loss-of-function animals. As a consequence of these extensive deaths, the affected animals die during embryogenesis (Hengartner *et al.*, 1992). These observations suggest that *ced-9* normally functions to prevent programmed cell death (*Figures 2 and 3*). Genetically, *ced-9* acts upstream of both *ced-3* and *ced-4*, as loss-of-function alleles of these genes can suppress the extensive cell-death phenotype of *ced-9* (lf) animals (Hengartner *et al.*, 1992) (*Figures 2 and 3*). Interestingly, embryos resulting from homozygous strong *ced-9* (lf) mutants can die at a very early developmental stage without any overt morphologic indication of programmed cell death, suggesting that *ced-9* might have an essential function in addition to its antiapoptotic role (Hengartner *et al.*, 1992). The fourth gene that affects all developmental cell death but that evaded discovery in the initial genetic screens is *egl-1* (egg laying defective) (Conradt and Horvitz, 1998) (*Figure 3*). Interestingly, it turned out that *egl-1* loss-of-function alleles are defective in all programmed cell deaths occurring during somatic development. *egl-1* function was determined to be upstream of the other known cell-death components because of the suppression of programmed cell death caused by the ectopic overexpression of *egl-1* in *ced-3* (lf), *ced-4* (lf), and *ced-9* (lf) mutants (Conradt and Horvitz, 1998; 1999).

Mosaic analysis suggests that *ced-3* and *ced-4* are likely to act cell-autonomously, indicating that they are needed to act within the cells that die (Yuan and Horvitz, 1990). This conclusion is further supported by experiments showing that *ced-3* and *ced-4* overexpression, in neuronal cells that usually do not die, is sufficient to induce programmed cell death (Shaham and Horvitz, 1996). In addition, these studies indicated that *ced-3* is likely to function downstream of *ced-4* because the induction of apoptosis by the overexpression of *ced-3* does not require *ced-4*, whereas, in the converse experiment, the induction of programmed cell death by *ced-4* overexpression requires the presence of *ced-3* (Shaham and Horvitz, 1996) (*Figure 2*).

The following statements of our general understanding of programmed cell death could be based on the studies described above. (i) The execution of programmed cell death during development, which, formally, may also be considered as a somewhat unusual terminal differentiation program, is subject to precise genetic control. (ii) The execution of the apoptotic fate is an active program that can be best described as cellular suicide; meaning a cell that senses that it is doomed to die participates actively in this process by inducing (or helping to induce) its own demise. (iii) Finally, all single cells may have the potential to undergo programmed cell death.

7. *egl-1*, *ced-3*, *ced-4*, and *ced-9* are functionally conserved throughout evolution

Positional cloning of the above-mentioned genes revealed that all these nematode cell-death genes have homologs in mammals that perform similar functions in the control of apoptosis. *ced-9* encodes a protein sharing 24% overall sequence identity with the mammalian Bcl-2 oncogene, which, like *ced-9*, negatively regulates cell death (Hengartner and Horvitz, 1994). As with *ced-9*, overexpression of *bcl-2* protects cells from dying, whereas *bcl-2* (lf) mutations make cells hypersensitive to death-inducing signals (for reviews, see Huang, 2000; Ferri and Kroemer, 2001). *ced-3* encodes a protein with similarity to a family of death-inducing proteases called caspases (Yuan *et al.*, 1993). CED-4 is related to mammalian Apaf-1, which in mammals acts together with caspase-9, caspase-3, and cytochrome *c* in the core apoptotic program (Yuan and Horvitz, 1992; Li *et al.*, 1997; Zou *et al.*, 1997). Finally, EGL-1 is related to the mammalian BH3-only-domain proteins, such as Bid, Bad, Bim, or Puma, all of which are implicated in proapoptotic signaling (Yuan and Horvitz, 1992; Yang *et al.*, 1995; Wang, K. *et al.*, 1996; Conradt and Horvitz, 1998; O'Connor *et al.*, 1998; Nakano and Vousden, 2001; Yu *et al.*, 2001; Wu and Deng, 2002).

Interestingly, of all the mentioned *C. elegans* cell-death genes, only one functional family member is encoded within the worm genome. *C. elegans* contains only one CED-9/bcl-2-like protein and only one CED-4/ apaf-1 look-alike. In contrast, *C. elegans* contains, besides CED-3, two additional caspases that do not seem to function in cell-death regulation (1998) (Shaham, 1998). Similarly, besides EGL-1, there is one additional protein encoded in the worm genome, containing a BH3 sequence motif (1998). The function of this protein is not known. It is currently still unknown whether cytochrome *c* and its release from mitochondria play a role in *C. elegans* cell-death regulation. The fact that there are two cytochrome *c* copies encoded in the worm genome and the fact that cytochrome *c* is an essential gene hindered genetic approaches to determine a potential proapoptotic function of cytochrome *c* in worm cell-death signaling. Similarly, it has not been analyzed whether the worm homolog of mammalian AIF (apoptosis-inducing factor) is implicated in worm programmed cell death. *C. elegans* also encodes two homologs of the IAP (inhibitor of apoptosis) protein family (Fraser *et al.*, 1999; Speliotes *et al.*, 2000). These proteins, which were initially found as bacolovirus proteins able to prevent cell death of host cells, are also implicated as antiapoptotic proteins in *Drosophila*. However, the two *C. elegans* IAP proteins do not seem to play a role in the regulation of programmed cell death. Indeed, *C. elegans* BIR-1 is involved in the regulation of cytokinesis and chromosome segregation similarly to the budding and fission yeast counterparts of this protein family (Fraser *et al.*, 1999; Speliotes *et al.*, 2000). In addition, the *C. elegans* genome does not encode any of the known components required for mammalian receptor-mediated, cell-death signaling. In summary, based on the finding that *C. elegans* has only one component of the core apoptotic pathway, *C. elegans* might contain the evolutionarily most simple cell-death pathway. Genome-sequencing and extensive EST projects with Protozoa and sponges will reveal whether these organisms also contain those four components of the core apoptotic pathway. It is also possible that there exist other, still unidentified components of the core programmed cell-death pathway in the worm. These putative genes might have eluded discovery because they are essential or because they act in redundant pathways.

8. EGL-1, CED-3, CED-4, and CED-9 are part of a molecular framework regulating programmed cell death

Biochemical studies indicate that the EGL-1, CED-9, CED-4, and CED-3 are part of a molecular pathway that regulates programmed cell death. The pattern of interactions between those proteins suggests a relatively simple molecular model (*Figure 4*). In cells that are supposed to survive, the proapoptotic activity of CED-4 and CED-3 is kept at bay by the binding of CED-4 to CED-9. CED-9 is localized at the outer mitochondrial membrane (Chen, F. *et al.*, 2000). At present, it is not clear whether CED-9 localization at the mitochondrial membrane is indeed necessary for antiapoptotic CED-9 activity. Likewise, it is unclear whether permeabilization of the outer mitochondrial membrane is necessary for the induction of programmed cell death, as is the case in mammalian cells (Hengartner, 2000). According to the model, apoptosis is initiated, at least during the execution of programmed cell death during development, by the transcriptional induction of the BH3-only-domain protein EGL-1 (Conradt and Horvitz, 1998; Conradt and Horvitz, 1999). Upon transcriptional induction, EGL-1 is supposed to bind to CED-9, thereby releasing CED-9-bound CED-4 (Conradt and Horvitz, 1998; del Peso *et al.*, 2000; Parrish *et al.*, 2000). The above model is supported by the *in vitro* interaction patterns of those proteins, as well as by the finding that *egl-1* is transcriptionally induced in many cells that are destined to die (Conradt and Horvitz, 1999). It should be noted that models describing *C. elegans* and mammalian cell-death activation differ with respect to CED-9 CED-4 interaction, as there is mounting evidence that the homologs of the two proteins Bcl-2 and Apaf-1 do not directly interact *in vivo* (Conus *et al.*, 2000). In addition, the order of protein function is supported by the fact that overexpression of both CED-3 and CED-4 can trigger programmed cell death in *C. elegans*, as well as upon transfection into mammalian cells (Shaham and Horvitz, 1996; Chinnaiyan *et al.*, 1997a,b; Wu, D. *et al.*, 1997). Upon liberation from CED-9, CED-4 translocates to the nuclear membrane *in vivo* (Chen *et al.*, 2000b). This translocation can be induced by *egl-1* overexpression, and it precedes the activation of programmed cell death; therefore, it might be an

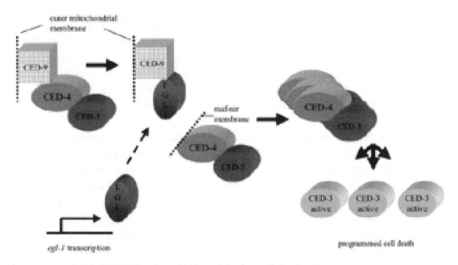

Figure 4. A molecular model for the activation of the core cell-death program.
The corresponding model is described in the text.

integral part of cell-death activation (Chen *et al.*, 2000b). Consequently, CED-4 is thought to confer CED-3 activation by its association with a homopolymeric complex that is needed to bring inactive CED-3 molecules into close proximity (via CED-4 binding), thereby leading to the activation of programmed cell death via CED-3 caspase activation (Yang, X. *et al.*, 1998).

9. Modulators and downstream components of the cell-death pathway

Very little is known about components that modulate the cell-death pathway or about targets of the cell-death pathway. *ced-8* causes an overall slowing down of the cell-death process (Ellis *et al.*, 1991; Stanfield and Horvitz, 2000). As compared to the wild-type animals, an equal number of programmed cell deaths occur, and these deaths are initiated at the same time as in wild-type animals. When *ced-8* is combined with very weak alleles of *ced-3* or *ced-4*, the weak cell-death defect of these two mutations is enhanced. *ced-8* encodes a protein similar to the human XK transmembrane protein, and worm *ced-8* does indeed localize to membranes (Stanfield and Horvitz, 2000). However, it is unclear how *ced-8* impinges on the regulation of the core apoptotic pathway. Another component that modulates programmed cell death was discovered via its interaction with *ced-4*. Upon overexpression of *mac-1*, which encodes an AAA type ATPase, the level of programmed cell death is reduced (Wu *et al.*, 1999b). Unfortunately, it could not be determined whether *mac-1* disruption leads to elevated levels of programmed cell death, as *mac-1* is essential for development. *nuc-1* encodes a worm DNAse II ortholog that was implicated in efficient DNA fragmentation during programmed cell death (Lyon *et al.*, 2000). *nuc-1* is also needed for the degradation of DNA in the digestive tract of *C. elegans* (Lyon *et al.*, 2000; Wu *et al.*, 2000). However, it seems that several other Dnases must also be involved in DNA degradation (see below). With respect to DNA fragmentation, corpses from *nuc-1* animals proceed to an intermediate stage that, in contrast to wild-type animals, allows the detection of apoptotic corpses due to accumulation of DNA 3′-hydroxyl ends via the Tunnel method. *nuc-1*-mediated DNA fragmentation is partially affected by corpse engulfment genes (see below), as Tunnel staining is reduced in *ced-1* and *ced-7* corpse engulfment-defective mutants. In contrast to the situation in mammals, no CPAN/CAP caspase-activated nuclease/caspase-activated DNAse nuclease activity has been defined in *C. elegans*.

For identification of further targets of *ced-3* caspase, a very elegant genetic screen was initiated. Assuming there might be redundancy in the cell-death pathway after the caspase step, Parrish *et al.* (2001) devised a sensitized genetic system for the efficient detection of mutations that enhance weak cell-death defects. As part of the experimental strategy, an activated *ced-3* caspase, as well as a GFP reporter, is expressed in six nonessential mechanosensory neurons. Enhancement of cell death can be easily detected by a reduced number of surviving mechanosensory cells (reduced number of GFP dots). One mutant (*cps-6*) identified is likely to encode the worm ortholog of endonuclease G (Parrish *et al.*, 2001). As with *ced-8* endonuclease G, inactivation enhances the weak cell-death defects of weak *ced-3* and *ced-4* mutations, suggesting that endonuclease G might act downstream of the *ced-3* caspase (Parrish *et al.*, 2001). It is not clear whether endonuclease G is cleaved by *ced-3*. Interestingly, worm endonuclease G localizes to mitochondria, like its mammalian otholog, which also has been implicated in programmed cell death (Li *et al.*, 2001; Parrish *et al.*, 2001). It will be interesting to see whether this genetic screen will uncover additional downstream components of the cell-death pathway.

10. The regulation of programmed cell death in specific cell types

In contrast to our advanced understanding of the molecular nature of the core apoptotic pathway, very little is known about the genes that commit specific subsets of cells to apoptotic death (recent developments concerning the regulation of programmed cell death in the germ line will be discussed below). In principle, there are two different classes of mutants that might affect cell type-specific deaths. Mutants might act indirectly, by affecting genes that are needed for specific cells to acquire their proper developmental cell fate. Alternatively, genes might be directly involved in the regulation of the cell-death machinery in specific target cells. The best-studied mutation that affects cell type-specific cell death is the *egl-1* gain-of-function mutation that causes the inappropriate death of both hermaphrodite HSN neurons. These two neurons are used in hermaphrodites for the innervation of vulval muscle cells to regulate egg laying. Consequently, and hence the name, egg laying is defective in *egl-1* (*gf*) animals (Desai *et al.*, 1988). In males, which do not need to lay eggs, the HSNs are also generated, but subsequently eliminated by programmed cell death. Interestingly, it turned out that all *egl-1* (gf) mutations are within a DNA binding site of the *tra-1* sex determination gene (Conradt and Horvitz, 1999). In hermaphrodites, TRA-1 acts at the *egl-1* promoter as a transcriptional repressor of the induction of programmed cell death of the HSN neuron by inhibiting the expression of *egl-1* transcription (Conradt and Horvitz, 1999). In males, TRA-1 does not bind to the transcriptional repressor; hence, HSNs die in male animals.

Another example of genes that are likely to control directly the induction of apoptosis in a specific subset of cells are the *ces* (cell-death specification) genes, which affect the apoptosis of the sister cells of the pharyngeal NSM neurons. Mutations in the two *ces* genes (*ces-1* and *ces-2*) define a linear pathway that leads to the cell type-specific activation of apoptosis (Ellis and Horvitz, 1991). Epistasis analysis reveals that within this pathway *ces-2* seems to be a negative upstream regulator of *ces-1*. The evidence for the epistatic relationship is derived from the analysis of various single and double mutants. *ces-1* gain-of-function mutants prevent death of the NSM sisters that usually die (as well as preventing death of two additional cells, the sisters of the I2 neurons). NSM sister cell survival is also triggered by the loss of function of *ces-2*. Since the cell survival caused by loss of *ces-2* function is suppressed by loss of *ces-1* function, *ces-1* probably acts downstream of *ces-2*. *ces-1* is likely to act upstream of the core apoptotic machinery since the observed cell death in *ces-1(lf)* is still readily suppressed by *ced-3* mutants (Ellis and Horvitz, 1991; *C. elegans* Sequencing Consortium, 1998). The *ces-2* gene encodes a putative bZIP transcription factor, suggesting that some nematode cell deaths might be regulated at a transcriptional level (Ellis and Horvitz, 1991; Metzstein *et al.*, 1996). *ces-1*encodes a Snail family zinc-finger protein, most similar in sequence to the *Drosophila* neuronal differentiation protein Scratch (Ellis and Horvitz, 1991; Metzstein and Horvitz, 1999). CES-2 may directly repress *ces-1* transcription by binding to a transcriptional element within the *ces-2* promoter (Ellis and Horvitz, 1991; Metzstein and Horvitz, 1999). Transcription of *ces-1* dependent genes may be required to prevent death of the NSM sister cells (Ellis and Horvitz, 1991; Metzstein and Horvitz, 1999). It is unclear whether *egl-1* transcription is inhibited by *ces-1* in NSM sister cells. Thus, a transcriptional cascade may be needed to control death of specific cells such as the NSM sister cells in *C. elegans*.

11. Germ-line cell death

The finding that programmed cell death occurs not only in the developmentally determined somatic cell lineage but also in the germ line of hermaphroditic worms has initiated a host of new discoveries. The *C. elegans* germ line is the only proliferating tissue in the adult animal. The distal tip cell generates germ cells and instructs them to divide mitotically until they reach the transition zone (Seydoux and Schedl, 2001). Here they initiate meiosis and proceed until diakinesis (Seydoux and Schedl, 2001). Upon fertilization in the spermatheca, the oozytes resume meiosis, giving rise to embryos (*Figure 1*). Gumienny *et al.* (1999) noticed the presence of about 0–4 germ-cell corpses at any given time in the adult hermaphroditic germ line. Germ-cell apoptosis is correlated with age, as the number of corpses increases with the age of the worm (Gumienny *et al.*, 1999). As a consequence, over 300 germ cells, corresponding to 50% of all germ cells produced, die through programmed cell death in an adult worm (Gumienny *et al.*, 1999). These deaths occur only in the female gonad in cells of the pachytene phase of meiosis I. Since these deaths occur independently of environmental stimuli, they were termed physiologic germ cell deaths. Interestingly, physiologic germ-cell death is dependent on the worm ras/map kinase pathway that is needed for the commitment of meiotic pachytene cells to enter the meiotic diplotene stage. In the absence of ras/map kinase signaling, germ-cell death cannot occur. It is unclear whether this dependency is due to a direct connection of map kinase signaling with proapoptotic genes. Alternatively, the effect of map kinase signaling might be indirect, as only cells committed to leave the pachytene via the activation of map kinase signaling have the capacity to apoptose (Gumienny *et al.*, 1999). Like somatic apoptosis, physiologic germ-cell death is dependent on *ced-3*and *ced-4*, and the engulfment of corpses requires the same set of genes as during somatic cell death. However, the somatic cell-death trigger *egl-1* is not required for physiologic germ-cell death (Gumienny *et al.*, 1999). Therefore, at least one additional, germ-line-specific apoptotic trigger has been postulated, but its identity has remained elusive.

In addition to physiologic germ-cell death, it was recently discovered that environmental stimuli can trigger programmed germ-cell death. Infection with a virulent form of the bacterium *Salmonella typhimurium* induces apoptosis in the *C. elegans* germ line (Aballay and Ausubel, 2001). *C. elegans* is usually fed on a lawn of *E. coli*, but when Aballay and Ausubel left worms with *S. typhimurium* as their only food source, they observed elevated levels of germ-cell death (Aballay and Ausubel, 2001). Bacteria-induced programmed germ-cell death requires all components of the core apoptotic pathway, including *egl-1*. Interestingly, worms mutant for *egl-1*, *ced-3*, *ced-4*, or *ced-9 (gf)* die earlier when fed on *S. typhimurium* than their wild-type counterparts (Aballay and Ausubel, 2001). It will be interesting to define the specific pathways that lead to bacteria-induced apoptosis.

Genotoxic stress, such as ionizing irradiation, can also lead to programmed cell death in *C. elegans* (Gartner *et al.*, 2000). Upon irradiation, germ cells activate checkpoint pathways, whose activation leads to a transient cell-cycle arrest and to programmed cell death (*Figure 5*). Interestingly, those two DNA damage responses are spatially separated. Whereas mitotic germ cells halt cell-cycle progression, meiotic pachytene cells undergo apoptosis (Gartner *et al.*, 2000). Cells outside the germ line show neither of these responses (Gartner *et al.*, 2000). Like *Salmonella*-induced germ-cell death, as well as physiologic germ-cell death, radiation-induced apoptosis appears to be restricted to the hermaphroditic germ line, requires *ced-3* and *ced-4*, and is also suppressed by *ced-9* loss-of-function mutations (Gartner *et al.*, 2000). In addition, radiation-induced apoptosis is partially dependent on *egl-1* (Gartner *et al.*, 2000). Consistent with the notion that double strand breaks may cause radiation-induced programmed cell death, the level of apoptosis is dramatically enhanced in mutants that are defective in double-strand break repair, such as *mre-11* and *rad-51* (Gartner *et al.*, 2000; Boulton *et al.*, 2002).

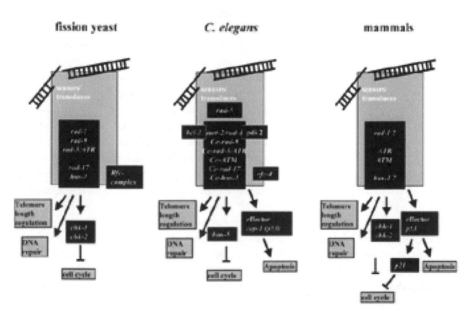

Figure 5. Simplified model of DNA damage checkpoint pathways in fission yeast, *C. elegans*, and mammals.
Simplified model of DNA damage checkpoint signaling in various organisms. Note that apoptosis does not occur in yeasts. Worm p53 affects programmed cell death, but not cell-cycle arrest. It has not been shown whether yeast or mammalian homologs of *rad-5* have a function in DNA damage checkpoint signaling.

Screens for mutants that are defective in DNA damage-induced apoptosis led to the discovery of *mrt-2*, *rad-5*, and *op241* (*Figure 5*). These mutants are also defective in cell-cycle arrest and render the worms hypersensitive to DNA damage (Gartner *et al.*, 2000). *mrt-2* was found to be required not only for DNA damage checkpoint but also for the regulation of telomere replication (Ahmed and Hodgkin, 2000; Gartner *et al.*, 2000). When propagated over several generations, *mrt-2* homozygous worms accumulate progressively shorter telomeres, leading to genomic instability and to the accumulation of chromosomal end-to-end fusion, and ultimately resulting in sterility (Ahmed and Hodgkin, 2000). Positional cloning of *mrt-2* revealed that this gene is the worm homolog of the *Schizzosaccharomyces pombe rad1* and the *Saccharomyces cerevisiae rad17* checkpoint genes. *rad1/rad17* has previously been shown to be involved in yeast DNA damage checkpoints (Ahmed and Hodgkin, 2000). In yeast, checkpoint signaling involves a group of proteins needed for the recognition of DNA damage and for the transduction of the damage signal to effect cell-cycle arrest (*Figure 5*) (for reviews, see Caspari and Carr, 1999; Foiani *et al.*, 2000; Murakami and Nurse, 2000; Rhind and Russell, 2000; Walworth, 2000; Abraham, 2001; Wahl and Carr, 2001). *Schizzosaccharomyces pombe* Rad9p, Rad1p, and Hus1p form a complex that structurally resembles a PCNA sliding DNA clamp, and has recently been shown, *in vivo*, to associate to DNA close to double-strand breaks in a Rad17p-dependent manner (Kondo *et al.*, 2001; Melo *et al.*, 2001). In addition, the *S. cerevisiae* homolog of mammalian ATM/ATR Mec1 also localizes to damaged DNA, and its kinase activity is activated in response to DNA damage. The DNA damage signal is then relayed via the scMec1p/spRad3p/ATM/ATR kinases to CHK1 and CHK2 kinases, which cause cell-cycle arrest via phosphorylation of key cell-cycle proteins. Given that *mrt-2* is involved in radiation-induced apoptosis, it was reasonable to assume that worm homologs of yeast checkpoint genes are also defective in DNA damage checkpoint responses. RNAi inactivation of these candidate worm checkpoint genes led to the inactivation of radiation-induced cell-cycle arrest (Boulton *et al.*, 2002) (*Figure 5*). However, defects in radiation in induced programmed cell death could be induced only at a low penetrance, probably because these checkpoint genes could be only partially inhibited in meiotic pachytene cells by RNAi (Boulton *et al.*, 2002). Generation of true checkpoint gene mutation will therefore be necessary to assign an unambiguous function to genes in radiation-induced programmed cell death.

rad-5 was the first conserved checkpoint gene whose function in the DNA damage checkpoint was defined in the *C. elegans* system (Hartman and Herman, 1982). *rad-5* is an essential gene, and the two known *rad-5* mutations are temperature-sensitive lethals (Hartman and Herman, 1982). Cloning of the *C. elegans rad-5* revealed that this gene is related to *S. cerevisiae tel-2*, an essential gene shown to be involved in telomere-length regulation (Ahmed *et al.*, 2001). *rad-5* function in telomere regulation is less clear, as telomere length in *rad-5* mutants fluctuates but does not become progressively either longer or shorter (Ahmed *et al.*, 2001). Interestingly, the checkpoint gene *rad-5* was found to be allelic with *clk-2* (Ahmed *et al.*, 2001). *clk* genes were found to slow the worm's development and to extend its lifespan (Lakowski and Hekimi, 1996; Benard *et al.*, 2001; Lim *et al.*, 2001). In *clk-2(qm37)*, this lifespan extension is only weak, and it is unclear whether this effect is merely an indirect consequence of the slow-growth phenotype of *clk-2 (qm37)* mutation.

A recent functional genomics-based approach has led to the identification of additional genes involved in worm checkpoint responses. Boulton *et al.* (2002) cloned all worm homologs of yeast checkpoint and DNA repair genes. Subsequently, all these genes were used for two-hybrid analysis to identify interaction partners (Boulton *et al.*, 2002). The potential checkpoint function of these interaction partners was than assessed by RNAi (*Figure 5*). This approach led to the identification of a worm homolog of the mammalian Bcl-3 oncogene needed for DNA damage signaling (Boulton *et al.*, 2002). It will be interesting to see whether the mammalian homolog has a similar defect in DNA damage signaling.

In addition to the checkpoint genes that affect both radiation-induced cell-cycle arrest and cell death, only one gene-product is known that affects only radiation-induced cell death. Although it initially evaded detection by conventional homology searches, bioinformatic approaches using generalized profiles revealed that the worm genome encodes for a distant homolog of the mammalian p53 tumor suppressor gene termed *cep-1* (*C. elegans* p53-like) (*Figure 5*) (Derry *et al.*, 2001; Schumacher *et al.*, 2001). However, sequence alignments revealed that many of the p53 residues that are implicated either in DNA binding or in oncogenesis are conserved in *cep-1* (Derry *et al.*, 2001; Schumacher *et al.*, 2001). Unlike mammalian p53 but like *Drosophila* p53, *cep-1* is not required for DNA damage-induced cell-cycle arrest (Derry *et al.*, 2001; Schumacher *et al.*, 2001). In addition to its function in cell-cycle arrest, Derry *et al.* (2001) showed an involvement of *cep-1* in the stress response and some involvement in meiotic chromosome segregation, as *cep-1*-mutant animals showed weak defects in meiotic chromosome segregation. It will be interesting to determine the transcriptional targets of *cep-1* that mediate programmed cell death. Furthermore, it will be important to determine why only germ cells are able to undergo programmed cell death in response to radiation.

The DNA damage checkpoint is activated not only after genotoxic insult but also upon meiotic defects, such as the accumulation of unprocessed meiotic recombination intermediates. As part of a normal meiotic division, double-strand breaks, which are needed for meiotic recombination, are generated by the meiotic endonuclease SPO-11 (for review, see Roeder, 1997). At a following step of meiotic recombination, double-strand breaks are resected to generate single-stand overhangs. These single-strand overhangs invade the homologous chromosome via a strand-exchange reaction that is mediated by the conserved RAD-51 strand-exchange protein (Roeder, 1997). Worms lacking RAD-51 are thought to accumulate unprocessed SPO-11-induced double-strand breaks (Rinaldo *et al.*, 1998; Takanami *et al.*, 1998). These breaks are recognized by the same checkpoint proteins as the proteins sensing radiation-induced double-strand breaks, and they result in elevated levels of germ-cell apoptosis (Gartner *et al.*, 2000). The observation that meiotic defects can lead to increased levels of germ-cell death can be employed in readily isolating mutations defective in meiotic chromosome pairing and recombination (Pawel Pasierbeck, Josef Loidl and Anton Gartner, unpublished observation).

The above data indicate conclusively that the DNA damage checkpoint pathways are conserved throughout evolution (*Figure 5*). However, the apoptotic response to DNA damage exists only in metazoans, since yeast does not have apoptosis. It is therefore conceivable that the p53 pathway is the link between the conserved DNA damage checkpoint pathway that already exists in yeasts and the metazoan specific apoptotic machinery. It will be of great interest to determine the molecular link between *cep-1* and the core apoptotic machinery, which is likely to involve transcriptional activation of *cep-1* target genes. Furthermore, additional genetic screens are likely to uncover novel potentially metazoan-specific DNA damage checkpoint genes.

12. Genes required for the engulfment of dying cells

The final executing steps in apoptosis are the engulfment by other cells and the degradation of dying cells (cell corpses). The engulfment process, which removes cell corpses before they can lyse and release harmful cytoplasmic contents, is highly efficient and lasts less than 1 h. Cell corpse engulfment is a multistep process that involves the recognition of the dying cell followed by the subsequent enclosure by expanding pseudopodia of the engulfing cell (Ellis *et al.*, 1991). The engulfment, observed by ultrastructural studies, can already occur even before any morphologic changes in the dying cell appear (Ellis *et al.*, 1991). In *C. elegans*, no professional phagocytes exist. Therefore, neighboring cells usually take up this function. Many cells have the potential to recognize and engulf apoptotic cells. During embryogenesis, dying cells are often engulfed by their sister cells, whereas postembryonic cell corpses are removed, usually by epithelial cells (Sulston and Horvitz, 1977; Hedgecock *et al.*, 1983). In contrast to the fixed fate of dying cells, engulfment involves stochastic events, as various cells can engulf a specific dying cell (Hoeppner *et al.*, 2001). In the *C. elegans* germ line, engulfment by gonadal sheath cells (these somatic cell surround the germ line) starts to occur immediately after the syncytial germ-line nuclei become cellularized during programmed cell death (Liu and Hengartner, 1998; Wu and Horvitz, 1998; Gumienny *et al.*, 1999).

So far, seven genes, *ced-1*, *ced-2*, *ced-5*, *ced-6*, *ced-7*, *ced-10*, and *ced-12*, have been found to be required for efficient engulfment of dying cells (Hedgecock *et al.*, 1983; Ellis *et al.*, 1991) (*Figure 3*). Mutations in any of these genes result in a persistent corpse phenotype. Dying cells in engulfment-defective worms display all morphologic changes indicative of apoptotic corpses, including the appearance of the highly refractive disks under Normarski optics. Corpses in engulfment-defective animals can stay for an extended period of time, disappear by secondary necrosis, or be phagocytized with delay (Hedgecock *et al.*, 1983; Ellis *et al.*, 1991; Wu and Horvitz, 1998).

Interestingly, two recent papers suggest that there is an unexpected link between the cell-death machinery and the corpse-engulfment machinery that can be best explained by a positive feedback regulation between cell-death and cell-engulfment signals (Hoeppner *et al.*, 2001; Reddien *et al.*, 2001). These studies took advantage of properties conferred by weak loss-of-function alleles of *ced-3* and *ced-4* that presumably result in threshold levels of caspase activity (Hoeppner *et al.*, 2001; Reddien *et al.*, 2001). These alleles result in only a weak cell-death defect, and approximately 80% of the programmed cell deaths that normally occur still happen. When the sequence of various cell-death stages was precisely followed by lineage analysis in these weak cell-death mutants, several unexpected observations were made (Hoeppner *et al.*, 2001; Reddien *et al.*, 2001). In many cases, the onset of death was later than normal, and the execution of cell death was slowed down as in the situation in *ced-8* mutants, as one would expect if there were only threshold levels of caspase activity. Similarly, in some instances, early stages of cell death occurred, but the cells eventually returned to normal nonapoptotic morphology and remained alive. In some cases, however, potential interactions between the cell-death and the cell-engulfment machinery were revealed. It was observed that corpses failed to be engulfed, indicating a caspase defect that might block the engulfment process. In other cases, corpses were engulfed at stages that occurred much earlier than in wild-type animals. In addition to these morphologic observations, double mutant analysis between weak *ced-3* and engulfment mutations also suggests an intricate link between the regulation of apoptosis and corpse engulfment, as engulfment mutations enhance the cell-death defects conferred by weak cell-death mutations (Hoeppner *et al.*, 2001; Reddien *et al.*, 2001). It will be interesting to clarify the pathways that communicate between the apoptosis and the engulfment pathways.

On the basis of genetic studies on the generation of many double-mutant combinations, engulfment genes can be placed into two distinct, partially redundant pathways involving the *ced-1*, *ced-6*, and *ced-7* and the *ced-2*, *ced-5*, *ced-10*, and *ced-12* group of engulfment genes (Ellis *et al.*, 1991). Various double-mutants between the two groups of mutants, but not within a single group, show a significantly elevated number of persistent cell corpses compared to single mutants. In addition to defects in corpse engulfment, *ced-2*, *ced-5*, *ced-10*, and *ced-12* also show a defect in the migrations of the gonadal distal tip cells (DTCs), resulting in abnormally shaped gonads that frequently include extragonadal bends or bends at inappropriate positions (Ellis *et al.*, 1991; Gumienny *et al.*, 2001). The migration defect seems to be specific to the DTC, as *ced-5* mutants appear to be normal in the migration of other cells, such as the HSN neurons.

All seven *C. elegans* engulfment genes have been cloned, and their homologs have been identified in vertebrate species, including humans. These studies have allowed us to begin to understand how engulfing cells recognize dying cells, and what the 'eat me' signals sensed by the dying cells are (*Figure 6*). Studies in vertebrates have identified a number of molecules that may participate in the recognition step of the engulfment process (for review, see Fadok *et al.*, 2001; Feller, 2001; Schlegel *et al.*, 2000; May and Machesky, 2001). At a certain stage, apoptotic cells induce the exposure of phosphatidylserins on the surface, an effect that is associated with a general loss of the phospholipid asymmetry. Further changes in surface carbohydrates and surface charges might also be involved in the recognition by the phagocytes, and a number of mammalian receptors have been thought to interact with ligands on the apoptotic cell surface. The *C. elegans* CED-1 protein bears structural similarities to many transmembrane receptors, the most similar of which is the human SREC (scavenger receptor from endothelial cells) that contains 16 EGF-like motifs (Zhou *et al.*, 2001b). Moreover, CED-1 contains a NPXY motif

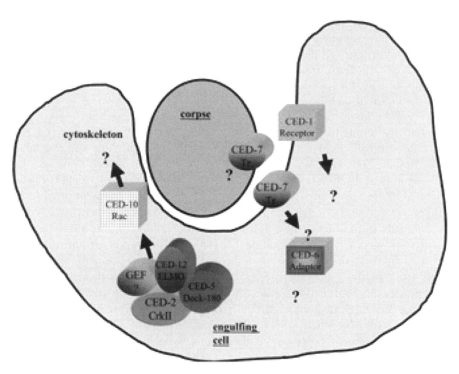

Figure 6. Simplified model of corpse engulfment in *C. elegans.*

and a YXXL motif (Zhou *et al.*, 2001b). These motifs are very common in the cytoplasmic domains of transmembrane receptors, where they often direct phosphorylation of tyrosine residues and the binding to adapter proteins. Consistent with CED-1's being a receptor, it is indeed localized to the cell surface, particularly on cell types that can function as engulfing cells, and the presence of dying cells induces CED-1 clustering around these cells (*Figure 6*) (Zhou *et al.*, 2001b). CED-1 clustering is independent of its cytoplasmic domain but dependent on the function of CED-7 (Zhou *et al.*, 2001b). It is not known what the actual substrate of the CED-1 receptor is. *ced-7* is the only engulfment gene known to play a role also in dying cells (Wu and Horvitz, 1998). CED-7 is similar to members of the ABC transporter superfamily that effect ATP-dependent translocation of specific substrates across cellular membranes (*Figure 6*) (Wu and Horvitz, 1998). Given these findings, two possible models for CED-1 and CED-7 function have been suggested. First, CED-7 may facilitate physical contact between dying cell and engulfing cell by exporting unknown adhesion molecules to the cell surface of dying and engulfing cells. In a second alternative model, CED-7 may function differently in dying and engulfing cells (Wu and Horvitz, 1998). In engulfing cells, it might be required for the clustering of CED-1. In dying cells, it might induce the exposure of apoptotic surface markers signaling the presence of cell corpses to neighboring cells that might be recognized by CED-1. CED-7 is most similar to the subfamily ABC-1, and ABC-1 family members have indeed been reported to promote the redistribution of phosphatidylserins to the outer layer of the plasma membrane (Luciani and Chimini, 1996). This model is intriguing, especially since phosphatidylserins have been supposed to play a crucial role in the recognition of dying cells (Schlegel *et al.*, 2000; Fadok *et al.*, 2001; Feller, 2001; May and Machesky, 2001). *ced-6*, which acts downstream of *ced-1* and *ced-7*, encodes for an adapter protein whose primary sequence suggests that it might interact with a tyrosine kinase pathway (*Figure 6*) (Liu and Hengartner, 1998). *ced-6* is functionally conserved, as human *ced-6* can complement *ced-6* mutations (Liu and Hengartner, 1999). It will be interesting to determine those signaling molecules that interact with CED-6.

ced-2, *ced-5*, *ced-10*, and *ced-12* encode homologs of mammalian CrkII, DOCK180, Rac, and ELMO, respectively (Wu and Horvitz, 1998; Reddien and Horvitz, 2000; Gumienny *et al.*, 2001; Wu *et al.*, 2001b; Zhou *et al.*, 2001a). Recent studies suggest that this subgroup of engulfment genes defines a conserved Rac signaling pathway that regulates reorganization of the actin cytoskeleton of the engulfing cell during corpse engulfment.

The small GTPase *ced-10*/Rac belongs, together with Rho and Cdc42, to a subgroup of the Ras-GTPase superfamily. The Rac GTPase has been implicated in the control of actin polymerization needed for membrane ruffling, cell motility, the formation of lamellipodia, and phagocytosis (Tosello-Trampont *et al.*, 2001). Genetic studies in the worm and biochemical analysis in mammalian cells suggest that CED-2/Crk, CED-5/DOCK180, and CED-12/ELMO function as upstream activators of CED-10/Rac (*Figure 6*) (Gumienny *et al.*, 2001; Wu *et al.*, 2001b). Human CrkII is a cytoskeleton-associated adapter protein consisting of a SH2 (Src-homology-2) domain followed by two C-terminal SH3 (Src-homology-3) domains, and it has been implicated in integrin-mediated signaling and cell movement (Matsuda *et al.*, 1991). CrkII localizes to focal adhesions with p130[CAS] and DOCK180, which was isolated on the basis of its physical interaction with the first SH3 domain of CrkII (Hasegawa *et al.*, 1996; Erickson *et al.*, 1997). CED-12 shares homology with members of the gene family ELMO, defined by three human orthologs, ELMO1, 2, and 3, and one ortholog in *Drosophila* (Gumienny *et al.*, 2001; Zhou *et al.*, 2001a). Both DOCK180 and ELMO1, under the control of a *C. elegans* heat-shock promoter, can rescue the migration defect of the DTC, but not the engulfment defect in *ced-5*-mutant animals. Detailed interaction analysis suggests an evolutionarily conserved trimeric complex between CED-12/ELMO, CED-5/DOCK180, and CED-2/CrkII (*Figure 6*) (Gumienny *et al.*, 2001). How does this putative ELMO/DOCK180/CrkII complex activate the Rac GTPase? Ras-related GTP-binding proteins function as molecular switches that cycle between GTP-bound 'on' and GDP-bound 'off' states regulating downstream effector molecules. Exchange of the bound GDP is facilitated by guanidine-nucleotide-exchange factors (GEFs). Several experiments have been done to determine the possible Rac-GEF activity of the ELMO:DOCK180:CrkII complex. However, it is unlikely that the ELMO:DOCK180:CrkII complex acts as a GEF (Gumienny *et al.*, 2001). *In vitro* assays with purified components did not reveal an associated GEF activity. Furthermore there is no Rac-GEF domain in ELMO, DOCK180, and CrkII. However, coimmunoprecipitation experiments with overexpressed human ELMO1and DOCK180 revealed a Rac GEF activity (Gumienny *et al.*, 2001). It is therefore likely that apoptotic signals induce the formation of an active trimeric ELMO/DOCK180/CrkII complex, which, in turn, becomes associated with a yet to be identified Rac-GEF activity, leading to Rac activation (Gumienny et al., 2001).

13. Potential functions of programmed cell death in *C. elegans*

Given the fact that *C. elegans* mutants defective in apoptosis remain viable and do not seem to have any gross deleterious effects in development, fertility, and behavior, we must ask what the biologic function is of programmed cell death in *C. elegans* (Ellis and Horvitz, 1986). If we consider developmental programmed cell death only, careful analysis reveals that mutant animals develop more slowly, have a slightly reduced fertility, and are affected in complex behavioral tasks such as chemotaxis. Although this lack of an overall phenotype provides advantages for the propagation of mutant animals, it leaves us with the question of the function of cell death during the development of *C. elegans*. The lack of gross alterations in the behavior of animals containing about 12% more cells due to the absence of apoptotic death seems especially surprising, as many of the 'undead cells,' although they never further divide, terminally differentiate (Ellis and Horvitz, 1986). Some of them, at least partially, take over the fate of cells closely related to them by lineage An illustrative example of such a process is provided by the 'undead' sister of the M4 neuron (Avery and Horvitz, 1987). The M4 neuron is needed for the proper operation of the pharynx in young animals, and its killing (as in laser microsurgery) leads to the starvation of operated animals. However, if the M4 cell is ablated in animals defective for apoptosis, its undead sister cell can take over its function and rescue the animals from death by starvation (Avery and Horvitz, 1987). As a matter of fact, the best case for the importance of programmed cell death during worm development might be made by evolutionary considerations. Morphogenetic modeling, as accomplished by dooming specific cells to undergo apoptosis, might be an efficient evolutionary strategy to remodel morphologic structures. Indeed, the death of a single cell can account for major anatomical changes. For example, the gross difference between the one-armed gonad of *Panagrellus revidus* and the two-armed gonad of *C. elegans* can be explained by the death of the posterior distal tip cell in *P. revidus* (DTCs direct the formation of gonadal arms) (Sternberg and Horvitz, 1981).

In view of the importance of programmed cell death occurring in the germ line, the various germ-line intrinsic cell-death pathways have to be considered separately. Spontaneous germ-cell death was supposed to occur because dying germ cells might function as nurse cells for oocytes, a hypothesis supported by the following observations (Gumienny et al., 1999). Within the germ line, mitotic and meiotic germ-cell nuclei are only partially enclosed by a plasma membrane and are thus part of a syncytium. Until the onset of diakinesis, these nuclei are characterized by a high nuclear to cytoplasmic volume ratio, a ratio that is reversed when oozytes are formed. After dying, apoptotic cells separate from the germ-line syncytium and extrude most of their cytoplasm to be used by growing oocytes in the germ-line syncytium. Only oocytes have to accumulate a high volume of cytoplasm; accordingly, male meiotic cells do not undergo apoptosis. Alternatively, germ-cell apoptosis could also be explained by developmental checkpoints that lead to the demise of naturally occurring compromised germ cells. Similarly, simple alterations in the metabolism of germ cells could also lead to programmed cell death. Indeed, this hypothesis is supported by the fact that a large number of genetic loci lead to excessive germ-cell apoptosis (S. Milstein, Anton Gartner, Pawel Pasierbeck and Michael Hengartner, unpublished observation). Besides the above-considered basal level of germ-cell death, germ-cell apoptosis can also be initiated by bacterial infection as well as by meiotic defects and genotoxic stress. The best case for biologic significance can be made for bacterially induced cell death, as worms defective in programmed cell death have a dramatically reduced live span as compared to their wild-type counterparts (Aballay and Ausubel, 2001). As to DNA damage-induced programmed cell death, it seems that the induction of programmed cell death by DNA damage checkpoints has only a minor biologic function. *mrt-2*, *rad-5*, and *cep-1* checkpoint mutants, as well as the general cell-death mutants *ced-3* and *ced-4*, are equally defective in inducing DNA damage-induced programmed cell death, but in contrast to the strong radiation sensitivity of *mrt-2* mutants, *rad-5* radiation sensitivity is very low, and *cep-1* as well as *ced-4* and *ced-3* animals are almost as radiation resistant as wild-type animals (Gartner *et al.*, 2000; Ahmed *et al.*, 2001; Schumacher *et al.*, 2001). Given that *cep-1* animals as well as *ced-3* and *ced-4* animals are able to undergo checkpoint induced cell cycle arrest, whereas *rad-5* animals are defective in both DNA damage-induced programmed cell death and cell-cycle arrest, it seems that DNA damage checkpoint-induced DNA repair (defective in *mrt-2*) is the most important damage response, and damage-induced apoptosis is the least important damage response (Gartner *et al.*, 2000; Ahmed *et al.*, 2001; Schumacher *et al.*, 2001). It is unclear why *C. elegans* evolved damage-induced programmed cell death. It is likely that the biologic significance of this response is to eliminate meiotic cells with defects in the regulation of meiotic recombination, a suggestion supported by the finding that checkpoint mutants, as well as the *cep-1* mutants, have a weak meiotic chromosome segregation defect (Derry *et al.*, 2001).

14. Conclusion

C. elegans has proved to be an invaluable model system for the study of programmed cell death. However, many key problems concerning the regulation of programmed cell death still have to be solved. It will be interesting to determine further key targets of the *ced-3* caspase. The various upstream pathways that lead to the activation of the cell-death machinery upon developmental and environmental signals have to be further elucidated. In addition, some emphasis has to be placed on the further elaboration of the corpse-engulfment and degradation pathway. It is thus likely that studies using *C. elegans* will continue to shed light on the conserved mechanisms of programmed cell death that may guide studies in vertebrate apoptosis.

Acknowledgments

Work performed in the Gartner laboratory is supported by the Max Planck Society (Erich Nigg) and by DGF grants DFG 703/1 and DFG 703/2 to Anton Gartner.

10. | Apoptosis in *Drosophila*

Kristin White

1. Introduction

Genetically tractable model systems have proven invaluable in identifying molecules important for apoptosis, and for understanding how these molecules interact to regulate apoptosis during development, and in response to stress. Two widely used models, the nematode *C. elegans* and the fruit fly *Drosophila melanogaster*, demonstrate distinct advantages for these studies. Genetic screens in *C. elegans* identified the ced-3 caspase as an essential effector of apoptosis, and the ced-4/Apaf-1 adapter proteins and the ced-9 and egl-1 bcl-2 family members as critical regulators of caspase activation (reviewed in Chapter 9). Work in a second genetic system, *Drosophila*, has uncovered a different caspase regulatory mechanism. As we describe below, developmental cell death in flies is regulated largely by the anti-apoptotic IAP proteins, and the proapoptotic Reaper (Rpr), Grim, and Hid proteins. The identification of different pathways in these two systems probably reflects differences in the relative importance of these two regulatory systems in worm and fly apoptosis, as well as distinct biases of the screens used to uncover these pathways. These findings also highlight the advantages of looking at the same pathways in multiple model systems.

2. Genetic screens uncover genes important in developmental apoptosis

Early studies on developmental cell death in flies revealed that a significant number of cells die during development, and that the death of these cells resembles apoptosis as described in mammalian cells, both ultrastructurally and biochemically (Abrams *et al.*, 1993). A simple assay for apoptosis was developed that facilitated the examination of cell death in *Drosophila* embryos. This assay was based on the finding that acridine orange specifically stained apoptotic cells in living embryos.

This allowed the overall pattern of apoptosis to be rapidly assessed in large numbers of embryos.

Using this assay, genetic screens were undertaken for mutations that perturbed the overall level of apoptosis in the embryo (White *et al.*, 1994). A collection of genomic deletions representing about half of the genome was screened for defects in apoptosis. In these screens, large numbers of embryos were collected from parents heterozygous for a genomic deletion, and defects in apoptosis were analyzed in the 25% of the progeny homozygous for the deletion. Although most of these deletions are large, deleting many functions that are required for normal embryonic development, the large supply of maternal gene products is generally sufficient to allow embryonic development to proceed long enough to assess apoptosis.

The majority of deletion strains examined in this screen produced no quantitative defects in apoptosis in homozygous embryos. Embryos homozygous for a number of genomic deletions were found to have a significant increase in the overall level of apoptosis. This may reflect the loss of antiapoptotic functions, such as DIAP1 (see below). In addition, general defects in development can also result in widespread apoptosis (see, for example, Pazdera *et al.*, 1998).

Three overlapping genomic deletions completely eliminated apoptosis in homozygous embryos (White *et al.*, 1994). These homozygous embryos died at the end of embryogenesis, and contained more cells than wild-type embryos. It is very likely that the presence of extra cells in these embryos is a direct result of the lack of apoptosis. However, since the deletions are relatively large and eliminate a number of genes, embryonic lethality cannot be attributed directly to the lack of apoptosis.

Attempts were made to generate single-gene mutations that showed the same cell death-defective phenotype as the deletion. This proved unsuccessful, as the loss of more than one gene in the interval is necessary to see the cell death-defective phenotype. Four genes, reaper (rpr) (White *et al.*, 1994), head involution defective (hid) (Grether *et al.*, 1995), grim (Chen *et al.*, 1996), and the newly identified sickle (skl) (Christich *et al.*, 2002; Srinivasula *et al.*, 2002; Wing *et al.*, 2002), have now been identified in this genomic region. These genes lie clustered together, within about 250 kb. The protein products of all four genes share a short stretch of homology at the N-terminus, which is important for their function. Otherwise the genes are fairly dissimilar. Deletions that remove only one of these genes have no or very weak effects on embryonic apoptosis, while the removal of two has stronger effects (Grether *et al.*, 1995; Chen *et al.*, 1996; Peterson *et al.*, 2002). Deletion of the genes hid, rpr, and grim together results in a complete block in embryonic apoptosis. A role for skl in developmental apoptosis remains to be demonstrated.

3. How Rpr, Hid, Grim, and Skl initiate apoptosis

Overexpression of Rpr, Hid, Grim, or Skl is sufficient to induce ectopic apoptosis in some tissues of the developing fly and in mammalian and insect cultured cells (Grether *et al.*, 1995; Chen *et al.*, 1996; Pronk *et al.*, 1996; White *et al.*, 1996; Evans *et al.*, 1997; Claveria *et al.*, 1998; McCarthy and Dixit, 1998; Haining *et al.*, 1999). Apoptosis is induced on overexpression of Rpr, Hid, or Grim in embryos homozygous for the deletion that removes all three of these genes (Grether *et al.*, 1995;

Chen *et al.*, 1996; White *et al.*, 1996). This demonstrates that each gene is sufficient to induce apoptosis in the absence of the others. Expression of Skl appears to be a less potent inducer of cell death, although it can enhance death induced by Rpr or Grim (Christich *et al.*, 2002; Srinivasula *et al.*, 2002; Wing *et al.*, 2002).

Analysis of mutant embryos lacking rpr, hid, and grim suggests that these proteins are likely to act as initiators rather than effectors of apoptosis. When these mutant embryos are exposed to high doses of x-rays, some apoptosis is induced (White *et al.*, 1994). Thus, at least some of the effector machinery is present in these embryos. Caspase activation is essential for the proapoptotic activity of Rpr, Hid, Grim, and Skl (Grether *et al.*, 1995; Chen *et al.*, 1996; White *et al.*, 1996; Srinivasula *et al.*, 2002; Wing *et al.*, 2002). Coexpression of the broad-spectrum caspase inhibitor p35 can block cell death induced by ectopic expression of all four genes.

How do Rpr, Hid, Grim, and Skl activate caspases to initiate apoptosis? The first clue came from experiments in the laboratory of Lois Miller, demonstrating that Rpr, Hid, and Grim were able to bind to IAPs (inhibitor of apoptosis proteins), a family of antiapoptotic proteins (Vucic *et al.*, 1997; 1998). Originally identified in baculovirus, antiapoptotic IAPs are found in insects and mammals (reviewed in Chapter 3). IAPs are characterized by the presence of one to three copies of a conserved Zn binding motif called the baculovirus IAP repeat or BIR domain. In addition, many IAPs also contain a C-terminal ring-finger domain. Both of these motifs appear to be essential for IAP function. IAPs exert their antiapoptotic activity by inhibiting caspases.

Three IAPs DIAP1, DIAP2, and Deterin, have been characterized in *Drosophila* (Hay *et al.*, 1995; Jones *et al.*, 2000). Overexpression of DIAP1 or DIAP2 has been shown to suppress apoptosis induced by overexpression of Rpr, Hid, or Grim in flies and in cultured cells (Hay *et al.*, 1995; Vucic *et al.*, 1997, 1998). In cultured cells, apoptosis induced by overexpressed Rpr can be suppressed by overexpression of Deterin as well (Jones *et al.*, 2000). DIAP1 can also partially inhibit apoptosis resulting from Skl expression in insect cells (Srinivasula *et al.*, 2002; Wing *et al.*, 2002).

A role for DIAP2 or Deterin in inhibiting developmental apoptosis has not been demonstrated. However, decreased levels of DIAP1 enhance killing by ectopically expressed Rpr, Grim, and Hid (Hay *et al.*, 1995; Lisi *et al.*, 2000). This suggests that DIAP1 is necessary to prevent apoptosis. Strong support for this hypothesis comes from the phenotype of embryos homozygous for DIAP1 mutations. Loss of DIAP1 function results in massive apoptosis early in embryogenesis (Wang, S.L. *et al.*, 1999; Goyal *et al.*, 2000; Lisi *et al.*, 2000). Importantly, the apoptosis resulting from the absence of DIAP1 is not dependent on Rpr, Grim, or Hid function. DIAP1 is therefore required to inhibit cell death, and Rpr, Hid, and Grim inhibit DIAP1 function.

Direct physical interaction between Rpr, Grim, Hid, or Skl and DIAP1 is likely to be required for inhibition of DIAP1 antiapoptotic activity. Mutant alleles of DIAP1 have been identified that block apoptosis induced by ectopic expression of Rpr and Hid (Goyal *et al.*, 2000). These mutations result in DIAP1 proteins that can no longer be bound by Rpr or Hid, supporting a role for direct physical interactions between the proteins.

The *Drosophila* caspases Dronc, Dcp-1, and drICE are found to bind to DIAP1 (Kaiser *et al.*, 1998; Hawkins *et al.*, 1999; Meier *et al.*, 2000). DIAP1 binds to the proform of Dronc and Dcp-1, and to the cleaved, activated form of drICE. Diap1 binding inhibits the activity of these caspases. Thus, DIAP1 is probably present in many or all cells, blocking caspase activation and activity. Supporting this idea is the finding that the *Drosophila* Apaf-1 homolog, Ark, is required for apoptosis resulting from decreased levels of DIAP1 (Rodriguez *et al.*, 2002; Zimmermann *et al.*, 2002). Apaf-1 in mammalian cells promotes the activation of certain caspases (see Chapter 7). If DIAP1 is required to block the activity of caspases activated by Ark, the absence of Ark could lower the general level of caspase activity in the cell, decreasing the requirement for caspase inhibition.

Rpr, Grim, Hid, and Skl bind to the second BIR domain of DIAP1 through the conserved 14-amino-acid stretch at their N-terminus (Vucic *et al.*, 1998; Wang *et al.*, 1999c; Wu, J.W. *et al.*, 2001; Srinivasula *et al.*, 2002). Deletion of N-terminal sequences abrogates the proapoptotic activity of Hid, as well as binding of Hid to DIAP1 (Vucic *et al.*, 1998). Deletion of this N-terminal motif from Rpr and Grim does not completely eliminate the killing activity of Rpr and Grim. Surprisingly, both of these truncated proteins still bind to DIAP1 when the proteins are co-over-expressed (P. Bangs and K. White, unpublished observations; Vucic *et al.*, 1998). This suggests that additional domains of both Rpr and Grim may act as DIAP1 inhibitory motifs. Other differences between the mechanisms of Rpr and Grim killing and Hid killing are suggested by the finding that specific gain of function alleles of DIAP1 can inhibit Rpr and Grim killing, but not Hid killing, or can inhibit Hid killing but have no appreciable impact on Rpr or Grim killing (Lisi *et al.*, 2000). These mutant proteins might be predicted to show differential abilities to bind Rpr and Grim in comparison with Hid.

Obvious sequence homologs of Rpr, Grim, and Hid have not been identified in mammalian systems. However, two mammalian proteins have recently been shown to act as functional homologs of these proteins. SMAC/Diablo and Omi/HtrA2 are both mitochondrial proteins that can be released when apoptosis is induced (Du *et al.*, 2000; Verhagen *et al.*, 2000; Suzuki *et al.*, 2001; Hegde *et al.*, 2002; Martins *et al.*, 2002; van Loo *et al.*, 2002; Verhagen *et al.*, 2002). Both proteins potentiate caspase activation, and can bind to mammalian IAPs. Importantly, the first four amino acids of the processed forms of SMAC/Diablo and Omi/HtrA2 are homologous to the N-terminal IAP-inhibitory domains of Rpr, Grim, Hid, and Skl, and bind to the same structural domain of the IAPs. It is likely that this IAP inhibitory domain will also be found in other mammalian proteins.

4. Caspases are required for Rpr-, Grim-, Hid-, and Skl-induced apoptosis

Seven caspases have been identified in *Drosophila*. Dredd/Dcp-2 (Inohara *et al.*, 1997; Chen *et al.*, 1998), Dronc/Nc (Dorstyn *et al.*, 1999b), and Strica/Dream (Doumanis *et al.*, 2001) appear to be initiator caspases, as they contain long N-terminal prodomains. Dronc has a caspase recruitment or CARD domain, while the Dredd prodomain shows some similarity to the death effector domain (DED). The long serine/threonine-rich prodomain of Strica bears no obvious homology to other caspase prodomains. The caspases Dcp-1 (Song *et al.*, 1997), drICE (Fraser and Evan, 1997), Decay (Dorstyn *et al.*, 1999b), and Damm (Harvey *et al.*, 2001) have the shorter prodomains consistent with executioner caspases.

Dronc is expressed in many cells of the developing embryo (Dorstyn *et al.*, 1999b). Genetic reduction of Dronc indicates that Dronc mediates apoptosis initiated by Rpr and Hid (Quinn *et al.*, 2000). In the embryo, reduction of Dronc by the RNA interference technique (reviewed in Sharp, 1999) suggests that Dronc is required for most developmental apoptosis. Interestingly, Dronc expression is induced by the steroid hormone ecdysone, which is a key regulator of apoptosis during metamorphosis (Dorstyn *et al.*, 1999b). No mutant alleles of Dronc have been isolated, so the role of Dronc in ecdysone-induced death has not yet been assessed.

Ectopically expressed Dronc can induce cell death in yeast cultured cells and cells in the developing *Drosophila* eye (Dorstyn *et al.*, 1999b; Hawkins *et al.*, 2000; Meier *et al.*, 2000). Dronc has been shown to have an interesting substrate specificity: it is autoprocessed after a glutamate, but cleaves drICE after a more typical aspartate (Hawkins *et al.*, 2000; Meier *et al.*, 2000). Dronc activation appears to require the activity of the *Drosophila* Apaf-1 homolog, Ark (Quinn *et al.*, 2000; Dorstyn *et al.*, 2002).

Like Dronc, the long prodomain containing caspase Dredd physically interacts with Ark (Rodriguez *et al.*, 1999). Although mutations in Dredd suppress apoptosis induced by Rpr and Grim, apoptosis induced by Hid is not effectively suppressed (Chen *et al.*, 1998). This supports the idea that different caspases may function downstream of the apoptotic inducers. This specificity also appears to extend to the executioner caspases, as described below.

The pattern of Dredd expression is particularly interesting (Chen *et al.*, 1998). Dredd transcripts appear to accumulate in cells that are fated to die in the embryo. Surprisingly, this accumulation is dependent on the expression of the apoptotic inducers Rpr, Grim, and Hid, as embryos homozygous for a deletion of these genes do not show wild-type Dredd expression. The mechanism by which this regulation of Dredd expression occurs is not yet known.

A role for Dredd outside apoptosis has also been uncovered. The innate immune response in flies is mediated in part by NFκB-dependent transcription of antibacterial and antifungal peptides (Hoffmann and Reichhart, 1997). Mutations in Dredd suppress the antibacterial response (Elrod-Erickson *et al.*, 2000; Leulier *et al.*, 2000). Overexpression of an upstream mediator of this response, Imd, results in increased Rpr expression and apoptosis (Georgel *et al.*, 2001). It will be interesting to test whether Rpr itself might play a role in innate immunity.

The Strica prodomain is high in serine and threonine residues, and shares no obvious homology to other caspase prodomains (Doumanis *et al.*, 2001). It is widely expressed during development, and accumulates in some tissues fated to undergo apoptosis. Although its structure suggests that it is an apical caspase, no role has yet been defined for Strica in developmental apoptosis. One unique feature of Strica is that it can bind to both DIAP1 and DIAP2, while Dronc, Dcp-1, and drICE have only been found to bind to DIAP1.

Among the putative effector caspases, Dcp-1, drICE, and Decay most resemble mammalian caspase-3 (Kumar and Doumanis, 2000). Mutants in Dcp-1 have been characterized (Song *et al.*, 1997; McCall and Steller, 1998). Although these mutants show no detectable defects in embryonic apoptosis, Dcp-1 plays an important role in oogenesis. If the female germ line lacks Dcp-1, the transfer of nurse-cell contents to the developing egg is inhibited. This nurse-cell 'dumping' appears to be a modified form of apoptosis. Dcp-1-mutant animals also have other defects that may indicate a role for this caspase in other developmental events. Activation of Dcp-1 is facilitated by Rpr and Grim, but not Hid, expression, suggesting that this caspase may act selectively downstream of Rpr and Grim (Song *et al.*, 2000).

Specific roles for drICE, Decay, and Damm have not yet been identified. However, depletion of drICE from extracts of *Drosophila* S2 tissue-culture cells inhibits some apoptotic activity in these extracts (Fraser *et al.*, 1997). Expression of catalytically inactive, and presumably dominant negative, Damm can block apoptosis induced by ectopically expressed Hid, but not Rpr, again suggesting specificity in the downstream caspases (Harvey *et al.*, 2001).

5. Modifiers of caspase activation: the mitochondrial pathway

One provocative difference between the *C. elegans* and mammalian apoptotic pathways is the function of the ced-4/Apaf-1 caspase activator. In worms, the apoptosome contains the caspase ced-3, the adapter ced-4, and the bcl-2-like protein ced-9, with ced-9 inhibiting ced-4 activation of ced-3 (reviewed in Chapter 9). This inhibition is relieved by the proapoptotic activity of the BH3 protein, egl-1. The mammalian apoptosome also consists of a caspase (usually caspase-9), and an adapter (Apaf-1). However, Apaf-1-mediated activation is inhibited by the c-terminal WD repeats of Apaf-1. Binding of cytochrome *c* to this domain relieves this inhibition (reviewed in Chapter 7). Bcl-2 family members seem to act upstream, regulating the release of mitochondrial proteins, including cytochrome *c*.

Drosophila appears to share characteristics with both of these systems. The Apaf-1/ced-4 homolog, Ark, is clearly important for the full response to Rpr, Grim, and Hid activation, and interacts with at least two apical caspases, Dronc and Dredd (Kanuka *et al.*, 1999; Rodriguez *et al.*, 1999; Zhou *et al.*, 1999). One form of Ark contains a WD domain similar to Apaf-1, and this form binds to cytochrome *c*. Provocatively, a short form of Ark has been reported by one group, and this form lacks the WD repeats, making it more similar to ced-4 (Kanuka *et al.*, 1999). This suggests that Ark may be activated independently of cytochrome *c* binding. Cytochrome *c* appears to undergo conformational changes during apoptosis in *Drosophila* cells. However, it does not appear to be released in large amounts from the mitochondria, as in mammalian cells (Varkey *et al.*, 1999; Dorstyn *et al.*, 2002; Zimmermann *et al.*, 2002). Some data suggest that cytochrome *c* is not required for all apoptosis in *Drosophila* cells (Zimmermann *et al.*, 2002).

Ark is broadly expressed during development, with higher levels of expression in a limited number of cells in the embryo (Kanuka *et al.*, 1999; Rodriguez *et al.*, 1999; Zhou *et al.*, 1999). Ark-mutant embryos have reduced levels of cell death, particularly in the central nervous system (CNS) and epidermal regions, and the larval nervous systems are substantially larger than those of the wild type and have excess cells (Kanuka *et al.*, 1999; Rodriguez *et al.*, 1999; Zhou *et al.*, 1999). Adults show a range of phenotypes, including abnormal wings, extra bristles, extra photoreceptors, and male sterility. The CNS of the larvae is also enlarged, a phenotype similar to that seen in Rpr mutants (see below). These phenotypes suggest that Ark is required for some, but not all, developmental apoptosis in the fly. RNA interference experiments also suggest a role for Ark in stress-induced apoptosis in S2 cells (Zimmermann *et al.*, 2002). In this study, Rpr- and Grim-induced apoptosis was not affected by the loss of Ark.

It is not yet known how (or whether) the fly Bcl-2 family members regulate mitochondrial function or Ark activity. Two Bcl-2 family members have been described in *Drosophila*, Debcl/dBorg-1/dRob-1 (Brachmann *et al.*, 2000; Colussi *et al.*, 2000; Igaki *et al.*, 2000) and Buffy (Flybase, 1994; Brachmann *et al.*, 2000; Colussi *et al.*, 2000). Buffy has not been well characterized. Debcl contains BH1, BH2, and BH3 domains, as well as a C-terminal transmembrane domain. Overexpression studies indicate that Debcl is able to induce apoptosis in a variety of tissues, while loss of function studies indicate that Debcl is required for some developmental apoptosis.

There is some evidence that Debcl has both pro- and antiapoptotic activities. Expression of low levels of Debcl protects cells from apoptosis induced by serum withdrawal (Brachmann *et al.*, 2000). However, expression of Debcl in the eye sensitizes cells to apoptosis induced by UV (Brachmann *et al.*, 2000). Some Debcl overexpression phenotypes are suppressed by coexpression of the broad-spectrum caspase inhibitor p35, while others are not (Colussi *et al.*, 2000; Igaki *et al.*, 2000; Zimmermann *et al.*, 2002). This suggests that in some situations Debcl may induce cell death by both caspase-dependent and caspase-independent mechanisms.

The mechanism of Debcl function is unclear. Although Debcl can bind to a number of the antiapoptotic Bcl-2 family members from other species (Colussi *et al.*, 2000), there is no evidence for an antiapoptotic Bcl-2 family member in flies. Genetic data indicate that decreased DIAP1 function enhances the ability of Debcl to induce apoptosis, while decreased Ark function suppresses Debcl-induced apoptosis (Colussi *et al.*, 2000). This may reflect a role for Debcl in Ark-mediated facilitation of caspase activation, or it may simply reflect a more general ability of Ark and DIAP1 to modulate caspase activation in cells. Decreased levels of DIAP1 are found to enhance many apoptotic inducers, probably because more active caspases are present in these cells. Decreased Ark may simply turn the rheostat the other way.

6. How is the mitochondrial pathway activated?

Genetic data indicate that Ark is required for some developmental apoptosis. However, as described above, it may be that decreased Ark levels simply lower the amount of activated caspase in the cell. Expression of Rpr, Grim, Hid, or Skl and subsequent inhibition of DIAP1 would simply result in lower levels of active caspase in the absence of Ark, and decrease the chance that a cell would undergo apoptosis. This model is supported by the finding that decreased Ark levels suppress apoptosis resulting from decreased DIAP1 activity (Rodriguez *et al.*, 2002; Zimmermann *et al.*, 2002).

Alternatively, it is possible that Rpr, Grim, and Hid activate the mitochondrial pathway, either directly or indirectly, perhaps through a caspase-mediated step. It is possible that the *Drosophila* homolog of Scythe connects Rpr to the mitochondrial pathway. Scythe is a Rpr-binding protein identified in *Xenopus* extracts (Thress *et al.*, 1998). Rpr interaction with Scythe releases a proapoptotic factor, which can release cytochrome *c* from mitochondria (Thress *et al.*, 1999). Thus, Rpr binding to Scythe could result in Ark-mediated caspase activation through cytochrome *c* increases (*Figure 1*).

7. Moving upstream: regulation of Rpr, Hid, Grim, and Skl activity

Different upstream pathways regulate the proapoptotic activities of Rpr, Hid, Grim, and Skl. These pathways include steroid hormones, DNA damage, growth factor signaling, and many as yet uncharacterized developmental cues (*Table 1*). Both transcriptional and post-translational mechanisms of regulation have been uncovered for these genes.

In the developing embryo, the pattern of Rpr, Grim, and Skl expression corresponds to the pattern of cell death (White *et al.*, 1994; Chen *et al.*, 1996; Christich *et al.*, 2002; Srinivasula *et al.*, 2002; Wing *et al.*, 2002). This suggests that the expression of these genes is elaborately regulated by many of the pathways that regulate other aspects of development. There are both similarities and differences in the expression patterns of these three genes. For example, Rpr expression is upregu-

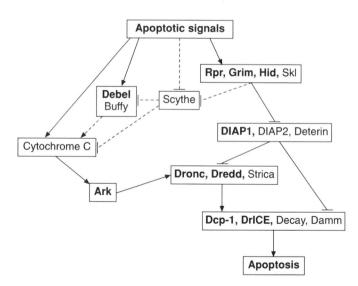

Figure 1. A speculative model for the interactions between proteins involved in apoptosis in *Drosophila*. The interactions between the apoptotic inducers Rpr, Grim, Hid, and Skl and DIAP1, and between DIAP1 and some of the caspases are supported by experimental data (see text). However, the mechanisms by which these inducers activate the mitochondria pathway are unknown. Genes shown to be involved in apoptosis *in vivo* are in bold. Speculative interactions are indicated by dotted lines.

Table 1. Signals that regulate apoptotic
functions in *Drosophila*

DNA damage	Rpr
	Skl
	Cytochrome C
	Debcl
Steroid hormones	Rpr
	Hid
	Dronc
	Damm
Developmental signals	Rpr
	Hid
	Grim
	Skl
	Dredd
Growth factors	Hid

lated when cells die due to defects in the normal developmental program, while Skl expression is not increased (Nordstrom *et al.*, 1996; Christich *et al.*, 2002). It is not yet clear whether developmental signals are integrated directly at the promoters of these genes, or whether the developmental regulation of their expression reflects the activity of a limited number of transcription factors, which in turn integrate several signaling pathways. Furthermore, it is not known whether these genes share regulatory sequences. Although the genes are clustered on the chromosome, they are spread over more than 100 kb (Adams *et al.*, 2000). Detailed molecular studies will reveal whether this region contains regulatory sequences that are shared by more than one gene.

Although the developmental pathways that regulate Rpr expression during embryogenesis have not yet been described, two regulators have been identified that directly activate the Rpr promoter: the steroid hormone receptor and p53. Expression of Rpr is detected in many effete larval tissues, shortly before they undergo apoptosis. The death of these tissues depends on changes in the level of the steroid hormone ecdysone. The Rpr promoter contains a steroid hormone-binding site, which is sufficient to activate Rpr expression in response to ecdysone (Jiang *et al.*, 2000). Expression of Hid, Dronc, and Damm is also activated in response to steroid hormones, although the activation of Hid is likely to be indirect. It is interesting to note that Rpr expression is turned on in some cells by rising titers of ecdysone, and in some tissues by falling titers (Robinow *et al.*, 1997). Apoptosis is also limited to a subset of tissues that respond to the hormone (Robinow *et al.*, 1993).

Exposure to ionizing radiation induces the expression of Rpr and Skl, and subsequent apoptosis (Nordstrom *et al.*, 1996; Brodsky *et al.*, 2000; Christich *et al.*, 2002). Studies using a dominant negative form of *Drosophila* p53 indicate that this protein is required for DNA damage-induced cell death (Brodsky *et al.*, 2000;

Ollmann *et al.*, 2000). A radiation-responsive element in the Rpr promoter contains a p53-binding site. This strongly suggests that ionizing radiation activates apoptosis in flies by inducing Rpr expression in a p53-dependent manner. Characterization of p53-induced death in a Rpr mutant, as described below, indicates that there are other p53 targets (Peterson *et al.*, 2002).

Hid is more broadly expressed than Rpr, Grim, and Skl. Although Hid is a potent inducer of cell death when ectopically expressed, Hid mRNA is observed in cells that do not undergo apoptosis (Grether *et al.*, 1995). This suggests that Hid activity can be modulated post-transcriptionally. Hid has been shown to be activated when growth factor signaling through the Ras/Raf/MAPK pathway is inhibited (Bergmann *et al.*, 1998; Kurada and White, 1998; Bergmann *et al.*, 2002). MAPK activation through this pathway inhibits Hid transcription, and also appears to inhibit Hid function by direct phosphorylation of the protein.

8. What is the role of apoptosis in *Drosophila* development?

Apoptosis serves to remove excess and effete cells during development in *Drosophila*, just as it does in other animals (Jacobson *et al.*, 1997). Throughout development, excess cells are generated, and then eliminated by apoptosis. The functions of Rpr, Grim, and Hid are required for the majority of developmental apoptosis outside the female germ line (Foley and Cooley, 1998). Examination of the phenotypes of mutants that remove one or more of these genes allows the identification of cells that are normally eliminated by apoptosis, and provides clues as to how these deaths are regulated. Characterization of Hid and Rpr mutants has shown that these genes are responsible for different subsets of developmental deaths.

Embryos that are homozygous for a deletion that removes Rpr, Grim, and Hid lack all developmental apoptosis, and are lethal at the end of embryogenesis. The most obvious phenotypes in these embryos are a block in a morphogenetic movement required for proper head development, and a large number of excess cells in the central nervous system (White *et al.*, 1994; Zhou *et al.*, 1995; Nassif *et al.*, 1998). This latter defect indicates that in flies, as in mammals, excess neurons are generated during development, to be eliminated by apoptosis. The defects in head development may arise from excess cells interfering in tissue involution.

Animals mutant for Hid are viable to adult stages, although short-lived (Abbott and Lengyel, 1991; Grether *et al.*, 1995). They have increased numbers of midline glia cells at the end of embryogenesis, and excess interommatidial cells in the adult eye. The viability of both of these cell types is known to be regulated by the activity of the EGF receptor (EGFR) (Freeman, 1996; Scholz *et al.*, 1997; Stemerdink and Jacobs, 1997; Bergmann *et al.*, 1998; Kurada and White, 1998; Miller and Cagan, 1998; Sawamoto *et al.*, 1998; Bergmann *et al.*, 2002). EGFR, acting through the Ras/Raf/MAPK pathway, downregulates Hid activity, as described above. Cell survival is favored when cells are exposed to high levels of EGFR activity, and cells undergo Hid-induced apoptosis in the absence of this activity. Hid mutants also show other defects in development – for example, in the developing wing – that have not yet been ascribed to defects in cell death.

Mutants that selectively remove Rpr have recently been described (Peterson *et al.*, 2002). Again these mutants are viable, and show relatively specific defects in cell death. Surprisingly, despite the demonstration that the ecdysone receptor and p53 directly regulate Rpr expression (Brodsky *et al.*, 2000; Jiang *et al.*, 2000; Ollmann *et al.*, 2000), no defects have been detected in hormone- or p53-induced apoptosis in the Rpr mutant. Ecdysone-induced expression of other proapoptotic proteins, such as Hid, is likely to be sufficient for apoptosis in the absence of Rpr. Interestingly, apoptosis induced by ionizing radiation is reduced in the Rpr mutants. Taken together, these data indicate that Rpr function is important for the full response to DNA damage, but is not rate limiting in the context of p53 overexpression.

Two populations of cells in the nervous system are increased in Rpr mutants. One is a group of neurons that normally die during metamorphosis, but survive in the Rpr-mutant adult, and the other is a subset of neuroblasts, or neural stem cells. Some neuroblasts are normally destined to undergo apoptosis at the end of embryonic development. In the Rpr mutant, these deaths do not occur, and the neuroblasts continue to divide, giving rise to a very enlarged nervous system. It is not known why the elimination of these populations of cells is so dependent on Rpr activity.

The generation of excess cells and their elimination by apoptosis provides the organism with significant flexibility in development. Studies where normal development is perturbed illustrate this point dramatically. Disruptions both in organ patterning and in the total number of cells in a tissue or in the whole organism can be compensated by increased cell death, as described below.

Ectopic expression of cyclin E in the embryo results in an extra round of cell division, nearly doubling the cell density in the embryo (Li, Q. *et al.*, 1999). Despite the huge excess of cells, most of these embryos develop and hatch into larvae, although at a slower rate of development. Examination of cell death in these embryos shows that hyperplasia induced by cyclin E overexpression is largely counteracted by increased apoptosis. The excess cells are removed in a pattern-specific fashion rather than randomly, and the doomed cells express Rpr prior to their death. Increased cell number in developing adult structures can also be compensated for by increased apoptosis. Expression of activated Ras can result in increased cell division, followed by increased cell death (Karim and Rubin, 1998).

In both embryonic and larval tissues, disruptions in pattern can also be counteracted by increased apoptosis. This 'pattern repair' can involve changes both in levels of apoptosis and in cell division. A nice example of this is provided by studies examining the effects of misexpressing the anterior morphogen bicoid. Increased bicoid in the anterior of the embryo results in an expansion of embryonic head structures (Driever and Nüsslein-Volhard, 1988). Surprising, a large number of these bicoid-overexpressing embryos develop into normal adults. Analysis of cell death in these embryos shows increased death in the anterior regions of the embryos, and decreased death in the posterior regions (Namba *et al.*, 1997). The anterior death is accompanied by increased Rpr expression. Thus, defects in global patterning can be accomplished by modulation of the normal cell-death pathway.

Within a single developing tissue or organ, the regulation of cell number is essential to producing a properly proportioned structure. In the developing wing, ectopically generated death on one side of the wing is counteracted by increased death on the opposite side, and by increased division on the smaller side (Milán *et al.*, 1997). If the ectopic death is not too extensive, the wing develops into a correctly proportioned adult wing of normal size. As in embryos, if the number of cells in a developing tissue is reduced below a threshold – for example, by a cell-cycle block – the ability of the tissue to compensate is insufficient, and defects in adult structures occur (de Nooij and Hariharan, 1995).

9. Conclusions

Drosophila provides a powerful model system to investigate how apoptosis is regulated and executed during development. Genetic screens have proven to be invaluable for identifying genes important for this process. In addition, the genetic and molecular tools available in this system allow the validation of findings generated in other systems. For example, loss-of-function genetics can be used to test the importance of interactions detected in overexpression studies.

Studies on the developmental regulation of cell death in *Drosophila* are greatly aided by the substantial information available on the regulation of other aspects of development. There is likely to be substantial overlap between the regulation of cell division, cell differentiation, and cell survival. Understanding how these processes are coordinately regulated will be critical to full understanding of patterning and organogenesis. In addition, the contribution of apoptosis must be considered to fully understand the phenotypes resulting from mutations in developmentally important genes.

Many of the pathways that regulate apoptosis in flies are similar to those in mammalian systems. In both organisms, DNA damage induces p53-dependent apoptosis, and steroid hormones activate both apoptotic and differentiation pathways. The available genetics in flies allows candidate target genes to be tested for importance. In the future, rapid new strategies for producing knockouts should allow even multiple targets to be tested together. The study of apoptosis during *Drosophila* metamorphosis will produce a better understanding of how different tissues respond differently to the same hormonal signal. Some of this specificity is probably due to the impact of other signaling pathways on the cell. It is probable that similar processes regulate specificity in response to hormones in other system.

Our understanding of apoptosis during animal development is far from complete. Determining which cells die, and how these patterns of death are controlled will be a critical contribution of *in vivo* models such as flies and worms. The problem of how perturbations in development are compensated by altered patterns of apoptosis is a fascinating area for future work.

11. Cell-culture systems in apoptosis

Stefan Grimm

1. Introduction: advantages of cell culture to investigate apoptosis

Cell culture is an invaluable tool to study the behavior of cells, as it allows us to investigate a great number of them with similar genotype and phenotype. It thereby circumvents genetic variations between individuals and also the ethical questions of human and animal experiments. Consequently, the definition of media that support the growth of cells explanted from tissues was a major achievement in biology (Eagle, 1959). Even though it constitutes a reductionist view of the cells to observe them without supporting neighboring cells and without the extracellular matrix, many cells – though usually tumor cells – still reflect their behavior *in vivo* (Masters, 2000).

Since apoptosis is a rapid process that is usually concluded within a few hours, cell culture is an adequate instrument to investigate this phenomenon as it allows us to resolve the sequence of the individual steps. Especially GFP-fusion proteins and immunofluorescence helped to reveal the dynamic processes that lead to cell death (Goldstein *et al.*, 2000). Given the ease of studying cells in culture, even genes that have been isolated by *in vivo* systems, such as those of *Drosophila*, are subsequently analyzed *in vitro* (McCarthy and Dixit, 1998). This is now aided by a plethora of different readouts for apoptosis in cell culture, from the demonstration of the DNA ladder (Wyllie, 1980) or the TUNEL assay that detects the free DNA ends generated during apoptosis (Gavrieli *et al.*, 1992) to the exposure of surface markers in apoptotic cells (Koopman *et al.*, 1994). Powerful supplementary methods such as flow cytometry rely on cell culture and have immensely contributed to the elucidation of apoptosis (Darzynkiewicz and Traganos, 1998).

An important aspect of cell culture is the easiness with which the cells can be genetically manipulated. Nowadays, different transfection methods facilitate the introduction of genetic material into virtually every cell. This permits many genes to be overexpressed by transfected plasmids. In addition, techniques such as antisense cDNAs (Albert and Morris, 1994), RNAi (Elbashir *et al.*, 2001), or antisense oligonucleotides (Ghosh and Iversen, 2000) allow us specifically to inactivate single mRNAs in order to assess their contribution to cell-death regulation. Moreover, pharmacologic studies are possible in cell culture, as drugs can be titrated in microtiter plates. This is especially important for cell-death research, since many proapoptotic stimuli, such as H_2O_2, UV irradiation, or cytostatic drugs, can be applied in cell culture.

2. Cell-culture approaches to isolate genes in apoptosis

Apoptosis is genetically determined. Even if the original stimulus is not encoded by genes, as in 'programmed cell death' (see Introduction), but rather constitutes an accident, the ensuing signaling pathway is executed by genes and their proteins. As the process of apoptosis is considered very complex, it is expected that there are still novel genes to be detected that promise to reveal the inner workings of cells undergoing apoptosis. Consequently, the isolation of genes is a major effort in apoptosis research. This review will concentrate on the use of cell-culture techniques to determine genes involved in apoptosis. It will present some of them and their relevance for apoptosis induction.

A number of methods, chiefly the two-hybrid assay in yeast (Fields and Song, 1989), allow us to clone genes according to their protein association. As the signaling pathway for apoptosis is in many cases mediated by protein–protein interactions, such techniques have turned out to be very powerful (Boldin *et al.*, 1995; Chinnaiyan *et al.*, 1995; Stanger *et al.*, 1995). However, this overview will focus solely on mammalian cell culture and its use in apoptosis research.

Apoptosis constitutes a degenerative process that eventually results in the disappearance of the affected cells. Given this transient nature of apoptosis, three basic strategies have emerged to isolate genes involved in cell death (*Figure 1*). The first is an approach that correlates apoptosis with the transcriptional up- or downregulation of specific genes. This requires the determination of differentially regulated genes during the process of apoptosis. The other two approaches are functional expression assays that allow us to clone genes on the basis of their activity for apoptosis. The first strategy uses the inactivation of endogenous genes (loss of function) or the overexpression of inhibitors of apoptosis to select those cells that are thereby protected against a proapoptotic stimulus. Rather than a selection, the alternative approach employs multiple transfections in parallel and therefore constitutes a screen for genes that have the dominant activity to induce apoptosis upon overexpression. All three strategies will now be described in more detail.

2.1 Correlative strategy

This approach is based on the earlier observation that transcription- or translation-inhibitors can reduce apoptosis (Lockshin, 1969). This led to the prediction that apoptosis-inducing proteins do not reside in healthy cells and that only an apoptosis stimulus causes the synthesis of novel proteins that induce cell death. Intuitively legitimate as this may be, there are probably more instances where the mediators of an apoptotic signal are already present in the cell and only have to be activated upon apoptosis induction (Jacobson *et al.*, 1994; Schulze-Osthoff *et al.*, 1994).

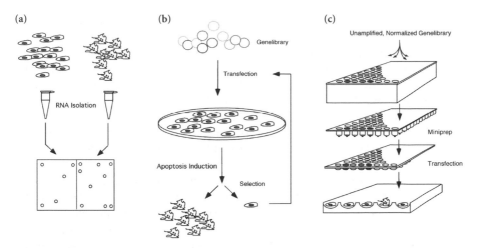

Figure 1. Different assays to determine apoptosis-associated genes.
(a) The correlative approach. RNA is isolated from healthy and apoptotic cells and the differentially expressed genes are compared by subtractive library construction, differential display, and SAGE or gene microarrays. (b) Selection strategy to isolate apoptosis genes. An expression library is introduced in a batch format in cells, and a selection is established by the application of a proapoptotic stimulus. From the cells that survived the induction of cell death, the transfected DNA is isolated and enriched by several rounds of amplification in bacteria and selection for survival. (c) A genetic screen to isolate apoptosis genes. Every single cDNA of an unamplified and normalized library is introduced into a separate population of cells by an individual transfection. The proapoptotic effect of the genes is then detected by an appropriate assay for cell death.

Hence, the determination of genes that are differentially expressed between healthy and apoptotic cells might be an adequate experiment only for some inducers of cell death. While the first attempts of this strategy were performed with subtractive cDNA libraries (Ishida *et al.*, 1992), current experiments involve DNA microarrays to investigate many genes in parallel (*Figure 1a*). Several studies have followed this road and come up with a number of differentially regulated genes. One such experiment was performed with irradiation as the cell-death inducer (Voehringer *et al.*, 2000), and another with kainic acid or potassium withdrawal as the proapoptotic stimulus in neurons (Chiang, L.W. *et al.*, 2001). Hundreds of genes have been found to be associated with the process of apoptosis, only some of them, such as caspase-3, being established players in apoptosis. Interestingly, Chiang and colleagues observed discrete waves of differentially expressed genes, making it possible that earlier genes influence the expression of the late genes. This suggests that a very complex transactivation pattern is responsible for apoptosis induction.

While all these experiments allow us to correlate different gene activities with the induction of apoptosis, their exact contribution to apoptosis induction has to be elucidated in every case. Starting with a similar approach to hunt for differentially expressed genes, Michael Green and colleagues were able to attribute to the upregulation of a lipocalin the cell death of a mouse pro-B cell line induced by interleukin (IL)-3 deprivation (Devireddy *et al.*, 2001). Lipocalins are a group of small, mostly secreted proteins with various functions, ranging from the transport of lipophilic molecules, prostaglandin synthesis, and the regulation of cell homoeostasis to the modulation of the immune response (Flower, 1996). Transcending the mere correlation of its upregulation, these workers demonstrated that this lipocalin is necessary for apoptosis induction in their system, since antisense oligonucleotides abrogated the effect. Furthermore, they showed that this molecule induces apoptosis by an autocrine loop that is established in B cells by IL-3 deprivation. How the signal of the proapoptotic lipocalin is transmitted in the cell is not yet known, although a connection to the dephosphorylation of Bad, a Bcl-2-like protein, and its subsequent activation was established.

Another interesting gene isolated with the correlative strategy is par-4, which was found to be upregulated in apoptotic prostate cells (Sells *et al.*, 1994). Par-4 features a death domain and a leucin zipper, which is known as a protein–protein interaction domain (Landschulz *et al.*, 1988). Subsequent investigations showed that par-4 is strongly expressed in a number of neurodegenerative diseases, such as Alzheimer's, Parkinson's, and Huntington's diseases, that are characterized by an excess of apoptosis induction. Par-4 could mediate apoptosis by its interaction with a number of apoptosis-regulating proteins, among them Bcl-2, protein kinase C zeta and caspase-8 (Mattson *et al.*, 1999). It has also been suggested that par-4 functionally associates with NF-κB, an inducible transcription factor that can both repress and mediate proapoptotic stimuli (Baichwal and Baeuerle, 1997).

Other investigators have used apoptosis induced by specific transcription factors for the analysis with DNA microarrays or SAGE (serial analysis of gene expression). Chiefly, p53, one of the most important tumor suppressor genes that can lead to apoptosis, was studied (Kannan *et al.*, 2001). Vogelstein and colleagues were able to derive a general concept of p53-induced apoptosis from an overview of p53-target genes (Polyak *et al.*, 1997). They found a striking pattern of genes that can cause oxidative stress, among them a galectin family member and a NADPH oxidoreductase-related gene. In addition, they could repress apoptosis with pyrrolidine dithiocarbamate (PDTC) and diphenyleneiodonium chloride (DPI), two antioxidants. Consequently, these workers postulated that reactive oxygen species are the determining factor in p53-activated cell death (Polyak *et al.*, 1997). Other studies by Arnold Levine's group, also aiming at the transcription targets of p53, isolated some, but not all, genes of the Vogelstein group (Zhao *et al.*, 2000). These scientists also determined cell-specific effects, the temporal succession of gene activation, and various p53 stimuli. Over 100 genes were found to be up- or downregulated. They belong to different classes such as growth factors, extracellular matrix proteins, and adhesion molecules, but they also included genes, such as Fas and DR5, known to be involved in apoptosis. These results point to a much more complex regulation of p53-induced apoptosis.

To sum up, the correlative strategy has already yielded a number of attractive genes whose transcriptional expression correlates with and is sometimes even responsible for the induction of apoptosis. The use of microarrays allows us to detect global changes in the expression pattern. It is therefore likely that array techniques will speed up the search for genes whose expression is changed during apoptosis induction. Nevertheless, individual genes that are transcriptionally altered during cell death have to be functionally verified for their role in apoptosis.

2.2 Selection strategies

Apoptosis allows researchers to select cells that withstand a proapoptotic stimulus in order to determine gene functions in cell death. As mammalian cells are permanently diploid (or polyploid, like many tumor cells), recessive gene activities are difficult to investigate. Transfection experiments to inactivate endogenous genes (and their proteins) have been established to tackle this problem. They comprise both antisense cDNAs against cellular genes and cDNA fragments encoding proteins that can act in a dominant-negative manner. By isolating transfected plasmids from the cells that survived a proapoptotic stimulus, one can determine genes that mediate a signal for apoptosis (*Figure 1b*).

This approach was pioneered by Adi Kimchi and her group (Deiss and Kimchi, 1991). They transfected an antisense cDNA library into HeLa cells and isolated cells that survived the exposure to interferon-γ (a strategy called 'technical knockout'). These workers expected to identify genes that mediate the signal for apoptosis and to define rate-limiting steps in this process (Levy-Strumpf and Kimchi, 1998). Five so-called DAP-genes (DAP: Death Associated Proteins) were determined that conferred resistance to interferon-γ. They turned out to be completely divergent and, partially, even localized to different compartments of the cell. Thioredoxin was the first gene described (Deiss and Kimchi, 1991). It encodes a thiol protein that exerts a potent reducing activity and might be able to keep in check the reactive oxygen species that are frequently formed in apoptosis. It also promotes DNA binding of transcription factors, such as NF-κB, AP-1, and p53, all of which have been implicated in apoptosis induction. Moreover, thioredoxin has an inhibitory effect on the TNF-induced activation of ASK1, possibly by direct binding to this MAPK kinase kinase (Ichijo *et al.*, 1997).

The other isolated genes from the screen include a calcium/calmodulin-regulated kinase with a death domain (DAP-kinase), a nucleotide-binding protein (DAP-3), a homolog of the translation initiation factor eIF4G (DAP-5), a small proline-rich cytoplasmic protein (DAP-1), and the lysosomal protease cathepsin D (Levy-Strumpf and Kimchi, 1998). The diversity of these genes makes it already evident how disparate their contribution must be and consequently how the complex signal for apoptosis is regulated. A surprising finding was that the lysosomal aspartyl protease, cathepsin D, plays a role in apoptosis induction and even mediates Fas- and TNF-induced apoptosis (Deiss *et al.*, 1996). Cathepsin D was one of the first proteases, besides the established cysteine proteases of the caspase family, found to contribute to apoptosis induction (see Chapter 13). This was even more astonishing as lysosomal proteases, unlike caspases, are known to have a wide range of substrates. How, then, could it induce such a defined cellular outcome as apoptosis? Subsequent studies revealed that cathepsin D accumulates in dying cells and might therefore change its subcellular localization and hence its substrates, a fact that could explain the involvement of this house-keeping gene in the induction of apoptosis (Deiss *et al.*, 1996).

The aforementioned DAP-kinase turned out to be especially interesting, as it is the founding member of a small kinase family that regulates cell death (Kogel *et al.*, 2001). This is a rare example in which a kinase is involved in apoptosis induction; usually, proapoptotic signals are mediated by direct protein–protein contact rather than by phosphorylation (see Chapter 5). The DAP kinase has been shown to be a tumor-suppressor gene (Inbal *et al.*, 1997): it is downregulated in highly metastasis-forming cells, and, upon reconstitution, it reduces the ability of cells to establish metastasis. This effect could be mediated by a signaling pathway involving the tumor suppressors p19ARF and p53 (Raveh *et al.*, 2001). The tumor-suppressor activity of the DAP kinase is a case in point of how apoptosis induction protects the organism against malignant cells.

One gene from this screen, DAP-3, is a supposed nucleotide-binding protein, as it harbors a 'P-loop' motif. This domain is necessary for its functioning since a point mutation in this domain abrogated its apoptosis induction. While one report localized this protein to mitochondria (Berger *et al.*, 2000), another study has recently shown that DAP-3 can bind to the TRAIL receptor on the plasma membrane (Miyazaki and Reed, 2001). This is a TNFR family member that has attracted much attention, as it induces apoptosis in a tumor-specific way (Walczak *et al.*, 1999). DAP-3 also binds to the adapter protein FADD and links it to the TRAIL receptor in a GTP-dependent manner. Consequently, DAP-3 might be an assembling factor in the TRAIL receptor complex.

DAP-5 was isolated as a cDNA fragment in a sense orientation that presumably encodes a dominant-negative version of this gene. It is homologous to the translation initiation factor eIF4GI but is missing its N-terminus, which otherwise associates with the cap-binding protein eIF4A. Interestingly, eIF4GI is cleaved by caspase-3. It is therefore possible that DAP-5, like the cleaved form of eIF4GI, works to initiate a cap-independent translation during apoptosis induction. This special form of protein synthesis can lead to the generation of both pro- and anti-apoptotic genes, such as XIAP and Apaf-1, whose mRNAs have IRES (internal ribosome entry sites). These results indicate another layer of cell-death regulation by the translation of specific genes (Holcik *et al.*, 2000).

An interesting aspect of this work was the finding that all genes except thioredoxin can also elicit a proapoptotic signal upon overexpression. This and all other results with DAP genes underscore the validity of this cloning approach. While an ordering of the various genes from this screen along a signaling pathway is still lacking, it is tempting to speculate that DAP-3, which seems to be localized on the membrane, is the gene most proximal to the signal created by interferon-γ. From there, the signal spreads to different organelles, including lysosomes, and activates caspases. Eventually, we will learn how all the genes from this screen contribute to apoptosis induction and how they can be integrated into a signaling pathway leading to cell death.

In an independent experiment, Gabig and colleagues determined an antisense construct to an endogenous gene by a similar expression-cloning assay, using IL-3 deprivation as apoptosis inducer in myloid cells. The corresponding gene, called Requiem (Gabig *et al.*, 1994), encodes a zinc-finger protein. When downregulated, it protects cells against IL-3 deprivation. From its primary sequence, Requiem is supposed to encode a transcription factor, but little else is known about this gene.

Using a retroviral insertion mutagenesis, Sarit Larisch and colleagues discovered ARTS, a septin-like protein whose downregulation protects cells against TGF-β apoptosis (Larisch et al., 2000). Until then, septins were only known to be part of the cytoskeleton and, possibly, to aid in the assembly of signaling complexes involving protein kinases. ARTS, in contrast, is an alternatively spliced isoform that lacks the septin-like C-terminus. Interestingly, this protein is localized to mitochondria and translocates to the nucleus upon apoptosis induction. While its exact function in apoptosis in this organelle remains to be determined, it is an additional protein that is released from mitochondria after apoptosis induction (see Chapter 7).

In a variation of the original antisense approach, Luciano D'Adamio and colleagues (Vito et al., 1996) transfected a random primed cDNA library into mouse T-cell hybridoma cells, and selected those that survived T-cell receptor cross-linking. This approach is based on preceding experiments that led to the isolation of growth-suppressor genes (Gudkov et al., 1994). Since the library in this experiment was generated in a random way, it could be expected to contain inhibitors of apoptosis that might function not only via antisense but also by truncated proteins that could fortuitously constitute dominant-negative variants. Of the six isolated genes, three have been characterized to date. ALG-3 and ALG-4 do encode proteins and act in a dominant-negative manner by interfering with their endogenous proteins. ALG-2, however, was expressed in an antisense orientation and downregulates the endogenous protein. It encodes a Ca^{2+}-binding protein that is also involved in glucocorticoid-mediated apoptosis, as its antisense can repress cell death by this hormone. ALG-3 encodes a truncated version of presenilin-2. Mutations in this gene are responsible for some familial cases of Alzheimer's disease, which is characterized by an aberrant cell death in neurons (Mattson, 2000). It is a conserved membrane protein that contributes to the γ-secretase activity necessary to liberate β-amyloid from its precursor protein. The localization of its protein to the endoplasmic reticulum (ER) – which is a Ca^{2+} store for apoptosis induction (see Chapter 6) – and the Ca^{2+}-binding properties of ALG-2 suggest that these two genes regulate the concentration of this proapoptotic ion. Both genes also mediated the proapoptotic effects of the Fas receptor.

ALG-4 is the gene responsible for upregulating the mRNA of the Fas-ligand. The dominant-negative clone that was isolated diminished the transcriptional activation of the Fas-ligand promoter. This effect seems to be accomplished by inhibiting (or activating, in the case of wild-type ALG-4) the inducible transcription factor NF-κB, which is a known transcriptional regulator of the Fas ligand promoter (Lacana and D'Adamio, 1999). The molecular cause of this effect is currently unknown.

Selection strategies can also be applied to dominant activities in apoptosis. In view of the observation that most apoptosis repressors are activated upon overexpression, several laboratories have performed experiments solely to isolate inhibitors of apoptosis. To this end, cDNA expression libraries were transfected into cells, a proapoptotic stimulus was applied, and the cells surviving this signal were selected. After several rounds of plasmid extraction from surviving cells and amplification in bacteria, genes could be isolated that repress TNF-mediated cell death. Two of these genes turned out to be the known apoptosis inhibitors Bcl-2 and Bcl-X (Jäättelä *et al.*, 1995). In a similar experiment, researchers tried to clone genes that impaired amyloid-β–induced apoptosis and discovered a secreted polypeptide, called 'humanin'. Interestingly, this gene seems to be specific in its cell-death repression for β-amyloid or other Alzheimer's disease genes (such as presenilins), since cell death caused by etoposide, huntingtin, and SOD mutants were not affected by humanin (Hashimoto *et al.*, 2001). A functional explanation for this effect is still lacking, but its investigation is certainly underway.

Other investigators discovered TIAF1 in a functional expression assay. TIAF is an apparently nuclear protein that features a Wilms' tumor protein family domain. Its expression inhibits TGF-β–induced apoptosis but also cell death caused by the TNF receptor and the overexpression of a variety of proapoptotic molecules. This is evidently achieved by downregulating IκB, the specific inhibitor of NF-κB, an effect that would result in the constitutive activation of its antiapoptotic effects (Chang *et al.*, 1998).

Gerry Nolan and coworkers used another selection strategy. Aiming at pharmacologically relevant gene targets in apoptosis, they did not concentrate on cDNAs that express endogenous proteins; instead, they used a retroviral gene library encoding an artificial permutation of random short peptides (Xu *et al.*, 2001). A peptide that inhibited taxol-induced cell death was isolated and shown to upregulate the multiple-drug resistance transporter (ABCB1). This artificial construct was used to isolate interacting proteins that might explain the peptide's antiapoptotic activity. Two subunits of the proteasome were determined, suggesting that the observed effect is regulated by this protein degradation complex.

In summary, a number of different stimuli have been applied to apoptosis induction to select genes or genetic elements (both antisense and fragments of cDNAs) that repress apoptosis. The information that could be inferred from these results is at the moment too complex to draw a coherent picture. Many approaches have yielded only individual genes that still require integration into a larger network of interacting proteins for apoptosis signaling. However, some systems, such as the one established by Adi Kimchi, which, it should be noted, is focused on only one single apoptosis inducer, already allow us to grasp the complexity of the signaling events in apoptosis.

2.3 Screening strategy

Another approach to clone apoptosis-relevant genes in cell culture was established in the laboratory of Philip Leder (Grimm and Leder, 1997). As many gene activities are redundant, knockouts (loss-of-function) cannot always be expected to yield a cellular effect. In addition, as the selection strategy uses cell death as the final readout, some genes that are part of a larger network of apoptosis regulators, and therefore are less effective mediators, might not get detected. Moreover, the investigators were motivated by a conspicuous correlation: almost every gene that transmits an apoptosis signal also has the dominant capacity to induce cell death upon overexpression. This activity is even conserved across species boundaries. It might be explained by the observation that apoptosis is mediated by specific protein–protein interactions that are induced when one of the partners is overexpressed. At first, it seemed impractical to use this dominant effect for isolating apoptosis genes because their activity leads to the degenerative process of cell death, instead of cell survival, and therefore amplification of the responsible genes, as in the case of the selection procedure. However, a genetic screen to isolate such dominant, apoptosis-inducing genes could be developed (*Figure 1c*). It is based on iterative transfections of very small expression plasmid pools into human embryonic kidney cells (HEK). In order to decrease the number of transfections, the cDNAs are taken from an unamplified, normalized gene library. The subsequent microscopic inspection of the apoptotic phenotype facilitates detection of cell death.

Since cell death can be induced by many different stimuli, the specificity of the signal that can lead to apoptosis in HEK cells has been characterized. Several control experiments with dominant-negative mutants of known genes, constitutively active oncogenes, and various cytostatic drugs could not generate apoptotic cells, indicating that induction of cell death needs a defined signal; an unspecific insult of the cells does not suffice (Albayrak *et al.*, submitted). This is also supported by a theoretical consideration: given the severe consequences, committing suicide cannot be a minor decision for a cell. Therefore, defined checks have to be overcome in a cell to induce apoptosis. The specificity of the proapoptotic signal is also underscored by the finding that adenine nucleotide translocase (ANT)-1, a component of the 'permeability transition' (PT-) pore, can potently induce apoptosis. In contrast, ANT-2, a gene that is 90% identical to ANT-1, is completely inactive in cell death induction (Bauer *et al.*, 1999).

After carrying out the screen for about 40% of a gene library, more than 70 positive genes were found (about 0.07% of all cDNAs). This high count correlates with the expected, and already experimentally shown, complex regulation of apoptosis induction, and it reflects the number of sensors that exist in the cell for this program. Among the isolated genes are several genes already known as apoptosis inducers, including ZIP kinase (Kawai *et al.*, 1998), NIP3 (Chen *et al.*, 1997), FADD (Chinnaiyan *et al.*, 1995), PERP (Attardi *et al.*, 2000), and CIDE-A and CIDE-B (Inohara *et al.*, 1998), which therefore serve as positive controls for the screen (Albayrak *et al.*, submitted).

The first gene isolated by this screen and further characterized was NDF (<u>N</u>eu <u>D</u>ifferentiation <u>F</u>actor) (Grimm and Leder, 1997; Grimm *et al.*, 1998). NDF is a growth factor that is first synthesized as a membrane-spanning precursor protein, from which the secreted growth factor moiety is proteolytically released. Interestingly enough, NDF can, when overexpressed, also function as an oncogene (Krane and Leder, 1996). This is probably achieved by activating receptors on neighboring cells, which subsequently proliferate and eventually undergo malignant transformation. One interpretation of NDF's proapoptotic activity is that it is a safeguard against its activation as an oncogene. Experiments indicated that its apoptosis-inducing domain is indeed encoded by the cytoplasmic moiety and therefore is physically separable from its extracellular growth factor domain, which mediates the oncogenic activity. Hence, only NDF-overexpressing cells undergo apoptosis. This finding is – much like the results with the DAP kinase – another example of how apoptosis contributes to the protection against malignant cells.

Nevertheless, extensive apoptosis was found in tumors caused by NDF in transgenic mice (Grimm *et al.*, 1998; Weinstein *et al.*, 1998). The activity of NDF in apoptosis induction therefore still exists in malignant cells but must be repressed by additional mutations in order for the tumor to thrive.

ANT-1, a component of the mitochondrial 'permeability transition' (PT) pore, was also isolated with this screen (Bauer *et al.*, 1999). The PT pore is a protein complex between the inner and outer mitochondrial membrane that has been implicated in apoptosis induction by a wide variety of reagents (Crompton, 1999). In apoptosis induction, it is converted into an unspecific pore by an unknown mechanism. This leads to the collapse of the membrane potential, a block of the respiratory chain, and to a release of apoptogenic proteins from mitochondria. Until now, ANT-1 was known only as a transport molecule for ATP and ADP. However, recent results have demonstrated that its transport activity is independent of its activity to induce apoptosis. Apoptosis induced by ANT-1 is probably mediated by specific protein–protein interactions that titrate out repressors of the PT pore. Hence, these investigations could contribute to a mechanistic understanding of how this pore and its components are activated for apoptosis induction (Bauer *et al.*, 1999). ANT-1 also constitutes an example of a gene from this screen that can be integrated into a pathologic context: it is overexpressed in dilated cardiomyopathy (DCM), a prevalent degenerative heart disease of unknown cause (Dorner *et al.*, 1997) that is characterized by an excess of apoptosis induction in the heart muscle. For this reason, it could be postulated that this pathologic apoptosis induction is caused by overexpressed ANT-1.

A dominant activity exerted by overexpression of genes is used in this screen. The examination of such dominant gene activities is becoming increasingly prevalent. For example, a similar approach uses microarrays that have been spotted with plasmid DNAs. Cells laid upon these arrays take up and express the DNA. A pilot experiment indicated that this transfection method might be suitable to determine apoptosis-inducing genes (Ziauddin and Sabatini, 2001). This would allow high-thoughput screening for cell-death inducers, an aim that is also pursued with the original screen (see below).

While the correlation and the selection strategy are focused on specific external stimuli for apoptosis, in the screening assay the transfected genes themselves induce apoptosis. What consequences does this have for the kind of genes that can be expected from this assay? For one, genes can be isolated whose endogenous counterparts mediate a genuine proapoptotic stimulus: upon their inactivation, less apoptosis is induced. As such a reduced cell death can lead to tumorigenesis, they are also candidates for tumor suppressors. In fact, two of the genes that have previously not been linked to apoptosis are only recently discovered tumor-suppressor genes (Albayrak *et al.*, submitted; Schoenfeld *et al.*, submitted). The cellular counterparts of the second group of genes from the screen might not themselves transmit a proapoptotic signal. Nevertheless, when overexpressed, they activate and therefore define sensors that mediate induction of cell death. These sensors can be expected to be protein complexes, as apoptosis seems to be induced – at least in most cases – by direct protein–protein interactions (see Chapter 5). To this class belong also those genes that are upregulated or constitutively active in diseases, and that therefore constitute potential drug targets (such as ANT-1 in DCM). Two genes whose dominant alleles are responsible for degenerative diseases have been uncovered by the screen (Albayrak *et al.*, submitted). As the expression of the genes isolated by this screen is sufficient for apoptosis induction, it is possible to detect genes outside the main apoptosis-signaling pathways. Hence, a subset of the genes might induce apoptosis only under specific circumstances. An example is NDF, which acquires its proapoptotic activity when it is activated as an oncogene. This category also includes genes that are operative only in certain cell lines, such as the transformed tumor cells used in the screen. Some examples of tumor-specific apoptosis-inducing genes have been described previously, such as apoptin (Danen-Van Oorschot *et al.*, 1997), mda-7 (Su *et al.*, 1998), and the TRAIL receptor (Walczak *et al.*, 1999).

Even though the screen still leads to individual proapoptotic genes, the long-term goal of this work is the same as with other assays for apoptosis genes: to establish a network of functionally interacting proteins – defined by the isolated genes – that eventually leads to the activation of caspases and to apoptosis induction. In the case of the screen, this concept calls for a complete coverage of apoptosis-inducing genes. Such a compendium of apoptosis inducers would indicate the sensors involved. In addition, the sheer number of genes from the screen makes it imperative to add additional information (such as expression data, disease-relevance, and sequence comparisons). An inventory of apoptosis genes would thereby allow researchers to focus on those that can be easiest integrated into a known physiologic or pathologic context in which their signal becomes relevant to induction of cell death. To implement a saturation screen, the assay has already gone through different stages to increase the throughput (*Figure 2*). While first done by isolating plasmid DNA in single reaction tubes (Grimm and Leder, 1997), it can now be performed in 96-well microtiter plates (Neudecker and Grimm, 2000). This allows workers to reduce the number of clones to single genes per transfection. Because an individual transfection is performed for every cDNA clone in a library, the process requires thousands of transfections in parallel. Thus, it was only a logical step to use robotic systems in the screen. Reporter assays to detect apoptosis and a transfection robot, especially tailored to multiple simultaneous transfections, were also developed (Kachel and Grimm, 2002).

3. Conclusions

Cell-culture studies with the correlative, the selection, and the screen approaches have already led to the description of a number of genes involved in apoptosis. However, most of them are still isolated entities, and such concise stories as derived from the work with *C. elegans* (Chapter 10) cannot yet be told about the genes isolated in cell culture. Clearly, additional efforts by the investigators are required to integrate these apoptosis genes into a signaling network. However, some general statements can already be deduced from the available data.

Interestingly, many inhibitors or mediators determined by the presented methods can also function with stimuli other than the ones that were originally used for their isolation. This argues for a very redundant signaling during apoptosis. Hence,

(a)

Unamplified, Normalized Library

Plate
Bacteria
and Repeat

Miniprep

Transfection

(b)

Unamplified, Normalized Genelibrary

Miniprep

Transfection

(c)

Figure 2. Development of the screen for dominant, apoptosis-inducing genes.
(a) The original version of the screen comprised the transfection of small plasmid pools of at least 20 different genes and the further investigation of single clones from positive pools. (b) The next format was implemented with 96-well plates, employing a special DNA isolation protocol that yielded transfection-pure plasmid DNA. (c) The robotic version of the screen uses a transfection robot tailored to the screen that allows researchers to perform a high-throughput assay.

we must revise the notion of a linear pathway emanating from one stimulus and progressing in a straight succession of sequentially activated apoptosis mediators down to caspases. It is more likely that nature uses many, partially redundant signaling pathways that feed back to further mediators of apoptosis. This cross-talk might stabilize the system and prevent mistakes in the activation of this deadly program.

While many chapters in this book focus on several gene families involved in apoptosis (such as the TNFR or the Bcl-2 family), these families are underrepresented in our description of cell-culture systems in apoptosis. Rather than a shortcoming, this could be an advantage, as the defined conditions of cell culture seem to allow researchers to clone genes outside the standard apoptosis pathways. As every gene from cell culture characterizes a signaling pathway, it is possible to use these genes to describe these unique signals and the cellular processes they activate.

All in all, cell culture has already made valuable contributions to apoptosis research and greatly deepened our understanding of cell death. With all the information gained from this work, it is safe to say that cell-culture systems can be expected to keep many researchers in the field of apoptosis busy for some time to come.

12. Caspase-independent cell death

Marcel Leist and Marja Jäättelä

1. Introduction

Caspase-mediated apoptosis is the death program of choice in many developmental and physiologic settings (Kerr *et al.*, 1972; Strasser *et al.*, 2000; Kaufmann and Hengartner, 2001). It would, however, be very dangerous for the organism to depend on a single protease family for clearance of unwanted and potentially dangerous cells. Indeed, the exclusive role of caspases in the execution of programmed cell death (PCD) has been challenged recently (Borner and Monney, 1999; Kitanaka and Kuchino, 1999; Johnson, 2000; Wang, 2000; Kaufmann and Hengartner, 2001; Leist and Jäättelä, 2001a). Since the first reports on caspase-independent PCD in the late 1990s, over 200 papers have been published on the topic. Now our understanding of the molecular control of alternative death pathways is growing, like that of the molecular anatomy of apoptosis at the time of the discovery of caspases less than a decade ago. Here, we review recently discovered triggers and molecular regulators of alternative cell-death programs and discuss the implications of the death mode for the surrounding tissue and the potential of caspase-independent PCD signaling pathways as therapeutic targets for the treatment of cancer and neurodegenerative disorders.

2. Four patterns of death: from apoptosis to necrosis

The unclear definition of the alternative death pathways has been the major obstacle to elucidating them. If PCD is used as a synonym of apoptosis and defined by caspase activation (Samali *et al.*, 1999), alternative caspase-independent PCD pathways are evidently not possible. In contrast, the classification that we use here takes into account the implications of the death mode for the surrounding tissue and leaves space for different mechanistic observations and alternative interpretations. PCD is simply defined as cell death that is dependent on signals or activities within the dying cell (Lockshin and Zakeri, 2001). According to the morphology of dying cells, PCD can be further divided into apoptosis, apoptosis-like, and necrosis-like PCD (Kitanaka and Kuchino, 1999; Leist and Jäättelä, 2001a) (*Figure 1*). Apoptosis is defined here by chromatin condensation to compact and apparently simple geometric figures (stage 2 chromatin condensation), phosphatidylserine exposure, cytoplasmic shrinkage, plasma membrane blebbing, and formation of apoptotic bodies (Kerr *et al.*, 1972; Woo *et al.*, 1998; Susin *et al.*, 2000; Leist and Jäättelä, 2001a). Moreover, apoptosis-like PCD is characterized by chromatin condensation and display of phagocytosis recognition molecules before the lysis of the plasma membrane. Chromatin condenses, however, to lumpy masses that are less compact than in apoptosis (stage 1 chromatin condensation) (Woo *et al.*, 1998; Susin *et al.*, 2000; Leist and Jäättelä, 2001a). Any degree and combination of other apoptotic features can be found. Necrosis-like PCD is used here to define PCD in the absence of chromatin condensation, or at best with chromatin clustering to loose speckles (Leist *et al.*, 1997; Vercammen *et al.*, 1998a,b; Mateo *et al.*, 1999; Holler *et al.*, 2000; Sperandio *et al.*, 2000). Varying degrees of apoptosis-like cytosolic features may occur before the lysis (Mateo *et al.*, 1999; Holler *et al.*, 2000). Unlike necrosis,

Shapes of death/chromatin condensation

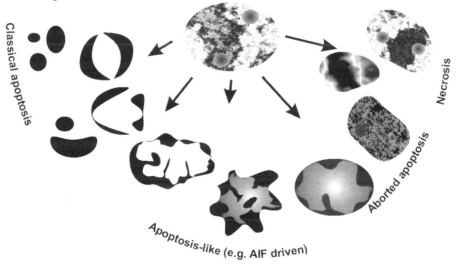

Figure 1. Nuclear alterations in different forms of PCD.
The use of chromatin condensation as a criterion to distinguish apoptosis from apoptosis-like PCD has been inconsistent in the scientific literature, and the potential for overlapping definitions and errors is large. Electron-microscopic examples of classic apoptosis and apoptosis-like PCD (Leist and Jäättelä, 2001a) or these schematic drawings might provide a general guideline. The control chromatin is speckled, showing areas of eu- and heterochromatin, and mostly one or several more condensed micronuclei (top middle). Caspase-dependent chromatin compaction and fragmentation to crescent- or spherical-shaped masses at the nuclear periphery is shown on the left. Caspase-independent chromatin margination triggered directly by microinjection of AIF or in a number of models of apoptosis-like death is shown at the bottom. Many intermediate forms and also transitions to necrosis are possible. Necrotic morphology is also observed in models where caspases are inhibited before apoptosis is completed (aborted apoptosis).

necrosis-like PCD is the result of active cellular processes that can be intercepted by, for instance, oxygen-radical scavengers (Schulze-Osthoff *et al.*, 1992; Vercammen *et al.*, 1998a,b), inhibition of poly(ADP) ribose polymerase (PARP) (Ha and Snyder, 1999) or mutations in intracellular signaling molecules (Holler *et al.*, 2000). A subgroup of necrotic PCD models is often classified as 'aborted apoptosis'; that is, a standard apoptosis program is initiated, then blocked at the level of caspase activation, and finally terminated by alternative, caspase-independent routes (Nicotera *et al.*, 1999). Autophagy is characterized by the formation of large lysosome-derived cytosolic vacuoles (Bursch *et al.*, 1996; Chi *et al.*, 1999; Elliott *et al.*, 2000), and dark cell death in specialized cells, such as chondrocytes (Roach and Clarke, 2000) or neurons (Turmaine *et al.*, 2000), usually lacks chromatin condensation and can thus be classified as necrosis-like PCD. Accidental necrosis, characterized by a rapid lysis of plasma membrane and organelle swelling, is the conceptual counterpart to PCD, since it is prevented only by removal of the stimulus. It occurs after exposure to high concentrations of detergents, oxidants, and ionophores, or high intensities of pathologic insult (Nicotera *et al.*, 1999). Finally, it should be noted that in tissue-culture conditions that lack the phagocytizing cells, the plasma membrane of cells dying either by classic apoptosis or apoptosis-like PCD will eventually break. Such loss of the cellular permeability barrier followed by passive changes in cell organelles is often confusingly referred to as secondary necrosis.

3. Death programs beyond caspases

In its classic form, apoptosis occurs almost exclusively when caspases, particularly caspase-3, are activated (Woo *et al.*, 1998; Susin *et al.*, 2000). The unexpected ability of certain cells to survive the activation of proapoptotic caspases (Lacana *et al.*, 1997; Wright *et al.*, 1997; Jäättelä *et al.*, 1998; De Maria *et al.*, 1999; Zeuner *et al.*, 1999; Harvey *et al.*, 2000; Foghsgaard *et al.*, 2001; Hoeppner *et al.*, 2001; Los *et al.*, 2001; Reddien *et al.*, 2001) demonstrates, however, the remarkable plasticity of the cellular death program, and argues against the idea that caspases alone are sufficient for the induction of mammalian PCD. Furthermore, recent evidence indicates a diversification of the apoptosis program in higher eukaryotes with respect to the necessity and role of caspases. Signals emanating from the established key players of apoptosis, including death receptors and caspases themselves, may result in necrosis-like PCD (Leist *et al.*, 1997; Vercammen *et al.*, 1998a,b; Khwaja and Tatton, 1999; Leist *et al.*, 1999; Holler *et al.*, 2000), and apoptosis-like PCD characterized by chromatin condensation and phosphatidylserine exposure is not necessarily accompanied by effector caspase activation (Berndt *et al.*, 1998; Lavoie *et al.*, 1998; Monney *et al.*, 1998; Mathiasen *et al.*, 1999; Roberts *et al.*, 1999; Nylandsted *et al.*, 2000b; Gingras *et al.*, 2002). Other important apoptosis hallmarks, such as detachment, shrinkage, and zeiosis, can also be present in cells dying in a caspase-independent manner (McCarthy *et al.*, 1997; Berndt *et al.*, 1998; Nylandsted *et al.*, 2000b; Foghsgaard *et al.*, 2001; Joza *et al.*, 2001; Gingras *et al.*, 2002).

Contrary to earlier expectations, the inhibition of caspase activation does not necessarily protect against cell death stimuli. Instead, it may reveal, or even enhance, underlying caspase-independent death programs. These programs may take the form of apoptosis-like (Deas *et al.*, 1998; Luschen *et al.*, 2000; Foghsgaard *et al.*, 2001; Joza *et al.*, 2001; Volbracht *et al.*, 2001b), or necrosis-like (Xiang *et al.*, 1996; Leist *et al.*, 1997; McCarthy *et al.*, 1997; Vercammen *et al.*, 1998a,b; Chautan *et al.*, 1999; Khwaja and Tatton, 1999; Xue *et al.*, 1999; Holler *et al.*, 2000; Matsumura *et al.*, 2000) PCD. In many experimental apoptosis models, including those triggered by death receptors (Vercammen *et al.*, 1998a,b; Holler *et al.*, 2000; Matsumura *et al.*, 2000), cancer drugs (Amarante-Mendes *et al.*, 1998), growth-factor deprivation (Xue *et al.*, 1999), staurosporine (Deas *et al.*, 1998), anti-CD2 (Deas *et al.*, 1998), oncogenes (McCarthy *et al.*, 1997), colchicine (Volbracht *et al.*, 2001b), GD3 (Simon *et al.*, 2001), or expression of Bax-related proteins (Xiang *et al.*, 1996; McCarthy *et al.*, 1997), the existence of backup death pathways has been uncovered following inhibition of caspase activity by pharmaceutical pancaspase inhibitors. However, several lines of evidence support the relevance of such 'second-line' mechanisms also for normal physiology and pathology. In addition to pharmacologic inhibitors, caspase pathways can be inactivated by other factors such as mutations (Chautan *et al.*, 1999), energy depletion (Leist *et al.*, 1997), nitra-tive/oxidative stress (Leist *et al.*, 1999), other proteases that are activated simulta-neously (Chua *et al.*, 2000; Lankiewicz *et al.*, 2000; Reimertz *et al.*, 2001), members of the 'inhibitor of apoptosis protein' (IAP) family (Jäättelä, 1999; Strasser *et al.*, 2000), defective release of Smac/Diablo (Deng *et al.*, 2002), or an array of viral pro-teins that can silence caspases (Strasser *et al.*, 2000). Thus, it is not surprising that the list of model systems where PCD is not accompanied by the effector caspase activation is growing (*Table 1*). This is especially evident in cancer cells, which often harbor defects in classic apoptosis pathways (Jäättelä, 1999).

Upon caspase inhibition, the alternative death pathways surface also *in vivo*. They are involved in processes such as the negative selection of lymphocytes (Smith *et al.*, 1996; Doerfler *et al.*, 2000), cavitation of embryoid bodies (Joza *et al.*, 2001), embryonic removal of interdigital webs (Chautan *et al.*, 1999), tumor necrosis fac-tor (TNF)-mediated liver injury (Kunstle *et al.*, 1999), and the death of chondro-cytes controlling the longitudinal growth of bones (Roach and Clarke, 2000). These examples may represent just the tip of the iceberg with regard to the complexity of death signaling *in vivo*. And the overlapping death pathways initiated by a single stimulus seem rather to be the rule than the exception (Holler *et al.*, 2000; Charette *et al.*, 2001; Joza *et al.*, 2001). The examination of potential crossovers of death pathways that lead eventually to different phenotypic outcomes may offer a chance to understand which events do determine commitment to death, and which ones are instead involved in upstream signaling or downstream execution.

4. Signaling in caspase-independent PCD

Several molecular mediators of classic caspase-mediated apoptosis pathways have been characterized during the last decade (Mattson, 2000; Strasser *et al.*, 2000; Kaufmann and Hengartner, 2001), whereas the description of most alternative death routines has remained limited to the phenomenological level. But recent mechanistic findings have opened a new era in this field. Like classic apoptosis,

Table 1. PCD models not accompanied by effector caspase activation

Stimulus	Cell type	Morphology	Rescued by	Reference
Adenoviral E4orf	Fibroblast p53-null fibroblast	Apoptosis-like	Bcl-2	Lavoie et al., 1998
Inhibition of the ubiquitin pathway	Fibroblast	Apoptosis-like	Bcl-2	Monney et al., 1998
CD4/CXCR4 receptor antibodies	CD4+ T cells	Apoptosis-like	Stromal cell growth factor	Berndt et al., 1998
Camptothecin	Liver cancer	Apoptosis-like	Cysteine cathepsin inhibitors	Roberts et al., 1999
Vitamin D compounds	Breast cancer	Apoptosis-like	Bcl-2, calpain inhibitors Calbindin	Mathiasen et al., 1999 I. S. Mathiasen and M. Jäättelä, unpublished
Antigen receptor cross-linking	B-cell lymphoma	Apoptosis-like	Inhibition of calpains and cysteine cathepsins	Katz et al., 2001
Sigma-2 receptor agonists	Breast cancer	Apoptosis-like		Crawford and Bowen, 2002
Hsp70 depletion	Various cancers (not fibroblast or immortalized epithelium)	Apoptosis-like	Cysteine cathepsin inhibitors (not by Bcl-2/Bcl-X$_L$)	Nylandsted et al, 2000a,b J. Nylandsted and M. Jäättelä, unpublished
Intracellular Acidification	Bladder cancer, fibroblast, leukemia	Necrosis-like	Inhibition of SAPK (not by Bcl-2)	Zanke et al., 1998
Oncogenic Ras	Glioblastoma gastric cancer	Necrosis-like autophagy	(not by Bcl-2)	Chi et al., 1999
Bin1	Liver cancer (not fibroblast)	Necrosis-like autophagy	SV40 large T antigen, serine protease inhibitor (not by Bcl-2)	Elliott et al., 2000
IGF1R	Kidney epithelium	Necrosis-like vacuolar degeneration	Actinomycin D Cycloheximide (not by Bcl-X$_L$)	Sperandio et al., 2000

CXCR4, chemokine coreceptor; SAPK, stress-activated protein kinase.

alternative death programs can be mediated by proteases and switched on by mitochondrial alterations or death receptors.

4.1 Noncaspase proteases as mediators of PCD

The most extensive evidence linking noncaspase proteases with PCD originates from studies of calpains; cathepsins B, D, and L; and granzymes A and B (Kitanaka and Kuchino, 1999; Johnson, 2000; Leist and Jäättelä, 2001[Q1]). These proteases cooperate often with caspases in classic apoptosis, but recent data suggest that they can also trigger PCD and bring about many of the morphologic changes characteristic of apoptosis in a caspase-independent manner (Johnson, 2000; Wang, 2000; Leist and Jäättelä, 2001a) (*Table 2*). As noncaspase proteases have only recently attracted broader interest among cell-death researchers, this list is likely to present only a fraction of all PCD-mediating proteases.

Cathepsins

The cathepsin protease family comprises cysteine, aspartate, and serine proteases (Johnson, 2000; Turk *et al.*, 2001). So far, the cysteine cathepsins B and L and the aspartate cathepsin D have been most clearly linked to PCD. Most cathepsins mature in the endosomal-lysosomal compartment. They can be activated by autoproteolysis in acidic pH or proteolysis by other proteases (for example, cathepsin D can activate cathepsins B and L). Furthermore, ceramide specifically binds to and promotes the proteolytic activation of cathepsin D, possibly linking sphingomylinase-mediated ceramide production and cathepsins to a common PCD pathway (Heinrich *et al.*, 1999). Until recently, the function of cathepsins was presumed to be limited to the disposal of proteins in the lysosomal compartment and degradation of extracellular matrix once secreted. During the past few years, however, many of them have been assigned specific functions, as, for example, in bone remodeling, hair follicle morphogenesis, antigen presentation, and PCD (Reinheckel *et al.*, 2001; Turk *et al.*, 2001). Genetic evidence for the role of cysteine cathepsins in PCD is provided by studies showing resistance to TNF-induced liver apoptosis in mice that lack cathepsin B (Guicciardi *et al.*, 2001), and massive PCD in the brains of mice that lack the cysteine cathepsin inhibitor, cystatin B (Pennacchio *et al.*, 1998).

Cathepsins participate in both caspase-dependent and -independent PCD induced by a variety of stimuli, including death receptors, camptothecin, B-cell receptors, bile salt, oxidants, and retinoids (Roberts *et al.*, 1997, 1999; van Eijk and de Groot, 1999; Guicciardi *et al.*, 2000; Foghsgaard *et al.*, 2001; Katz *et al.*, 2001; Roberg, 2001; Zang *et al.*, 2001). Cathepsins translocate from lysosomes to the cytosol and/or nucleus before the appearance of gross morphologic changes indicative of PCD. Notably, the inhibition of cathepsin activity protects cells from PCD without preventing the release of cathepsins from the lysosomes (Foghsgaard *et al.*, 2001). These data indicate that the release of cathepsins is not merely a sign of final organelle disintegration in the dying cell, and suggest that cathepsins have to escape the lysosomal compartment to trigger PCD. The latter hypothesis is further supported by the data showing that microinjection of cathepsin D (K. Roberg, personal communication), as well as limited disruption of lysosomes, triggers

Table 2. Characteristics of noncaspase proteases involved in programmed cell death

Protease/antiprotease	External stimuli	Cellular localization	Substrates	Effects
Cathepsin B and L/Cystatins (stefins)-cysteine proteases	TNF, TRAIL, camptothecin, Hsp70 depletion, B-cell antigen receptor, laser-beam-triggered microcavitation	Lysosomal or extracellular, translocates to the cytosol in apoptotic cells	Bid, PARP, procasp-1, -2, -6, - 7, 11, cathepsin C, Extracellular matrix	Cytochrome c release, chromatin condensation, blebbing, PS exposure, degradation of extracellular matrix
Cathepsin D -Aspartyl protease	TNF, interferon-γ, oxidative stress, retinoid CD437, ceramide photosensitizers	Lysosomal or extracellular translocates to the cytosol in apoptotic cells	Cystatins, Procathepsins	Cytochrome c release, activation of cathepsin B
Calpains/Calpastatin -cysteine proteases	Irradiation, ionophores, vitamin D compounds, TGF-β, β-lapacphone, dexamethasone, etoposide neurotoxins	Mainly cytosolic	Bax, Bcl-X_L, fodrin, procasp-3, -9, and -12, gelsolin, FAK, actin, c-fos, c-jun, c-mos, p53, cyclin D, etc.	Cytochrome c release, PS exposure, activation of procaspase-12
Granzymes/Serpin PI-9 -Serine proteases	Cytotoxic T cells, perforin-assisted entry to target cell	Cytotoxic granules in T cells	procasp-3, -6, -7, 8, -9, ICAD, lamins, histones, Bid, SET, DNA-PK_{CS}	Cytochrome c release, DNA breaks and fragmentation, nuclear condensation
AP24/Serpins -Serine protease	Death receptors, UV, DNA-damaging drugs	Cytosolic, translocates into nucleus in dying cells		DNA fragmentation
Omi (htra2)/serpins -Serine protease		Mitochondrial, translocates into cytosol in dying cells		Inhibition of IAPs, Cell rounding and shrinkage
Other serine proteases/ Serpins	Bin-1 tumor-suppressor, Death receptors		LEI (elastase)	L-DNase II-mediated DNA fragmentation

TGF, transforming growth factor; UV, ultraviolet light; FAK, focal adhesion kinase; PARP, poly (ADP) ribose polymerase; ICAD, inhibitor of caspase-activated DNase; SET, nucleosome assembly protein; DNA-PK_{CS}, catalytic subunit of DNA-dependent protein kinase.

apoptosis dependent on cathepsin activity (Kagedal *et al.*, 2001). It should also be noted that in some cells cathepsins may be pivotal for survival, as the cysteine cathepsin inhibitor, CATI-1, kills leukemia and lymphoma cells (Zhu and Uckun, 2000).

Calpains

Calpains are cysteine proteases that reside in the cytosol in an inactive zymogen form (Johnson, 2000; Wang, 2000). Two forms of calpains, μ-calpain and m-calpain, are ubiquitously expressed in human cells and have been linked to differentiation and PCD. The active forms of the enzymes consist of a variable large subunit (80 kDa) and a common small subunit (30 kDa). Activation of calpains requires an elevation in intracellular calcium $[Ca^{2+}]_i$. Proteolytic cleavage and association with membrane phospholipids may further contribute to their activation, possibly by lowering the $[Ca^{2+}]_i$ requirement. Calpain activity is controlled by calpastatin, a natural inhibitor that can be inactivated by calpain- or caspase-mediated cleavage. Calpains are activated by various stimuli (see *Table 2*) that increase the $[Ca^{2+}]_i$ and they can participate in PCD signaling upstream or downstream of caspases (Leist *et al.*, 1998; Waterhouse *et al.*, 1998; Nakagawa and Yuan, 2000; Choi *et al.*, 2001; Varghese *et al.*, 2001). Furthermore, calpains can mediate apoptosis-like PCD, even in the absence of caspase activation (Mathiasen *et al.*, 1999; I. S. Mathiasen and M. Jäättelä, unpublished). For example, EB1089/seocalcitol, a vitamin D analog currently on phase III clinical trials for the treatment of cancer, induces calcium- and calpain-dependent apoptosis-like PCD in breast cancer cells without triggering detectable caspase activation.

Serine proteases

The most prominent components of cytotoxic granules of cytotoxic lymphocytes are the pore-forming protein, perforin, and the serine proteases, granzyme A and granzyme B (Johnson, 2000; Trapani *et al.*, 2000). Upon activation, cytotoxic lymphocytes release their granular contents. Subsequently, target cells take up granzymes by receptor-mediated endocytosis and subsequent perforin-mediated release from endosomes into the cytosol or by diffusion via perforin-generated pores in the plasma membrane. Studies employing mice lacking either granzyme A or B have demonstrated that granzyme B is required for the granule-induced rapid caspase-mediated apoptosis (Johnson, 2000). Granzyme B cleaves its substrates after aspartate residues and can thus directly activate caspases. However, in the presence of caspase inhibitors, granzyme B triggers a slower necrosis-like PCD (Talanian *et al.*, 1997). Granzyme A is a trypsin-like protease that cleaves its substrates after lysine or arginine residues. Death induced by granzyme A is associated with DNA single-strand breaks created by a granzyme A-activated DNase (Beresford *et al.*, 2001).

Other serine proteases that have been associated with cell death include apoptotic protease 24 (AP24), which mediates DNA fragmentation in TNF-, UV light-, and chemotherapy-induced PCD of some cancer cells (Wright *et al.*, 1997; 1998), and the recently identified omi/htra2 (Suzuki *et al.*, 2001), which is released from mitochondria into the cytosol during apoptosis, and can mediate caspase-independent PCD dependent on its serine protease activity and/or contribute to the caspase activation by counteracting members of IAP family. A family of protease inhibitors called serpins inhibits the activity of serine proteases. Interestingly, the serine-protease-mediated inactivation of a serpin, leukocyte elastase inhibitor (LEI), transforms LEI into an endonuclease, L-DNase II (Torriglia *et al.*, 1998). L-Dnase II translocates to the nucleus in various PCD models; it can induce pyknosis and DNA degradation *in vitro*. Thus, the transformation of LEI to L-DNase II may act as an important switch of protease and nuclease pathways during caspase-independent PCD.

The definition of the role of the individual proteases in the complex process of PCD still requires much careful work. The dependence on certain proteases may be extremely cell-type and stimulus specific, and may depend on the relative expressions, activations, and inactivations of proteases and protease inhibitors (*Table 2* and *Figure 2*). Genetic approaches need to be combined with meticulous pharmacologic titration of inhibitors (Foghsgaard *et al.*, 2001), since it turns out that pan-caspase inhibitors, as well as many active site inhibitors of other proteases, are highly unspecific at concentrations widely used to test their role in PCD (Schotte *et al.*, 1998; Waterhouse *et al.*, 1998; Johnson, 2000; Foghsgaard *et al.*, 2001).

4.2 Death receptors as triggers of alternative PCD

The best-studied members of the death-receptor family are TNF receptor 1 (TNF-R1), Fas (also known as CD95 or Apo-1), and the receptors for TNF-related apoptosis-inducing ligand (TRAIL). Whereas it has long been known that TNF-induced death can take the shape of either apoptosis or necrosis (Laster *et al.*, 1988), the ability of the Fas ligand (FasL) and TRAIL to induce necrosis-like PCD has been

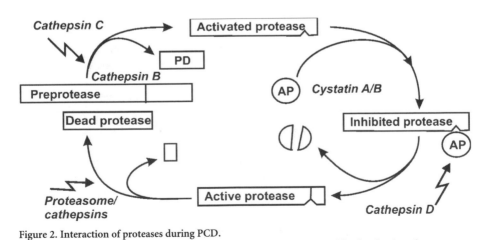

Figure 2. Interaction of proteases during PCD.
Frequently, an inactive or weakly active zymogen (preprotease) is activated by the cleaving of a prodomain (PD) (the first level of protease family interaction; examples shown in italics). Intracellular protease activity is prevented by specific antiproteases (AP), and ultimate activation requires inactivation of AP (often by proteolysis, the second level of protease family interaction). Further proteolysis can lead to inactivation of the active protease and eventually degradation (the third level of protease family interaction). The balance of all players in this circle determines which proteases dominate the death process. Pharmacologic inhibition of one protease easily shifts the balance to another pathway.

described only recently (Leist *et al.*, 1997; Kawahara *et al.*, 1998; Vercammen *et al.*, 1998a,b; Leist *et al.*, 1999; Holler *et al.*, 2000). In activated primary T cells, this caspase-independent necrosis-like PCD seems, at least in some cases, to be the dominant mode of death (Holler *et al.*, 2000). This may explain why inhibition of caspase activity in mouse T cells *in vivo* does not induce the lymphadenopathy and/or autoimmune disease usually manifested in mice with inactivating mutations in Fas or Fas ligand (Smith *et al.*, 1996).

Except for the dependence on reactive oxygen species (ROS) and, in some cases, serine protease activity, necrotic signaling pathways have remained ambiguous until recently (Denecker *et al.*, 2001). Novel data demonstrate now that TNF, FasL, and TRAIL can trigger caspase-8-independent necrosis-like PCD that is dependent on the Fas-associated death domain (FADD) protein and the kinase activity of the receptor-interacting protein (RIP) (Holler *et al.*, 2000). The dependence of RIP-mediated necrotic PCD on proteases remains to be studied. Interestingly, some TNF-resistant cells are sensitized to TNF-induced necrosis-like PCD upon inhibition of caspases, suggesting that caspases act as survival factors that directly inhibit the TNF-induced necrotic pathway (Khwaja and Tatton, 1999). Death receptors can also trigger caspase-independent apoptosis-like PCD. In immortalized epithelial cells, activated Fas has been reported to recruit Daxx from the nucleus to the receptor complex, and to trigger its binding with apoptosis signal-regulating kinase 1 (Ask1) (Charette *et al.*, 2000; Ko *et al.*, 2001). Others have, however, have failed to detect Daxx in the cytosol and have suggested that Daxx enhances Fas-induced caspase-dependent death from its nuclear localization (Torii *et al.*, 1999). Thus, Daxx may stimulate Fas-induced death by two independent mechanisms, the caspase-independent pathway being evident only when caspase activation is defective (Charette *et al.*, 2000) and enough Ask1 is available (Ko *et al.*, 2001). In addition to a caspase-dependent proapoptotic function that depends on its kinase activity, Ask1 possesses a caspase-independent killing function that is independent of its kinase activity and is activated by interaction with Daxx (Charette *et al.*, 2001). Ask1 has also been found to be essential for TNF-triggered apoptosis of primary fibroblasts, but its activation by TNF appears to require ROS (Tobiume *et al.*, 2001) instead of Daxx (Yang *et al.*, 1997). In TNF-treated fibrosarcoma cells cysteine cathepsins act as dominant execution proteases and bring about apoptosis-like morphologic changes (Foghsgaard *et al.*, 2001). Whether Ask1 and cathepsins act on the same signaling pathway is as yet unknown.

The picture described above suggests a complexity of death-receptor-induced apoptotic and necrotic signaling networks that far exceeds that of the simple linear pathway originally suggested by the discovery of the receptor-triggered caspase cascade (*Figure 3*).

4.3 Mitochondrial control of caspase-independent PCD

Analogous to classic apoptosis, mitochondrial membrane permeabilization (MMP) controlled by Bcl-2 family proteins resides at the heart of several alternative death pathways. The prevailing theory suggests that 'multidomain' proapoptotic Bcl-2 family members (such as Bax and Bak) are the actual pore-forming effector molecules required for MMP (Cheng *et al.*, 2001; Wei *et al.*, 2001). Bax

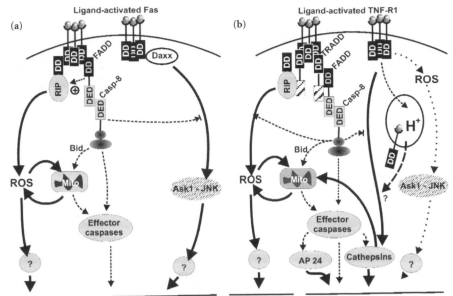

Figure 3. Multiple death pathways triggered by death receptors.
Death-receptor signaling is initiated by ligand-induced receptor trimerization. (a) Receptor death domains (DD) of Fas then recruit FADD, RIP, and/or Daxx to the receptor complex. Caspase-8 becomes activated after recruitment to FADD via death effector domain (DED) interaction, and triggers effector caspases, either directly or through a Bid-mediated mitochondrial pathway (Strasser *et al.*, 2000) (dashed lines). RIP initiates a caspase-independent (solid lines) necrotic pathway mediated by the formation of reactive oxygen species (ROS). Daxx activates the Ask1-JNK kinase pathway, leading to caspase-independent apoptosis. (b) Tumor necrosis factor receptor-1 (TNF-R1) signaling differs from that of Fas in the following steps: (i) Binding of FADD and RIP to the receptor complex requires the adapter protein TRADD. (ii) Binding of Daxx to TNF-R1 has not been demonstrated, and the Ask1-JNK pathway is activated by ROS (dotted line; caspase involvement unclear). (iii) The RIP-mediated necrotic pathway is inhibited by caspase-8. (iv) TNF-R1 can initiate a caspase-independent direct cathepsin B-mediated pathway. (v) Cathepsin B can enhance the mitochondrial death pathway. (vi) The final execution of the death – that is, phosphatidylserine exposure, chromatin condensation, and loss of viability – is brought about by effector caspases, the serine protease AP24, or cathepsin B in a cell-type-specific manner.

and/or Bak can be activated transcriptionally or by conformational change induced by cleavage or binding to an activated BH3-only Bcl-2 family member (*Figure 4*). Antiapoptotic Bcl-2 proteins (such as Bcl-2 and Bcl-X_L) oppose the MMP, probably by heterodimerization with Bax-like proteins, whereas 'BH3-only' Bcl-2 family members either oppose the inhibitory effect of Bcl-2-like proteins (Bad, Bim, Noxa, PUMA, etc.) or activate Bax-like proteins by direct binding (truncated Bid). The pathways upstream of MMP are numerous and with a few exceptions caspase-independent (Heibein *et al.*, 2000; Strasser *et al.*, 2000; Choi *et al.*, 2001; Ferri and Kroemer, 2001; Kaufmann and Hengartner, 2001; Stoka *et al.*, 2001) (*Figure 4*).

MMP does not necessarily result in an irreversible mitochondrial dysfunction and cell death. Initially, the pore-forming proapoptotic Bcl-2 family members Bax and Bak induce only outer membrane permeability but leave intact inner membrane energization, protein import function, and the ultrastructure of mitochon-

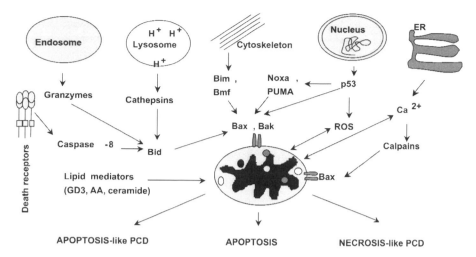

Figure 4. Caspase-independent signaling pathways leading to MMP.
The pore-forming proteins Bax and/or Bak can be activated by BH3-only proteins. Death receptors can activate caspase-8, which cleaves and activates Bid. Granzymes released from the granules of cytotoxic T cells and natural killer cells can be taken up by the target cells through perforin-assisted diffusion or endocytosis. Once released to the cytosol of the target cell by the action of perforin, granzyme B may also cleave and activate Bid. TNF and TRAIL, as well as various oxidants, detergents, and chemotherapeutic drugs, can induce the release of active cathepsins from the lysosomal compartment, and the cathepsin-mediated cleavage of Bid has been held to mediate cathepsin-induced MMP. Disruption of the cytoskeleton leads to the release of the BH3-only proteins Bim and Bmf. DNA damage induced by radiation or various chemotherapeutic drugs induces a p53-mediated transcription of genes encoding Bax, BH3-domain only proteins (Noxa or PUMA), and proteins involved in ROS generation. ER stress results in the release of Ca^{2+}, which may cause direct mitochondrial damage or activate Bax via calpain-mediated cleavage. Various death stimuli trigger the production of lipid second messengers that are involved in MMP and mitochondrial damage. Depending on the stimulus and the cell type, as well as the metabolic status of the cell, MMP leads to either caspase-mediated apoptosis or caspase-independent PCD.

dria (Von Ahsen *et al.*, 2000b). Recent data indicate that MMP prompts several caspase-dependent and -independent death pathways (Leist and Jäättelä, 2001a) (*Figure 5*). The apoptosome-caspase pathway leading to classic apoptosis is initiated by the MMP-dependent release of cytochrome *c* from the mitochondrial intermembrane space. Together with other essential factors (such as ATP), it triggers assembly of the apoptosome complex, which forms the template for efficient caspase processing. As a further safeguard mechanism, caspase-inhibitory factors (IAPs and XIAP) have to be removed by additional proteins (DIABLO/SMAC or Omi/HtrA2) released from mitochondria before the execution caspases can become fully active and produce the typically apoptotic morphology (Strasser *et al.*, 2000; Kaufmann and Hengartner, 2001).

The second mitochondrial death pathway leads to necrotic PCD, without necessarily activating caspases. A prominent example is TNF induced necrosis-like PCD mediated by mitochondria-derived ROS (Schulze-Osthoff *et al.*, 1992). Intracellular control of this pathway is indicated by its susceptibility to attenuation by antioxidants (Vercammen *et al.*, 1998a,b; Schulze-Osthoff *et al.*, 1992).

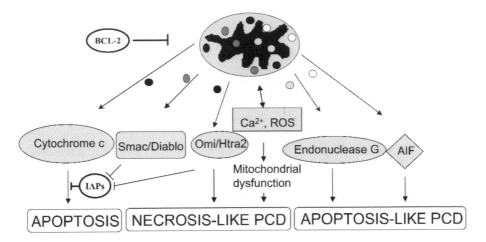

Figure 5. Death pathways downstream of MMP.
Mitochondrial damage leads to the release of numerous mitochondrial proteins that may trigger the execution of PCD. Cytochrome *c* release triggers caspase activation and classic apoptosis. Smac/Diablo and Omi/htra2 assist cytochrome *c*-induced caspase activation by counteracting IAPs. AIF triggers a caspase-independent death pathway, culminating in the DNA fragmentation and stage 1 chromatin condensation characteristic of apoptosis-like PCD. Endonuclease G cleaves DNA and induces stage 1 chromatin condensation. The serine protease activity of Omi/htra2 mediates caspase-independent cellular rounding and shrinkage without changes in the nuclear morphology. Ca^{2+} and ROS can lead to severe mitochondrial dysfunction and necrosis-like PCD. All these mitochondrial events are at least partially inhibited by Bcl-2.

A third distinct pathway from mitochondria is the release of the apoptosis-inducing factor (AIF) from the intermembrane space (Susin *et al.*, 1999; Suter *et al.*, 2000; Braun *et al.*, 2001). Recent genetic evidence indicates that this factor controls PCD in early development; that is, all the hallmarks of early morphogenetic death, including cytochrome *c* release, are prevented by deletion of AIF (Joza *et al.*, 2001). AIF induces caspase-independent formation of large (50-bp) chromatin fragments, whereas oligonucleosomal DNA fragments are generated only when caspase-activated DNase (CAD) is activated (Susin *et al.*, 1999; Strasser *et al.*, 2000). This biochemical difference is reflected by slight morphologic differences in the shape of the condensed chromatin (*Figure 1*).

Finally, endonuclease G or a serine protease, Omi/htra2, released from mitochondria during PCD can contribute to the caspase-independent death signaling downstream of MMP (Li *et al.*, 2001; Suzuki *et al.*, 2001). Extramitochondrial expression of Omi/htra2 induces caspase-independent PCD, and endonuclease G causes caspase-independent DNA fragmentation in isolated nuclei. However, direct evidence connecting endonuclease G and PCD-associated DNA fragmentation in mammalian cells is still lacking.

Often, more than one pathway seems to be activated simultaneously (Jäättelä *et al.*, 1998; Susin *et al.*, 1999; Mattson, 2000; Suter *et al.*, 2000). The cell fate and death mechanism are then determined by the relative speed of each process in a given model system, and by the antagonists of the individual pathways differentially expressed in different cell types. AIF, caspases, and ROS can feed back on mitochondria, causing enough structural and functional damage to trigger the release of other death factors, independently of the upstream signals (Nicotera *et al.*, 1999; Susin *et al.*, 1999; Mattson, 2000; Strasser *et al.*, 2000).

Defects in any step of the cytochrome *c* or AIF pathways can switch apoptosis or apoptosis-like PCD to death with a necrotic morphology (Leist *et al.*, 1997; McCarthy *et al.*, 1997; Daugas *et al.*, 2000). This death would still fulfill the criteria of PCD, as it is blocked by the antiapoptotic oncogenes Bcl-2 or Bcr-Abl (Amarante-Mendes *et al.*, 1998; Daugas *et al.*, 2000; Single *et al.*, 2001) or by the deletion of proapoptotic Bax (Miller *et al.*, 1997). Moreover, caspase inhibition changes the mode of death, but not its extent, once the signal has arrived at mitochondria (Xiang *et al.*, 1996; Hirsch *et al.*, 1997; Leist *et al.*, 1997; McCarthy *et al.*, 1997; Miller *et al.*, 1997; Amarante-Mendes *et al.*, 1998; Nicotera *et al.*, 1999; Daugas *et al.*, 2000). Thus, it seems that in many models of cell death the master controllers of PCD operate at the mitochondrial level, while the decision on the form of death is taken at the level of caspase activation (Nicotera *et al.*, 1999).

There are, however, certain cases where Bcl-2 expression is not protective, and where mitochondria may not have a regulatory role (Chi *et al.*, 1999; Schierle *et al.*, 1999; Elliott *et al.*, 2000; Finn *et al.*, 2000; Nylandsted *et al.*, 2000b; Sperandio *et al.*, 2000). Although the alternative control mechanisms are not well characterized, emerging candidates include different chaperone systems, such as heat-shock proteins (Jäättelä *et al.*, 1998; Charette *et al.*, 2000; Nylandsted *et al.*, 2000b) or ORP150 (Tamatani *et al.*, 2001). Organelles that have not received much attention until recently, such as the endoplasmic reticulum and lysosomes, might also take an essential role in the control of death (Mattson, 2000; Ferri and Kroemer, 2001; Leist and Jäättelä, 2001a) (*Figure 4*).

5. The significance of the program: removal of corpses

Death is not the only important endpoint of PCD. A much less complicated machinery would be sufficient to permeabilize the plasma membrane. The classic apoptosis program is, in fact, optimized to ensure that signals for phagocytosis are displayed well before cellular constituents might be released (Savill and Fadok, 2000; Strasser *et al.*, 2000). In extreme cases, there is even a feedback control of phagocytosis on the death program itself to ensure that death occurs only when phagocytosis has been initiated (Hoeppner *et al.*, 2001; Reddien *et al.*, 2001). Does this also apply to caspase-independent programs? A dominant uptake signal in mammalian cells is the translocation of phosphatidylserine to the outer leaflet of the plasma membrane. Also this 'eat-me' indicator is uncoupled from caspase activation in many model systems (Berndt *et al.*, 1998; Mateo *et al.*, 1999; Fröhlich and Madeo, 2000; Hirt *et al.*, 2000; Foghsgaard *et al.*, 2001), and nonapoptotically dying eukaryotic cells can be efficiently phagocytized (Hirt *et al.*, 2000). Mechanisms that can lead to the translocation of phosphatidylserine and phagocytosis in cells undergoing caspase-independent death include disturbances of cellular calcium homeostasis and protein kinase C activation (Hirt *et al.*, 2000; Volbracht *et al.*, 2001b). Noncaspase cysteine proteases might be involved not only in the alternative death execution, but also in alternative phagocytosis signal pathways. For instance, cathepsin B activity is required for the translocation of phosphatidylserine in TNF-challenged tumor cells (Foghsgaard *et al.*, 2001), and, in the apoptosis-like death of platelets, calpain inhibitors selectively block phagocytosis signals (Wolf *et al.*, 1999). Finally, genetic analysis in *C. elegans* has shown that the same phagocytosis-recognition molecules are involved in removing corpses produced by caspase-dependent apoptosis and caspase-independent necrosis (Chung *et al.*, 2000).

6. Complex control of tumor cell death

Paradoxically, the cell proliferation induced by enhanced activity of oncoproteins (such as Myc, E1A, E2F, and CDC25) or inactivation of tumor suppressor proteins (retinoblastoma protein, for example) is often associated with caspase activation and accelerated apoptosis (Schmitt and Lowe, 1999). The coupling of cell division to cell death has thus been proposed to act as a barrier that must be circumvented for cancer to occur (Jäättelä, 1999; Schmitt and Lowe, 1999). Indeed, high expression of the antiapoptotic proteins (Bcl-2, Bcl-x$_L$, survivin, and Bcr-Abl) and/or inactivation of the proapoptoic tumor-suppressor proteins (p53, p19^{ARF}, and PTEN) controlling caspase-dependent apoptosis pathways are often seen in human tumors (Jäättelä, 1999; Schmitt and Lowe, 1999).

6.1 Alternative death pathways in cancer

Despite showing severe defects in classic apoptosis pathways, cancer cells have not lost the ability to commit suicide. On the contrary, spontaneous apoptosis is common in aggressive tumors, and most of them respond to therapy (Kerr *et al.*, 1994). One explanation may be that defects in the signaling pathways leading to caspase activation may still allow caspase-independent death pathways to execute tumor cell death.

The alternative death pathways may also be enhanced by transformation (*Figure 6*). For example, oncogenic Ras can induce caspase- and Bcl-2-independent autophagic death (Chi *et al.*, 1999), and tumor-associated Src family kinases are involved in caspase-independent cytoplasmic execution of apoptotic programs induced by the adenovirus protein E4orf4 (Lavoie *et al.*, 2000; Gingras *et al.*, 2002). Furthermore, a transformation-associated caspase-, p53-, and Bcl-2-independent, apoptosis-like death program can be activated in tumor cell lines of different origins by depletion of a 70-kDa heat-shock protein (Hsp70) (Nylandsted *et al.*, 2000a,b). This death is preceded by a translocation of active cysteine cathepsins from lysosomes to cytosol, and inhibitors of their activity partially protect against death. Interestingly, cysteine cathepsins, as well as other noncaspase proteases, are highly expressed in aggressive tumors (Duffy, 1996).Therefore, expression of protease inhibitors may increase a cancer cell's chances of survival by impairing alternative death routes (Alexander *et al.*, 1996; Foghsgaard *et al.*, 2001; Leist and Jäättelä, 2001a).

Alternative death pathways can also function at an initial stage of tumorigenesis to limit tumor formation. Bin1, a tumor-suppressor protein that is often missing or functionally inactivated in human cancer, can activate a caspase-

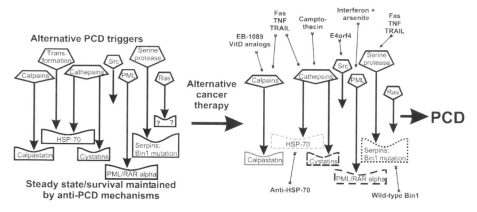

Figure 6. Alternative death pathways as regulators of tumor cell survival and as putative targets for cancer therapy.

(*Left*) Transformation is associated with upregulation of proteins that sensitize cells to caspase-independent PCD. As a defense line, death-promoting proteins are inactivated or expression of survival proteins is enhanced. Analogous changes in proteins regulating caspase-dependent apoptosis have also been demonstrated in cancer. (*Right*) Strategies of cancer therapy aimed at facilitating alternative death pathways.

independent apoptosis-like death process that is blocked by a serine protease inhibitor or the simian virus large T antigen, but not by overexpression of Bcl-2 or inactivation of p53 (Elliott *et al.*, 2000). Similarly, the promyelocytic leukemia PML/RARα oncoprotein also inhibits caspase-independent PCD induced by the PML tumor-suppressor protein (Quignon *et al.*, 1998). Interestingly, the cytoplasmic apoptotic features induced by ectopic expression of PML can even be enhanced by pancaspase inhibitors (Quignon *et al.*, 1998). It should, however, be noted that PML/RARA can also interfere with caspase activation in some death models (Wang, Z.G. *et al.*, 1998).

6.2 Designing new therapies based on alternative PCD pathways

Experimental gene-therapy approaches point to alternative death pathways as promising targets for tumor therapy. For example, the expression of Bin1 tumor suppressor or depletion of Hsp70 results in effective induction of caspase-independent apoptosis-like PCD in tumor cells (Elliott *et al.*, 2000; Nylandsted *et al.*, 2000a,b). Remarkably, adenoviral transfer of antisense Hsp70 cDNA also efficiently eradicates glioblastoma, breast-cancer, and colon-cancer xenografts in mice (J. Nylandsted and M. Jäättelä, unpublished). The ineffective delivery of viral vectors into multiple tumor sites appears to be the major limitation for the usage of this gene therapy in the treatment of human cancer. Thus, clinical applications of this approach require further development of the delivery systems or other means to activate Bin1 or to neutralize the antiapoptotic effect of Hsp70. However, in the case of local inoperable tumors, such as glioblastoma, locoregional gene therapy may prove beneficial.

Even though the signaling pathways regulating alternative PCD are only beginning to emerge, potentially cancer-relevant drugs or drug targets engaging caspase-independent death routines already exist (*Figure 6*). For instance, the topoisomerase inhibitor, camptothecin, induces cathepsin D/B-mediated apoptosis-like PCD in hepatocellular carcinoma cells (Roberts *et al.*, 1999); activation of a thrombospondin receptor (CD47) by thrombospondin or agonistic antibodies initiates programmed necrosis in B-cell-chronic lymphoma cells (Mateo *et al.*, 1999); antibodies to CD99 trigger a rapid, apoptosis-like PCD in transformed T cells (Pettersen *et al.*, 2001); interferons and arsenite initiate a caspase-independent death pathway, possibly mediated by PML (Quignon *et al.*, 1998); EB 1089, a synthetic vitamin D analog presently in phase III trials for the treatment of cancer, kills breast-cancer cells in a caspase-independent manner (Mathiasen *et al.*, 1999); and treatment of breast-cancer cells with sigma-2 receptor agonists triggers apoptosis-like PCD independently of p53 or caspase activity (Crawford and Bowen, 2002).

7. Alternative cell death in the nervous system

Caspase-driven neuronal apoptosis strictly following the classic apoptosome pathway is best documented during development of the nervous system (Los *et al.*, 1999), where many superfluous cells are produced and turned over (Raff, 1992), and in *in vitro* cultures of cells derived from developing brain (Mattson, 2000). Evidence is scarce for adult neurons, and here caspase-dependent mechanisms may yield to alternative death pathways (Johnson *et al.*, 1999). Notably, a re-evaluation of cell death in caspase knockout mice showed that apoptosis is reduced during development, but cell death in many brain regions proceeds to the same extent in a caspase-independent necrosis-like PCD often characterized by cytoplasmic vacuoles (Oppenheim *et al.*, 2001). Cell suicide in the adult nervous system has serious implications for the whole organism, since turnover is classically very limited. Thus, a rapid caspase cascade, which is advantageous for efficient elimination of unwanted or rapidly replaceable cells, is dangerous in the developed brain and must be tightly controlled. For instance, neurons can survive cytochrome *c* release from mitochondria if they do not simultaneously receive a second signal leading to 'competence to die' (Deshmukh *et al.*, 2000). Neuronally expressed apoptosis inhibitor proteins (IAP, NAIP) buffer the caspase system, and need to be inactivated before classic apoptosis can be executed (Kaufmann and Hengartner, 2001). This buffering capacity may allow for localized caspase activation (Mattson, 2000) (within synapses or neurites, for example) or sequestration of active caspases (Stadelmann *et al.*, 1999), without a buildup of the death cascade affecting the entire neuron. Stressed neurons might also acquire a temporary resistance that allows them to withstand otherwise lethal insults, such as those by excitotoxins (Hansson *et al.*, 1999). Such circumstances favor activation of slow, caspase-independent elimination routines, where damaged organelles are digested within a stressed cell, and the chance for rescue and reversibility is maintained until the process is complete (Jellinger and Stadelmann, 2000; Yamamoto *et al.*, 2000; Xue *et al.*, 2001).

Although some caspase-dependent apoptosis might occur in adult brain (Mattson, 2000), at least part of PCD in chronic neurodegenerative disease follows alternative mechanisms and results in different morphologies (Miller *et al.*, 1997;

Colbourne *et al.*, 1999; Roy and Sapolsky, 1999; Stadelmann *et al.*, 1999; Fujikawa, 2000; Jellinger and Stadelmann, 2000; Sperandio *et al.*, 2000; Turmaine *et al.*, 2000) (*Figure 7*). Further variation is observed in acute insults, such as ischemia or traumatic brain injury. Here, neurons within one brain region are exposed to different intensities of stress that trigger different death programs. Some of the main excitotoxic processes, such as mitochondrial impairment and dissipation of cell membrane potential, differentially impair various secondary routines of PCD (Nicotera

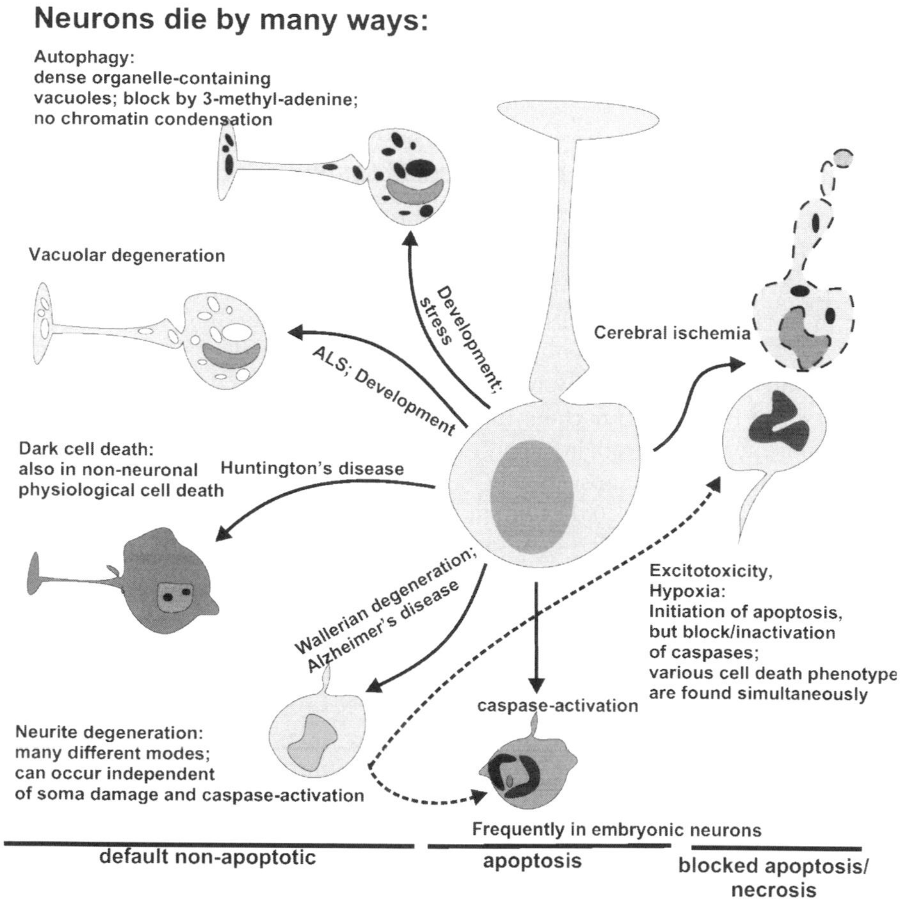

Neurons die by many ways:

Autophagy:
dense organelle-containing
vacuoles; block by 3-methyl-adenine;
no chromatin condensation

Vacuolar degeneration

Development; stress

ALS; Development

Cerebral ischemia

Dark cell death:
also in non-neuronal
physiological cell death

Huntington's disease

Wallerian degeneration;
Alzheimer's disease

Excitotoxicity,
Hypoxia:
Initiation of apoptosis,
but block/inactivation
of caspases;
various cell death phenotype
are found simultaneously

caspase-activation

Neurite degeneration:
many different modes;
can occur independent
of soma damage and caspase-activation

Frequently in embryonic neurons

default non-apoptotic | apoptosis | blocked apoptosis/necrosis

Figure 7. Developmental cell death occurs by caspase-dependent apoptosis or by morphologically and mechanistically distinct autophagy.
In various human diseases or animal models of them, the dominant form of neuronal disease is, for example, dark cell death in a Huntington's disease model, or vacuolar degeneration in a model of amyotrophic lateral sclerosis (ALS). Selective neurite degeneration occurs independently of caspase activation in different situations, and may eventually lead either to caspase-dependent apoptosis of cell bodies or to nonapoptotic death with irregular chromatin condensation. Excitotoxic death may take many forms and mechanisms depending on the intensity of insult, the age of the animal, and the brain region affected. It often results in mixed apoptotic and necrotic features, including cellular swelling, blebbing, nuclear pyknosis, display of phosphatidylserine, and some autophagic processes, such as uptake of mitochondria into lysosomes.

et al., 1999; Roy and Sapolsky, 1999; Fujikawa, 2000). For instance, rapid ATP depletion or disturbance of the intracellular ion composition impairs cytochrome *c*-induced caspase activation, and massive production of nitric oxide (NO) or calpain activation directly inactivates caspases (Nicotera *et al.*, 1999; Lankiewicz *et al.*, 2000). Accordingly, cell death has mixed features of apoptosis and necrosis, and might rely on either caspases or calpains as the dominant execution proteases (Wang, 2000; Volbracht *et al.*, 2001b), or the activation of PARP (Ha and Snyder, 1999) as a controller of programmed necrosis. Another group of proteases implicated as executors of ischemic death are the cysteine cathepsins (Yamashima, 2000). Possibly, they interact with calpains, and notably there is massive PCD in the brains of mice lacking the cathepsin inhibitor cystatin B (Pennacchio *et al.*, 1998).

The special shape of neurons (with projections up to 40 000 times longer than their cell bodies) allows degradative processes to be localized to a part of neurons and different death processes to be activated in different subsections of the cell (Nicotera *et al.*, 1999; Mattson, 2000). For instance, synaptic damage and neurite regression can occur by Bcl-2- and caspase-independent mechanisms (Sagot *et al.*, 1995; Finn *et al.*, 2000; Volbracht *et al.*, 2001b) and be initially reversible (Yamamoto *et al.*, 2000), whereas final elimination of cells may depend on caspases or proteasomal activities (Volbracht *et al.*, 2001b). The role of caspases as enhancers of the final phase of cell degeneration may apply to many common diseases. The longevity of neurons, combined with their dependence on effective intracellular transport, makes them sensitive to a slow form of death associated with the formation of intracellular polypeptide aggregates involving the amyloid-? precursor protein (APP), ataxins, presenilins, huntingtin, tau, and alpha-synuclein (Mattson, 2000). As most of these proteins are caspase targets (Wellington and Hayden, 2000) and become more toxic after cleavage, caspases might contribute to the accelerated death of neurons at the end of a caspase-independent degeneration phase, or vice versa, make neurons sensitive to alternative mechanisms without directly participating in death execution (Zhang *et al.*, 2000b).

8. Evolution of cell-death principles

The driving evolutionary pressures for the development of multiple cell-death programs have increased in parallel with the increased complexity and life span of organisms (Aravind *et al.*, 2001). But when in evolution did the caspase-independent death mechanisms arise? Caspase-coding sequences are absent from the known genomes of many nonanimal species (Aravind *et al.*, 2001). Nevertheless, such organisms – including plants and a number of single-celled eukaryotes – undergo PCD under conditions of stress (Ameisen, 1996; Fröhlich and Madeo, 2000) (see also Chapter 9). For instance, in yeast, this apoptosis-like death is associated with DNA-fragmentation, plasma membrane blebbing, phosphatidylserine exposure, and chromatin condensation (Fröhlich and Madeo, 2000), and can be selectively triggered or blocked by Bax-like or ced-9-related genes, respectively (see Chapter 8). Furthermore, programmed necrosis-like death is well characterized in caspase-deficient slime molds (Wyllie and Golstein, 2001).

The introduction of the caspases, and especially of the mitochondrial Ced-9/Bcl-2-related death switches (Ameisen, 1996; Aravind *et al.*, 2001), may represent a decisive refinement of the old caspase-independent death programs. The relative importance of different death mechanisms seems to have been optimized subsequently in various ways. One form of extreme specialization is exemplified by the somatic cell death in the nematode *Caenorhabditis elegans* (see Chapter 10). The requirements for PCD in *C. elegans* are adapted to its specific needs, and have diverged widely from those of mammals (Aravind *et al.*, 2001). Since the environmental pressure to provide a flexible death response is very low in this short-lived organism, evolutionary optimization has resulted in a single caspase-dependent apoptosis program. In contrast to mammals, control by mitochondrial proteins may play a minor role, and some degradative enzymes are supplied by the phagocytizing cell rather than by the dying cell itself (Strasser *et al.*, 2000; Hoeppner *et al.*, 2001; Kaufmann and Hengartner, 2001; Reddien *et al.*, 2001). Apoptosis in *C. elegans* is commonly cell-autonomous, that is, it is not signaled or controlled from outside, and the entire system of death receptors appears to be absent. In accordance with this minimalist program, somatic PCD is not essential for survival or development in *C. elegans* (Ellis and Horvitz, 1986). Vestiges of alternative apoptotic programs are, however, still found in the male linker cell, where a possibly Ced-3-independent PCD is triggered from outside (Ellis and Horvitz, 1986). The role of mitochondrial endonuclease G in caspase-independent degradation of DNA might also be conserved from worm to man (Li *et al.*, 2001; Parrish *et al.*, 2001).

The mammalian system of death programs could represent an opposite form of evolutionary direction, where, besides the multiple caspases, many other cysteine proteases and mitochondrial factors have taken additional roles in development and life (Los *et al.*, 1999; Strasser *et al.*, 2000). The essential nature of some factors (knockout lethality [Los *et al.*, 1999; Joza *et al.*, 2001] combined with the redundancy of others (difficulty with interpretation of knockouts [Los *et al.*, 1999]) has made the study of their specific role in PCD technically challenging. In addition, it has remained unclear which mechanisms are essential for commitment to death, and which ones merely determine the phenotypic outcome (Nicotera *et al.*, 1999).

9. Conclusion

The discovery and understanding of alternative death pathways will open new perspectives for the treatment of disease (*Figure 7*), and one of these therapies (vitamin D analogs) has already entered clinical phase III trials. New options and targets have emerged also for the prevention of death processes in neurodegenerative disease. Prominent examples of such targets that have reached the stage of clinical trials include PARP in necrosis and calpains in excitotoxicity (Ha and Snyder, 1999; Johnson *et al.*, 1999; Wang, 2000).

On a more general biologic level, the mode of cell death may differentially affect the surrounding tissue (Savill and Fadok, 2000). The important roles of caspase-independent/alternative death in the development of tumor immunity are just emerging (reviewed in Hirt *et al.*, 2000). Most recent evidence shows that the mode of cell demise controls the horizontal spread of oncogenic information (Bergsmedh *et al.*, 2001) and of infection (Boise and Collins, 2001). Since these processes can be favored by caspase activation, the classic apoptosis pathways can, in fact, be detrimental to the organism. This may explain the need for extremely tight control of caspase-activation by the cellular energy level (Leist *et al.*, 1997). The apparent paradox that death-bound ATP-depleted cells are not 'allowed' to activate caspases may then be explained by the fact that such cells would release activated caspases into the extracellular space upon premature lysis (Hentze *et al.*, 2001). Thus, non-apoptotic death may be not only a passive accidental event, but also, in some cases, a desirable death option for long-lived organisms having to deal with tumors, infections, and other nonlethal tissue insults throughout their life span.

Acknowledgments

We thank our colleagues for stimulating discussions, and the Danish Cancer Society, the German Research Council, and the Danish Medical Research Council for financial support.

References

Aballay, A. and Ausubel, F.M. (2001) Programmed cell death mediated by ced-3 and ced-4 protects *Caenorhabditis elegans* from *Salmonella typhimurium*-mediated killing. *Proc. Natl Acad. Sci. U. S. A.* **98**: 2735–2739.

Abbott, M.K. and Lengyel, J.A. (1991) Embryonic head involution and rotation of male terminalia require the *Drosophila* locus head involution defective. *Genetics* **129**: 783–789.

Abraham, R.T. (2001) Cell cycle checkpoint signaling through the ATM and ATR kinases. *Genes Dev.* **15**: 2177–2196.

Abrams, J.M., White, K., Fessler, L.I. and Steller, H. (1993) Programmed cell death during *Drosophila* embryogenesis. *Development* **117**: 29–43.

Acehan, D., Jiang, X., Morgan, D.G., Heuser, J.E., Wang, X. and Akey, C.W. (2002) Three-dimensional structure of the apoptosome. Implications for assembly, procaspase-9 binding, and activation. *Mol. Cell* **9**: 423–432.

Adams, J.M. and Cory, S. (2001) Life-or-death decisions by the Bcl-2 protein family. *Trends Biochem. Sci.* **26**: 61–66.

Adams, M.D., Celniker, S.E., Holt, R.A., *et al.* (2000) The genome sequence of *Drosophila melanogaster*. *Science* **287**: 2185–2195.

Ahmad, M., Srinivasula, S.M., Wang, L., Talanian, R.V., Litwack, G., Fernandes-Alnemri, T. and Alnemri, E.S. (1997) CRADD, a novel human apoptotic adaptor molecule for caspase-2, and FasL/tumor necrosis factor receptor-interacting protein RIP. *Cancer Res.* **57**: 615–619.

Ahmed, S., Alpi, A., Hengartner, M.O. and Gartner, A. (2001) *C. elegans* RAD-5/CLK-2 defines a new DNA damage checkpoint protein. *Curr. Biol.* **11**: 1934–1944.

Ahmed, S. and Hodgkin, J. (2000) MRT-2 checkpoint protein is required for germline immortality and telomere replication in *C. elegans*. *Nature* **403**: 159–164.

Akgul, C., Moulding, D.A., White, M.R. and Edwards, S.W. (2000) In vivo localisation and stability of human Mcl-1 using green fluorescent protein (GFP) fusion proteins. *FEBS Lett.* **478**: 72–76.

Albert, P.R. and Morris, S.J. (1994) Antisense knockouts: molecular scalpels for the dissection of signal transduction. *Trends Pharmacol. Sci.* **15**: 250–254.

Alcamo, E., Mizgerd, J.P., Horwitz, B.H., Bronson, R., Beg, A.A., Scott, M., Doerschuk, C.M., Hynes, R.O. and Baltimore, D. (2001) Targeted mutation of TNF receptor I rescues the RelA-deficient mouse and reveals a critical role for NF-kappa B in leukocyte recruitment. *J. Immunol.* **167**: 1592–1600.

Alderson, M.R., Tough, T.W., Davis-Smith, T., *et al.* (1995) Fas ligand mediates activation-induced cell death in human T lymphocytes. *J. Exp. Med.* **181**: 71–77.

Alexander, C.M., Howard, E.W., Bissell, M.J. and Werb, Z. (1996) Rescue of mammary epithelial cell apoptosis and entactin degradation by a tissue inhibitor of metalloproteinases-1 transgene. *J. Cell Biol.* **135**: 1669–1677.

Alimonti, J.B., Shi, L., Baijal, P.K. and Greenberg, A.H. (2001) Granzyme B induces BID-mediated cytochrome c release and mitochondrial permeability transition. *J. Biol. Chem.* **276**: 6974–6982.

Altucci, L., Rossin, A., Raffelsberger, W., Reitmair, A., Chomienne, C. and Gronemeyer, H. (2001) Retinoic acid-induced apoptosis in leukemia cells is mediated by paracrine action of tumor-selective death ligand TRAIL. *Nat. Med.* **7**: 680–686.

Amarante-Mendes, G.P., Finucane, D.M., Martin, S.J., Cotter, T.G., Salvesen, G.S. and Green, D.R. (1998) Anti-apoptotic oncogenes prevent caspase-dependent and independent commitment for cell death. *Cell Death Differ.* **5**: 298–306.

Ambrosini, G., Adida, C. and Altieri, D.C. (1997) A novel anti-apoptosis gene, survivin, expressed in cancer and lymphoma. *Nat. Med.* **3**: 917–921.

Ameisen, J.C. (1998) The evolutionary origin and role of programmed cell death in single celled organisms: a new view of executioners, mitochondria, host-parasite interactions, and the role of death in the process of natural selection. In *When Cells Die* (ed. Richard A. Lockshin), Wiley-Liss, Chichester, pp. 3–56.

Ameisen, J.C. (1996) The origin of programmed cell death. *Science* **272**: 1278–1279.

Ameisen, J.C., Idziorek, T., Billaut-Mulot, O., Loyens, M., Tissier, J.P., Potentier, A. and Ouaissi, M.A. (1995) Apoptosis in the unicellular eukaryote (*Trypanosoma cruzei*): implications for the evolutionary origin and role of programmed cell death in the control of cell proliferation, differentiation and survival. *Cell Death Differ.* **2**: 285–300.

Annaert, W. G., Becker, B., Kistner, U., Reth, M. and Jahn, R. (1997) Export of cellubrevin from the endoplasmic reticulum is controlled by BAP31. *J. Cell Biol.* **139**: 1397–1410.

Annis, M.G., Zamzami, N., Zhu, W., Penn, L.Z., Kroemer, G., Leber, B. and Andrews, D.W. (2001) Endoplasmic reticulum localized Bcl-2 prevents apoptosis when redistribution of cytochrome c is a late event. *Oncogene* **20**: 1939–1952.

Antonsson, B. (2001) Bax and other pro-apoptotic Bcl-2 family 'killer-proteins' and their victim, the mitochondrion. *Cell Tissue Res.* **306**: 347–361.

Antonsson, B., Conti, F., Ciavatta, A., *et al.* (1997) Inhibition of Bax channel-forming activity by Bcl-2. *Science* **277**: 370–372.

Antonsson, B. and Martinou, J.C. (2000) The Bcl-2 protein family. *Exp. Cell Res.* 256: 50–57.

Antonsson, B., Montessuit, S., Lauper, S., Eskes, R. and Martinou, J.C. (2000) Bax oligomerization is required for channel-forming activity in liposomes and to trigger cytochrome c release from mitochondria. *Biochem. J.* 345: 271–278.

Antonsson, B., Montessuit, S., Sanchez, B. and Martinou, J.C. (2001) Bax is present as a high molecular weight oligomer/complex in the mitochondrial membrane of apoptotic cells. *J. Biol. Chem.* 276: 11615–11623.

Aoki, H., Yoshimura, K., Kang, P.M., Hampe, J., Noma, T., Matsuzaki, M. and Izumo, S. (2002) Direct activation of mitochondrial apoptosis machinery by c-Jun N-terminal kinase in adult cardiac myocytes. *J. Biol. Chem.* 10: 10.

Apweiler, R., Attwood, T.K., Bairoch, A., *et al.* (2001) The InterPro database, an integrated documentation resource for protein families, domains and functional sites. *Nucleic Acids Res.* 29: 37–40.

Aravind, L., Dixit, V.M. and Koonin, E.V. (2001) Apoptotic molecular machinery: vastly increased complexity in vertebrates revealed by genome comparisons. *Science* 291: 1279–1284.

Aravind, L., Dixit, V.M. and Koonin, E.V. (1999) The domains of death: evolution of the apoptosis machinery. *Trends Biochem. Sci.* 24: 47–53.

Arch, R.H., Gedrich, R.W. and Thompson, C.B. (2000) Translocation of TRAF proteins regulates apoptotic threshold of cells. *Biochem. Biophys. Res. Commun.* 272: 936–945.

Arnoult, D., Tatischeff, I., Estaquier, J., *et al.* (2001) On the evolutionary conservation of the cell death pathway: mitochondrial release of an apoptosis-inducing factor during *Dictyostelium discoideum* cell death. *Mol. Biol. Cell.* 12: 3016–3030.

Ashkenazi, A. and Dixit, V.M. (1998) Death receptors: signalling and modulation. *Science* 281: 1305–1308.

Ashkenazi, A., Pai, R.C., Fong, S., *et al.* (1999) Safety and antitumor activity of recombinant soluble Apo2 ligand. *J. Clin. Invest.* 104: 155–162.

Asoh, S., Ohtsu, T. and Ohta, S. (2000) The super anti-apoptotic factor Bcl-xFNK constructed by disturbing intramolecular polar interactions in rat Bcl-xL. *J. Biol. Chem.* 275: 37240–37245.

Attar, R.M., Caamano, J., Carrasco, D., Iotsova, V., Ishikawa, H., Ryseck, R.P., Weih, F. and Bravo, R. (1997) Genetic approaches to study Rel/NF-kappa B/I kappa B function in mice. *Semin. Cancer Biol.* 8: 93–101.

Attardi, L.D., Reczek, E.E., Cosmas, C., Demicco, E.G., McCurrach, M.E., Lowe, S.W. and Jacks, T. (2000) PERP, an apoptosis-associated target of p53, is a novel member of the PMP-22/gas3 family. *Genes* 14: 704–718.

Attwood, T.K. (2000) The quest to deduce protein function from sequence: the role of pattern databases. *Int. J. Biochem. Cell Biol.* 32: 139–155.

Avery, L. and Horvitz, H.R. (1987) A cell that dies during wild-type *C. elegans* development can function as a neuron in a ced-3 mutant. *Cell* 51: 1071–1078.

Ayllon, V., Martinez, A.C., Garcia, A., *et al.* (2000) Protein phosphatase 1alpha is a Ras-activated Bad phosphatase that regulates interleukin-2 deprivation-induced apoptosis. *EMBO J.* 19: 2237–2246.

Baehrecke, E.H. (2000) Steroid regulation of programmed cell death during *Drosophila* development. *Cell Death Differ.* 7: 1057–1062.

Baichwal, V.R. and Baeuerle, P.A. (1997) Activate NF-kappa B or die? *Curr. Biol.* 7: 94–96.

Baldi, P., Chauvin, Y., Hunkapiller, T. and McClure, M.A. (1994) Hidden Markov models of biological primary sequence information. *Proc. Natl Acad. Sci. U. S. A.* 91: 1059–1063.

Banner, D.W., D'Arcy, A., Janes, W., Gentz, R., Schoenfeld, H.J., Broger, C., Loetscher, H. and Lesslauer, W. (1993) Crystal structure of the soluble human 55 kd TNF receptor-human TNF beta complex: implications for TNF receptor activation. *Cell* 73: 431–445.

Barcinski, M.A. and DosReis, G.A. (1999) Apoptosis in parasites and parasite-induced apoptosis in the host immune system: a new approach to parasitic diseases. *Braz. J. Med. Biol. Res.* 32: 395–401.

Barker, M.G., Brimage, L.J. and Smart, K.A. (1999) Effect of Cu,Zn superoxide dismutase disruption mutation on replicative senescence in *Saccharomyces cerevisiae*. *FEMS Microbiol. Lett.* 177: 199–204.

Barkett, M., Xue, D., Horvitz, H.R. and Gilmore, T.D. (1997) Phosphorylation of IkappaB-alpha inhibits its cleavage by caspase CPP32 in vitro. *J. Biol. Chem.* 272: 29419–29422.

Barry, M. and MacFadden, G. (1998) Apoptosis regulators from DNA viruses. *Curr. Opin. Immunol.* 10: 422–430.

Barry, M., Heibein, J.A., Pinkoski, M.J., Lee, S.F., Moyer, R.W., Green, D.R. and Bleackley, R.C. (2000) Granzyme B short-circuits the need for caspase 8 activity during granule-mediated cytotoxic T-lymphocyte killing by directly cleaving Bid. *Mol. Cell. Biol.* 20: 3781–3794.

Basañez, G., Nechushtan, A., Drozhinin, O., *et al.* (1999) Bax, but not Bcl-xL, decreases the lifetime of planar phospholipid bilayer membranes at sub-nanomolar concentrations. *Proc. Natl Acad. Sci. U. S. A.* 96: 5492–5497.

Basañez, G., Zhang, J., Chau, B.N., *et al.* (2001) Pro-apoptotic cleavage products of Bcl-xL form cytochrome c-conducting pores in pure lipid membranes. *J. Biol. Chem.* 276: 31083–31091.

Bateman, A., Birney, E., Durbin, R., Eddy, S.R., Howe, K.L. and Sonnhammer, E.L. (2000) The Pfam Protein Families Database. *Nucleic Acids Res.* 28: 263–266.

Baud, V. and Karin, M. (2001) Signal transduction by tumor necrosis factor and its relatives. *Trends Cell Biol.* 11: 372–377.

Bauer, M.K.A., Schubert, A., Rocks, O. and Grimm, S. (1999) Adenine nucleotide translocase-1, a component of the permeability transition pore, can dominantly induce apoptosis. *J. Cell Biol.* 147: 1493–1502.

Beg, A.A. and Baltimore, D. (1996) An essential role for NF-kappaB in preventing TNF-alpha-induced cell death. *Science* 274: 782–784.

Beg, A.A., Sha, W.C., Bronson, R.T., Ghosh, S. and Baltimore, D. (1995) Embryonic lethality and liver degeneration in mice lacking the RelA component of NF-kappa B. *Nature* 376: 167–170.

Benard, C., McCright, B., Zhang, Y., Felkai, S., Lakowski, B. and Hekimi, S. (2001) The *C. elegans* maternal-effect gene clk-2 is essential for embryonic development, encodes a protein homologous to yeast Tel2p and affects telomere length. *Development* **128**: 4045–4055.

Bentires-Alj, M., Dejardin, E., Viatour, P., Van Lint, C., Froesch, B., Reed, J.C., Merville, M.P. and Bours, V. (2001) Inhibition of the NF-kappa B transcription factor increases Bax expression in cancer cell lines. *Oncogene* **20**: 2805–2813.

Berard, M., Mondiere, P., Casamayor-Palleja, M., Hennino, A., Bella, C. and Defrance, T. (1999) Mitochondria connect the antigen receptor to effector caspases during B cell receptor-induced apoptosis in normal human B cells. *J. Immunol.* **163**: 4655–4662.

Beresford, P.J., Zhang, D., Oh, D.Y., Fan, Z., Greer, E.L., Russo, M.L., Jaju, M. and Lieberman, J. (2001) Granzyme A activates an endoplasmic reticulum-associated caspase-independent nuclease to induce single-stranded DNA nicks. *J. Biol. Chem.* **276**: 43285–43293.

Berger, T., Brigl, M., Herrmann, J. M., Vielhauer, V., Luckow, B., Schlondorff, D. and Kretzler, M. (2000) The apoptosis mediator mDAP-3 is a novel member of a conserved family of mitochondrial proteins. *J. Cell Sci.* **113**: 3603–3612.

Bergeron, L., Perez, G.I., Macdonald, G., *et al.* (1998) Defects in regulation of apoptosis in caspase-2-deficient mice. *Genes Dev.* **12**: 1304–1314.

Bergmann, A., Agapite, J., McCall, K. and Steller, H. (1998) The *Drosophila* gene hid is a direct molecular target of Ras-dependent survival signaling. *Cell* **95**: 331–341.

Bergmann, A., Tugentman, M., Shilo, B.Z. and Steller, H. (2002) Regulation of cell number by MAPK-dependent control of apoptosis: a mechanism for trophic survival signaling. *Dev. Cell* **2**: 159–170.

Bergsmedh, A., Szeles, A., Henriksson, M., Bratt, A., Folkman, M.J., Spetz, A.L. and Holmgren, L. (2001) Horizontal transfer of oncogenes by uptake of apoptotic bodies. *Proc. Natl Acad. Sci. U. S. A.* **98**: 6407–6411.

Bernardi, P. (1999) Mitochondrial transport of cations: channels, exchangers, and permeability transition. *Physiol. Rev.* **79**: 1127–1155.

Berndt, C., Mopps, B., Angermuller, S., Gierschik, P. and Krammer, P.H. (1998) CXCR4 and CD4 mediate a rapid CD95-independent cell death in CD4(+) T cells. *Proc. Natl Acad. Sci. U. S. A.* **95**: 12556–12561.

Berridge, M.J., Bootman, M.D. and Lipp, P. (1998) Calcium – a life and death signal. *Nature* **395**: 645–648.

Berridge, M.J., Lipp, P. and Bootman, M.D. (2000) The versatility and universality of calcium signalling. *Nat. Rev. Mol. Cell. Biol.* **1**: 11–21.

Bertin, J., Armstrong, R.C., Ottilie, *et al.* (1997) Death effector domain-containing herpesvirus and poxvirus proteins inhibit both Fas- and TNFR1-induced apoptosis. *Proc. Natl Acad. Sci. U. S. A.* **94**: 1172–1176.

Bertin, J. and DiStefano, P.S. (2000) The PYRIN domain: a novel motif found in apoptosis and inflammation proteins. *Cell Death Differ.* **7**: 1273–1274.

Beutner, G., Ruck, A., Riede, B. and Brdiczka, D. (1998) Complexes between porin, hexokinase, mitochondrial creatine kinase and adenylate translocator display properties of the permeability transition pore. Implication for regulation of permeability transition by the kinases. *Biochim. Biophys. Acta* **1368**: 7–18.

Birnbaum, M.J., Clem, R.J. and Miller, L.K. (1994) An apoptosis-inhibiting gene from a nuclear polyhedrosis virus encoding a polypeptide with Cys/His sequence motifs. *J. Virol.* **68**: 2521–2528.

Blanchard, H., Kodandapani, L., Mittl, P.R., Marco, S.D., Krebs, J.F., Wu, J.C., Tomaselli, K.J. and Grutter, M.G. (1999) The three-dimensional structure of caspase-8: an initiator enzyme in apoptosis. *Structure Fold Des.* **7**: 1125–1133.

Bodmer, J.L., Schneider, P. and Tschopp, J. (2002) The molecular architecture of the TNF superfamily. *Trends Biochem. Sci.* **27**: 19–26.

Bodmer, J.L., Burns, K., Schneider, P., *et al.* (1997) TRAMP, a novel apoptosis-mediating receptor with sequence homology to tumor necrosis factor receptor 1 and Fas(Apo-1/CD95) *Immunity* **6**: 79–88.

Boise, L.H. and Collins, C.M. (2001) *Salmonella*-induced cell death: apoptosis, necrosis or programmed cell death? *Trends Microbiol.* **9**: 64–67.

Boise, L.H., Noel, P.J. and Thompson, C.B. (1995) CD28 and apoptosis. *Curr. Opin. Immunol.* **7**: 620–625.

Boldin, M.P., Goncharov, T.M., Goltsev, Y.V. and Wallach, D. (1996) Involvement of MACH, a novel MORT1/FADD-interacting protease, in Fas/APO-1- and TNF receptor-induced cell death. *Cell* **85**: 803–815.

Boldin, M.P., Varfolomeev, E.E., Pancer, Z., Mett, I.L., Camonis, J.H. and Wallach, D. (1995) A novel protein that interacts with the death domain of Fas/APO1 contains a sequence motif related to the death domain. *J. Biol. Chem.* **270**: 7795–7798.

Borner, C. and Monney, L. (1999) Apoptosis without caspases: an inefficient molecular guillotine. *Cell Death Differ.* **6**: 497–507.

Bosch, T.C. and David, C.N. (1984) Growth regulation in *Hydra*: relationship between epithelial cell cycle length and growth rate. *Dev. Biol.* **104**: 161–171.

Bossy-Wetzel, E., Newmeyer, D.D. and Green, D.R. (1998) Mitochondrial cytochrome c release in apoptosis occurs upstream of DEVD-specific caspase activation and independently of mitochondrial transmembrane depolarization. *EMBO J.* **17**: 37–49.

Boulton, S.J., Gartner, A., Reboul, J., Vaglio, P., Dyson, N., Hill, D.E. and Vidal, M. (2002) Combined functional genomic maps of the *C. elegans* DNA damage response. *Science* **295**: 127–131.

Brachmann, C.B., Jassim, O.W., Wachsmuth, B.D. and Cagan, R.L. (2000) The *Drosophila* bcl-2 family member dBorg-1 functions in the apoptotic response to UV irradiation. *Curr. Biol.* **10**: 547–550.

Brady, H.J., Gil-Gomez, G., Kirberg, J., *et al.* (1996) Bax alpha perturbs T cell development and affects cell cycle entry of T cells. *EMBO J.* **15**: 6991–7001.

Brakebusch, C., Nophar, Y., Kemper, O., Engelmann, H. and Wallach, D. (1992) Cytoplasmic truncation of the p55 tumour necrosis factor (TNF) receptor abolishes signalling, but not induced shedding of the receptor. *EMBO J.* **11**: 943–950.

Braun, J.S., Novak, R., Murray, P.J., *et al.* (2001) Apoptosis-inducing factor mediates microglial and neuronal apoptosis caused by pneumococcus. *J. Infect. Dis.* **184**: 1300–1309.

Breckenridge, D.G., Nguyen, M., Kuppig, S., Reth, M. and Shore, G. (2002) Novel procaspase-8L isoform recruited to the BAP31 complex at the endoplasmic reticulum. *Proc. Natl Acad. Sci. U. S. A.* **99**: 4331–4336.

Brenner, C., Cadiou, H., Vieira, H.L., *et al.* (2000) Bcl-2 and Bax regulate the channel activity of the mitochondrial adenine nucleotide translocator. *Oncogene* **19**: 329–336.

Brewer, J.W. and Diehl, J.A. (2000) PERK mediates cell-cycle exit during the mammalian unfolded protein response. *Proc. Natl Acad. Sci. U. S. A.* **97**: 12625–12630.

Brierley, G.P., Baysal, K. and Jung, D.W. (1994) Cation transport systems in mitochondria: Na^+ and K^+ uniports and exchangers. *J. Bioenerg. Biomembr.* **26**: 519–526.

Brodsky, M.H., Nordstrom, W., Tsang, G., Kwan, E., Rubin, G.M. and Abrams, J.M. (2000) *Drosophila* p53 binds a damage response element at the reaper locus. *Cell* **101**: 103–113.

Broun, M., Sokol, S. and Bode, H.R. (1999) Cngsc, a homologue of goosecoid, participates in the patterning of the head, and is expressed in the organizer region of *Hydra*. *Development* **126**: 5245–5254.

Brunner, T., Mogil, R.J., LaFace, D., *et al.* (1995) Cell-autonomous Fas (CD95)/Fas-ligand interaction mediates activation-induced apoptosis in T-cell hybridomas. *Nature* **373**: 441–444.

Brustovetsky, N. and Klingenberg, M. (1996) Mitochondrial ADP/ATP carrier can be reversibly converted into a large channel by Ca^{2+}. *Biochemistry (Mosc.)* **35**: 8483–8488.

Brustovetsky, N. and Klingenberg, M. (1994) The reconstituted ADP/ATP carrier can mediate H^+ transport by free fatty acids, which is further stimulated by mersalyl. *J. Biol. Chem.* **269**: 27329–27336.

Bucher, P., Karplus, K., Moeri, N. and Hofmann, K. (1996) A flexible motif search technique based on generalized profiles. *Comput. Chem.* **20**: 3–23.

Buendia, B., Santa-Maria, A. and Courvalin, J.C. (1999) Caspase-dependent proteolysis of integral and peripheral proteins of nuclear membranes and nuclear pore complex proteins during apoptosis. *J. Cell Sci.* **112**: 1743–1753.

Burns, T.F. and el Deiry, W.S. (2001) Identification of inhibitors of TRAIL-induced death (ITIDs) in the TRAIL-sensitive colon carcinoma cell line SW480 using a genetic approach. *J. Biol. Chem.* **276**: 37879–37886.

Bursch, W., Ellinger, A., Kienzl, H., Torok, L., Pandey, S., Sikorska, M., Walker, R. and Hermann, R. S. (1996) Active cell death induced by the anti-estrogens tamoxifen and ICI 164 384 in human mammary carcinoma cells (MCF-7) in culture: the role of autophagy. *Carcinogenesis* **17**: 1595–1607.

Burtis, K.C., Thummel, C.S., Jones, C.W., Karim, F.D. and Hogness, D.S. (1990) The *Drosophila* 74EF early puff contains E74, a complex ecdysone-inducible gene that encodes two ets-related proteins. *Cell* **6**: 8–99.

Butt, A.J., Harvey, N.L., Parasivam, G. and Kumar, S. (1998) Dimerization and autoprocessing of the Nedd2(caspase-2) precursor requires both the prodomain and the carboxyl-terminal regions. *J. Biol. Chem.* **273**: 6763–6768.

Cai, J. and Jones, D.P. (1999) Mitochondrial redox signaling during apoptosis. *J. Bioenerg. Biomembr.* **31**: 327–334.

Cakouros, D., Daish, T., Martin, D., Baehrecke, E.H. and Kumar, S. (2002) Ecdysone-induced expression of the caspase DRONC during hormone dependent programmed cell death in *Drosophila* is regulated by Broad-Complex. *J. Cell Biol.* **57**: 985–996.

Calfon, M., Zeng, H., Urano, F., Till, J.H., Hubbard, S.R., Harding, H.P., Clark, S.G. and Ron, D. (2002) IRE1 couples endoplasmic reticulum load to secretory capacity by processing the XBP-1 mRNA. *Nature* 415: 92–96.

Campbell, R.D. (1976) Elimination by *Hydra* interstitial and nerve cells by means of colchicine. *J. Cell Sci.* 21: 1–13.

Carlberg, M., Dricu, A., Blegen, H., Kass, G. E., Orrenius, S. and Larsson, O. (1996) Short exposures to tunicamycin induce apoptosis in SV40–transformed but not in normal human fibroblasts. *Carcinogenesis* 17: 2589–2596.

Carratore, R.D., Della Croce, C., Simili, M., Taccini, E., Scavuzzo, M. and Sbrana, S. (2002) Cell cycle and morphologic alterations as indicative of apoptosis promoted by UV irradiation in *S. cerevisiae*. *Mutat. Res.* 513: 183–191.

Caspari, T. and Carr, A.M. (1999) DNA structure checkpoint pathways in *Schizosaccharomyces pombe*. *Biochimie* 81: 173–181.

Cecconi, F., Alvarez-Bolado, G., Meyer, B.I., Roth, K.A. and Gruss, P. (1998) Apaf1 (CED-4 homolog) regulates programmed cell death in mammalian development. *Cell* 94: 727–737.

C. elegans Sequencing Consortium (1998) Genome sequence of the nematode *C. elegans*: a platform for investigating biology. *Science* 282: 2012–2018.

Cha, S.S., Sung, B.J., Kim, Y.A., Song, Y.L., Kim, H.J., Kim, S., Lee, M.S. and Oh, B.H. (2000) Crystal structure of TRAIL-DR5 complex identifies a critical role of the unique frame insertion in conferring recognition specificity. *J. Biol. Chem.* 275: 31171–31177.

Chai, J., Du, C., Wu, J.W., Kyin, S., Wang, X. and Shi, Y. (2000) Structural and biochemical basis of apoptotic activation by Smac/DIABLO. *Nature* 406: 855–862.

Chai, J., Shiozaki, E., Srinivasula, S.M., Wu, Q., Datta, P., Alnemri, E.S., Shi, Y. and Dataa, P. (2001a) Structural basis of caspase-7 inhibition by XIAP. *Cell* 104: 769–780.

Chai, J., Wu, Q., Shiozaki, E., Srinivasula, S.M., Alnemri, E.S. and Shi, Y. (2001b) Crystal structure of a procaspase-7 zymogen: mechanisms of activation and substrate binding. *Cell* 107: 399–407.

Chandra, J., Zhivotovsky, B., Zaitsev, S., Juntti-Berggren, L., Berggren, P.O. and Orrenius, S. (2001) Role of apoptosis in pancreatic beta-cell death in diabetes. *Diabetes* 50 (Suppl 1): S44–S47.

Chang, H. Y. and Yang, X. (2000) Proteases for cell suicide: functions and regulation of caspases. *Microbiol. Mol. Biol. Rev.* 64: 821–846.

Chang, N.S., Mattison, J., Cao, H., Pratt, N., Zhao, Y. and Lee, C. (1998) Cloning and characterization of a novel transforming growth factor-beta1-induced TIAF1 protein that inhibits tumor necrosis factor cytotoxicity. *Biochem. Biophys. Res. Commun.* 253: 743–749.

Chang, S.H., Phelps, P.C., Berezesky, I.K., Ebersberger, M.L., Jr. and Trump, B.F. (2000) Studies on the mechanisms and kinetics of apoptosis induced by microinjection of cytochrome c in rat kidney tubule epithelial cells (NRK-52E). *Am. J. Pathol.* 156: 637–649.

Chao, D.T. and Korsmeyer, S.J. (1998) BCL-2 family: regulators of cell death. *Annu. Rev. Immunol.* 16: 395–419.

Charette, S.J., Lambert, H. and Landry, J. (2001) A kinase-independent function of ask1 in caspase-independent cell death. *J. Biol. Chem.* **276**: 36071–36074.

Charette, S.J., Lavoie, J.N., Lambert, H. and Landry, J. (2000) Inhibition of Daxx-mediated apoptosis by heat shock protein 27. *Mol. Cell. Biol.* **20**: 7602–7612.

Chautan, M., Chazal, G., Cecconi, F., Gruss, P. and Golstein, P. (1999) Interdigital cell death can occur through a necrotic and caspase- independent pathway. *Curr. Biol.* **9**: 967–970.

Chen, C., Edelstein, L.C. and Gelinas, C. (2000) The Rel/NF-kappaB family directly activates expression of the apoptosis inhibitor Bcl-x(L) *Mol. Cell. Biol.* **20**: 2687–2695.

Chen, F., Hersh, B.M., Conradt, B., Zhou, Z., Riemer, D., Gruenbaum, Y. and Horvitz, H.R. (2000) Translocation of *C. elegans* CED-4 to nuclear membranes during programmed cell death. *Science* **287**: 1485–1489.

Chen, G., Ray, R., Dubik, D., Shi, L., Cizeau, J., Bleackley, R.C., Saxena, S., Gietz, R.D. and Greenberg, A.H. (1997) The E1B 19K/Bcl-2–binding protein Nip3 is a dimeric mitochondrial protein that activates apoptosis. *J. Exp. Med.* **186**: 1975–1983.

Chen, M., Ona, V.O., Li, M., *et al.* (2000) Minocycline inhibits caspase-1 and caspase-3 expression and delays mortality in a transgenic mouse model of Huntington disease. *Nat. Med.* **6**: 797–801.

Chen, P., Nordstrom, W., Gish, B. and Abrams, J.M. (1996) grim, a novel cell death gene in *Drosophila*. *Genes Dev.* **10**: 1773–1782.

Chen, P., Rodriguez, A., Erskine, R., Thach, T. and Abrams, J.M. (1998) Dredd, a novel effector of the apoptosis activators reaper, grim and hid in *Drosophila*. *Dev. Biol.* **201**: 202–216.

Chen, Y., Molloy, S.S., Thomas, L., Gambee, J., Bachinger, H.P., Ferguson, B., Zonana, J., Thomas, G. and Morris, N.P. (2001) Mutations within a furin consensus sequence block proteolytic release of ectodysplasin-A and cause X-linked hypohidrotic ectodermal dysplasia. *Proc. Natl Acad. Sci. U. S. A.* **98**: 7218–7223.

Cheng, E.H., Kirsch, D.G., Clem, R.J., Ravi, R., Kastan, M.B., Bedi, A., Ueno, K. and Hardwick, J.M. (1997) Conversion of Bcl-2 to a Bax-like death effector by caspases. *Science* **278**: 1966–1968.

Cheng, E.H., Levine, B., Boise, L.H., Thompson, C.B. and Hardwick, J.M. (1996) Bax-independent inhibition of apoptosis by Bcl-XL. *Nature* **379**: 554–556.

Cheng, E.H., Nicholas, J., Bellows, D.S., Hayward, G.S., Guo, H.-G., Reitz, M.S. and Hardwick, J.M. (1997b) A Bcl-2 homolog encoded by Kaposi sarcoma-associated virus, human herpes virus 8, inhibits apoptosis but does not heterodimerize with Bax or Bak. *Proc. Natl. Acad. Sci. U. S. A.* **94**: 690–694.

Cheng, E.H., Wei, M.C., Weiler, S., Flavell, R.A., Mak, T.W., Lindsten, T. and Korsmeyer, S.J. (2001) BCL-2, BCL-X(L) sequester BH3 domain-only molecules preventing BAX- and BAK-mediated mitochondrial apoptosis. *Mol. Cell* **8**: 705–711.

Chen-Levy, Z. and Cleary, M.L. (1990) Membrane topology of the Bcl-2 proto-oncogenic protein demonstrated in vitro. *J. Biol. Chem.* **265**: 4929–4933.

Chi, S., Kitanaka, C., Noguchi, K., *et al.* (1999) Oncogenic Ras triggers cell suicide through the activation of a caspase-independent cell death program in human cancer cells. *Oncogene* **18**: 2281–2290.

Chiang, C.W., Harris, G., Ellig, C., *et al.* (2001) Protein phosphatase 2A activates the proapoptotic function of BAD in interleukin-3-dependent lymphoid cells by a mechanism requiring 14-3-3 dissociation. *Blood* **97**: 1289–1297.

Chiang, L.W., Grenier, J.M., Ettwiller, L., Jenkins, L.P., Ficenec, D., Martin, J., Jin, F., DiStefano, P.S. and Wood, A. (2001) An orchestrated gene expression component of neuronal programmed cell death revealed by cDNA array analysis. *Proc. Natl Acad. Sci. U. S. A.* **98**: 2814–2819.

Chicheportiche, Y., Bourdon, P.R., Xu, H., Hsu, Y.M., Scott, H., Hession, C., Garcia, I. and Browning, J.L. (1997) TWEAK, a new secreted ligand in the tumor necrosis factor family that weakly induces apoptosis. *J. Biol. Chem.* **272**: 32401–32410.

Chinnaiyan, A.M., Chaudhary, D., O'Rourke, K., Koonin, E.V. and Dixit, V.M. (1997a) Role of CED-4 in the activation of CED-3. *Nature* **388**: 728–729.

Chinnaiyan, A.M., O'Rourke, K., Lane, B.R. and Dixit, V.M. (1997b) Interaction of CED-4 with CED-3 and CED-9: a molecular framework for cell death. *Science* **275**: 1122–1126.

Chinnaiyan, A.M., O'Rourke, K., Tewari, M. and Dixit, V.M. (1995) FADD, a novel death domain-containing protein, interacts with the death domain of Fas and initiates apoptosis. *Cell* **81**: 505–512.

Chinnaiyan, A. M., O'Rourke, K., Yu, G. L., *et al.* (1996a) Signal transduction by DR3, a death domain-containing receptor related to TNFR-1 and CD95. *Science* **274**: 990–992.

Chinnaiyan, A.M., Tepper, C.G., Seldin, M.F., O'Rourke, K., Kischkel, F.C., Hellbardt, S., Krammer, P.H., Peter, M.E. and Dixit, V.M. (1996b) FADD/MORT1 is a common mediator of CD95 (Fas/APO-1) and tumor necrosis factor receptor-induced apoptosis. *J. Biol. Chem.* **271**: 4961–4965.

Chittenden, T., Flemington, C., Houghton, A.B., Ebb, R.G., Gallo, G.J., Elangovan, B., Chinnadurai, G. and Lutz, R.J. (1995) A conserved domain in Bak, distinct from BH1 and BH2, mediates cell death and protein binding functions. *EMBO J.* **14**: 5589–5596.

Choi, W.S., Lee, E.H., Chung, *et al.* (2001) Cleavage of Bax is mediated by caspase-dependent or -independent calpain activation in dopaminergic neuronal cells: protective role of Bcl-2. *J. Neurochem.* **77**: 1531–1541.

Chou, J.J., Li, H., Salvesen, G.S., Yuan, J. and Wagner, G. (1999) Solution structure of BID, an intracellular amplifier of apoptotic signaling. *Cell* **96**: 615–624.

Chou, J.J., Matsuo, H., Duan, H. and Wagner, G. (1998) Solution structure of the RAIDD CARD and model for CARD/CARD interaction in caspase-2 and caspase-9 recruitment. *Cell* **94**: 171–180.

Christensen, S.T., Chemnitz, J., Straarup, E.M., Kristiansen, K., Wheatley, D.N. and Rasmussen, L. (1998) Staurosporine-induced cell death in *Tetrahymena thermophila* has mixed characteristics of both apoptotic and autophagic degeneration. *Cell Biol. Int.* **22**: 591–598.

Christensen, S.T., Quie, H., Kemp, K. and Rasmussen, L. (1996) Insulin produces a biphasic response in *Tetrahymena thermophila* by stimulating cell survival and activating proliferation in two separate concentration intervals. *Cell Biol. Int.* **20**: 437–444.

Christensen, S.T., Sorensen, H., Beyer, N.H., Kristiansen, K., Rasmussen, L. and Rasmussen, M.I. (2001) Cell death in *Tetrahymena thermophila*: new observations on culture conditions. *Cell Biol. Int.* **25**: 509–519.

Christich, A., Kauppila, S., Chen, P., Sogame, N., Ho, S.I. and Abrams, J.M. (2002) The damage-responsive *Drosophila* gene sickle encodes a novel IAP binding protein similar to but distinct from reaper, grim, and hid. *Curr. Biol.* **12**: 137–140.

Chu, Z.L., McKinsey, T.A., Liu, L., Gentry, J.J., Malim, M.H. and Ballard, D.W. (1997) Suppression of tumor necrosis factor-induced cell death by inhibitor of apoptosis c-IAP2 is under NF-kappaB control. *Proc. Natl Acad. Sci. U. S. A.* **94**: 10057–10062.

Chua, B. T., Guo, K. and Li, P. (2000) Direct cleavage by the calcium-activated protease calpain can lead to inactivation of caspases. *J. Biol. Chem.* **275**: 5131–5135.

Chung, S., Gumienny, T.L., Hengartner, M.O. and Driscoll, M. (2000) A common set of engulfment genes mediates removal of both apoptotic and necrotic cell corpses in *C. elegans*. *Nat. Cell Biol.* **2**: 931–937.

Chuntharapai, A., Dodge, K., Grimmer, K., Schroeder, K., Marsters, S.A., Koeppen, H., Ashkenazi, A. and Kim, K.J. (2001) Isotype-dependent inhibition of tumor growth in vivo by monoclonal antibodies to death receptor 4. *J. Immunol.* **166**: 4891–4898.

Cikala, M., Wilm, B., Hobmayer, E., Bottger, A. and David, C.N. (1999) Identification of caspases and apoptosis in the simple metazoan hydra. *Curr. Biol.* **9**: 959–962.

Clapham, D.E. (1995) Calcium signaling. *Cell* **80**: 259–268.

Claveria, C., Albar, J., Serrano, A., Buesa, J., Barbero, J., Martinez, A.-C. and Torres, M. (1998) *Drosophila* grim induces apoptosis in mammalian cells. *EMBO J.* **17**: 7199–7208.

Cleary, M.L., Smith, S.D. and Sklar, J. (1986) Cloning and structural analysis of cDNAs for bcl-2 and a hybrid bcl-2/immunoglobulin transcript resulting from the t(14;18) translocation. *Cell* **47**: 19–28.

Clem, R.J., Cheng, E.H.Y., Karp, C.L., *et al.* (1998) Modulation of cell death by Bcl-XL through caspase interaction. *Proc. Natl Acad. Sci. U. S. A.* **95**: 554–559.

Clem, R.J., Sheu, T.T., Richter, B.W., He, W.W., Thornberry, N.A., Duckett, C.S. and Hardwick, J.M. (2001) c-IAP1 is cleaved by caspases to produce a proapoptotic C-terminal fragment. *J. Biol. Chem.* **276**: 7602–7608.

Cohen, G.M. (1997) Caspases: the executioners of apoptosis. *Biochem. J.* **326**: 1–16.

Cohen, P.T. (2002) Protein phosphatase 1 – targeted in many directions. *J. Cell Sci.* **115**: 241–256.

Colbourne, F., Sutherland, G.R. and Auer, R.N. (1999) Electron microscopic evidence against apoptosis as the mechanism of neuronal death in global ischemia. *J. Neurosci.* **19**: 4200–4210.

Coleman, M.L., Sahai, E.A., Yeo, M., Bosch, M., Dewar, A. and Olson, M.F. (2001) Membrane blebbing during apoptosis results from caspase-mediated activation of ROCK I. *Nat. Cell Biol.* **3**: 339–345.

Colussi, P.A., Harvey, N.L. and Kumar, S. (1998a) Prodomain-dependent nuclear localization of the caspase-2(Nedd2) precursor. *J. Biol. Chem.* **273**: 24535–24542.

Colussi, P.A., Harvey, N.L., Shearwin-Whyatt, L.M. and Kumar, S. (1998b) Conversion of procaspase-3 to an autoactivating caspase by fusion to the caspase-2 prodomain. *J. Biol. Chem.* **273**: 26566–26570.

Colussi, P.A. and Kumar, S. (1999) Targeted disruption of caspase genes in mice: what they tell us about the functions of individual caspases in apoptosis. *Immunol. Cell Biol.* **77**: 59–64.

Colussi, P.A., Quinn, L.M., Huang, D.C., Coombe, M., Read, S.H., Richardson, H. and Kumar, S. (2000) Debcl, a proapoptotic Bcl-2 homologue, is a component of the *Drosophila melanogaster* cell death machinery. *J. Cell Biol.* **148**: 703–714.

Conradt, B. and Horvitz, H.R. (1999) The TRA-1A sex determination protein of *C. elegans* regulates sexually dimorphic cell deaths by repressing the egl-1 cell death activator gene. *Cell* **98**: 317–327.

Conradt, B. and Horvitz, H.R. (1998) The *C. elegans* protein EGL-1 is required for programmed cell death and interacts with the Bcl-2–like protein CED-9. *Cell* **93**: 519–529.

Consortium, T.F.F. (1997) A candidate gene for familial Mediterranean fever. *Nat.Genet.* **17**: 25–31.

Consortium, T.I.F. (1997) Ancient missense mutations in a new member of the RoRet gene family are likely to cause familial Mediterranean fever. *Cell* **90**: 797–807.

Conus, S., Rosse, T. and Borner, C. (2000) Failure of Bcl-2 family members to interact with Apaf-1 in normal and apoptotic cells. *Cell Death Differ.* **7**: 947–954.

Cornillon, S., Foa, C., Davoust, J., Buonavista, N., Gross, J.D. and Golstein, P. (1994) Programmed cell death in *Dictyostelium*. *J. Cell Sci.* **107**: 2691–2704.

Costa, V. and Moradas-Ferreira, P. (2001) Oxidative stress and signal transduction in *Saccharomyces cerevisiae*: insights into ageing, apoptosis and diseases. *Mol. Aspects Med.* **22**: 217–246.

Cosulich, S.C., Worrall, V., Hedge, P.J., Green, S. and Clarke, P.R. (1997) Regulation of apoptosis by BH3 domains in a cell-free system. *Curr. Biol.* **7**: 913–920.

Crawford, K.W. and Bowen, W.D. (2002) Sigma-2 receptor agonists activate a novel apoptotic pathway and potentiate antineoplastic drugs in breast tumor cell lines. *Cancer Res.* **62**: 313–322.

Craxton, A., Otipoby, K.L., Jiang, A. and Clark, E.A. (1999) Signal transduction pathways that regulate the fate of B lymphocytes. *Adv. Immunol.* **73**: 79–152.

Crompton, M. (1999) The mitochondrial permeability transition pore and its role in cell death. *Biochem. J.* **341**: 233–249.

Crompton, M., Virji, S., Doyle, V., Johnson, N. and Ward, J.M. (1999) The mitochondrial permeability transition pore. Review. *Biochem. Soc. Symp.* **66**: 167–179.

Crook, N.E., Clem, R.J. and Miller, L.K. (1993) An apoptosis-inhibiting baculovirus gene with a zinc finger-like motif. *J. Virol.* **67**: 2168–2174.

Danen-Van Oorschot, A.A., Fischer, D.F., Grimbergen, J.M., *et al.* (1997) Apoptin induces apoptosis in human transformed and malignant cells but not in normal cells. *Proc. Natl Acad. Sci. U. S. A.* **94**: 5843–5847.

Darnay, B.G. and Aggarwal, B.B. (1997) Early events in TNF signaling: a story of associations and dissociations. *J. Leukoc. Biol.* **61**: 559–566.

Darzynkiewicz, Z. and Traganos, F. (1998) Measurement of apoptosis. *Adv. Biochem. Eng. Biotechnol.* **62**: 33–73.

Datta, S.R., Dudek, H., Tao, X., *et al.* (1997) Akt phosphorylation of BAD couples survival signals to the cell-intrinsic death machinery. *Cell* **91**: 231–241.

Datta, S.R., Katsov, A., Hu, L., Petros, A., Fesik, S. W., Yaffe, M.B. and Greenberg, M.E. (2000) 14–3–3 proteins and survival kinases cooperate to inactivate BAD by BH3 domain phosphorylation. *Mol. Cell* **6**: 41–51.

Daugas, E., Susin, S.A., Zamzami, N., *et al.* (2000) Mitochondrio-nuclear translocation of AIF in apoptosis and necrosis. *FASEB J.* **14**: 729–739.

Davis, M.C., Ward, J.G., Herrick, G. and Allis, C.D. (1992) Programmed nuclear death: apoptotic-like degradation of specific nuclei in conjugating *Tetrahymena*. *Dev. Biol.* **154**: 419–432.

De Giorgi, F., Lartigue, L., Bauer, M.K., Schubert, A., Grimm, S., Hanson, G.T., Remington, S.J., Youle, R.J. and Ichas, F. (2002) The permeability transition pore signals apoptosis by directing Bax translocation and multimerization. *FASEB J.* **16**: 607–09.

de Jong, D., Prins, F.A., Mason, D.Y., Reed, J.C., van Ommen, G.B. and Kluin, P.M. (1994) Subcellular localization of the bcl-2 protein in malignant and normal lymphoid cells. *Cancer Res.* **54**: 256–260.

De Maria, R., Zeuner, A., Eramo, A., *et al.* (1999) Negative regulation of erythropoiesis by caspase-mediated cleavage of GATA-1. *Nature* **401**: 489–493.

de Nooij, J.C. and Hariharan, I.K. (1995) Uncoupling cell fate determination from patterned cell division in the *Drosophila* eye. *Science* **270**: 983–985.

De Smaele, E., Zazzeroni, F., Papa, S., Nguyen, D.U., Jin, R., Jones, J., Cong, R. and Franzoso, G. (2001) Induction of gadd45beta by NF-kappaB downregulates pro-apoptotic JNK signalling. *Nature* **414**: 308–313.

Deas, O., Dumont, C., MacFarlane, M., *et al.* (1998) Caspase-independent cell death induced by anti-CD2 or staurosporine in activated human peripheral T lymphocytes. *J. Immunol.* **161**: 3375–3383.

Debatin, K.M., Fahrig-Faissner, A., Enenkel-Stoodt, S., Kreuz, W., Benner, A. and Krammer, P.H. (1994) High expression of APO-1 (CD95) on T lymphocytes from human immunodeficiency virus-1– infected children. *Blood* **83**: 3101–3103.

Degli-Esposti, M.A., Dougall, W.C., Smolak, P.J., Waugh, J.Y., Smith, C.A. and Goodwin, R.G. (1997a) The novel receptor TRAIL-R4 induces NF-kappaB and protects against TRAIL-mediated apoptosis, yet retains an incomplete death domain. *Immunity* **7**: 813–820.

Degli-Esposti, M.A., Smolak, P.J., Walczak, H., Waugh, J., Huang, C.P., DuBose, R.F., Goodwin, R.G. and Smith, C.A. (1997b) Cloning and characterization of TRAIL-R3, a novel member of the emerging TRAIL receptor family. *J. Exp. Med.* **186**: 1165–1170.

Degterev, A., Lugovskoy, A., Cardone, M., Mulley, B., Wagner, G., Mitchison, T.

and Yuan, J. (2001) Identification of small-molecule inhibitors of interaction between the BH3 domain and Bcl-xL. *Nat. Cell Biol.* **3**: 173–182.

Deiss, L.P., Galinka, H., Berissi, H., Cohen, O. and Kimchi, A. (1996) Cathepsin D protease mediates programmed cell death induced by interferon-gamma, Fas/APO-1 and TNF-alpha. *EMBO J.* **15**: 3861–3870.

Deiss, L.P. and Kimchi, A. (1991) A genetic tool used to identify thioredoxin as a mediator of a growth inhibitory signal. *Science* **252**: 117–120.

Del Carratore, R., Della Croce, C., Simili, M., Taccini, E., Scavuzzo, M. and Sbrana, S. (2002) Cell cycle and morphological alterations as indicative of apoptosis promoted by UV irradiation in *S. cerevisiae*. *Mutat. Res.* **513**: 183–191.

del Peso, L., Gonzalez, V.M., Inohara, N., Ellis, R.E. and Nunez, G. (2000) Disruption of the CED- 9:CED-4 complex by EGL-1 is a critical step for programmed cell death in *Caenorhabditis elegans*. *J. Biol. Chem.* **275**: 27205–27211.

del Peso, L., Gonzalez, V. M. and Nunez, G. (1998) *Caenorhabditis elegans* EGL-1 disrupts the interaction of CED-9 with CED-4 and promotes CED-3 activation. *J. Biol. Chem.* **273**: 33495–33500.

del Peso, L., Gonzalez-Garcia, M., Page, C., Herrera, R. and Nunez, G. (1997) Interleukin-3–induced phosphorylation of BAD through the protein kinase Akt. *Science* **278**: 687–689.

Delhase, M., Hayakawa, M., Chen, Y. and Karin, M. (1999) Positive and negative regulation of IkappaB kinase activity through IKKbeta subunit phosphorylation. *Science* **284**: 309–313.

Denecker, G., Vercammen, D., Steemans, M., *et al.* (2001) Death receptor-induced apoptotic and necrotic cell death: differential role of caspases and mitochondria. *Cell Death Differ.* **8**: 829–840.

Deng, Y., Lin, Y. and Wu, X. (2002) TRAIL-induced apoptosis requires Bax-dependent mitochondrial release of Smac/DIABLO. *Genes Dev.* **16**: 33–45.

Derry, W.B., Putzke, A.P. and Rothman, J.H. (2001) *Caenorhabditis elegans* p53: role in apoptosis, meiosis, and stress resistance. *Science* **294**: 591–595.

Desagher, S., Osen-Sand, A., Montessuit, S., Magnenat, E., Vilbois, F., Hochmann, A., Journot, L., Antonsson, B. and Martinou, J.C. (2001) Phosphorylation of bid by casein kinases I and II regulates its cleavage by caspase 8. *Mol. Cell* **8**: 601–611.

Desagher, S., Osen-Sand, A., Nichols, A., Eskes, R., Montessuit, S., Lauper, S., Maundrell, K., Antonsson, B. and Martinou, J.C. (1999) Bid-induced conformational change of Bax is responsible for mitochondrial cytochrome c release during apoptosis. *J. Cell Biol.* **144**: 891–901.

Desai, C., Garriga, G., McIntire, S.L. and Horvitz, H.R. (1988) A genetic pathway for the development of the *Caenorhabditis elegans* HSN motor neurons. *Nature* **336**: 638–646.

Deshmukh, M., Kuida, K., and Johnson, E.M. Jr. (2000) Caspase inhibition extends the commitment to neuronal death beyond cytochrome c release to the point of mitochondrial depolarization. *J. Cell Biol.* **150**: 131–143.

Deuschle, K., Funck, D., Hellmann, H., Daschner, K., Binder, S. and Frommer, W.B. (2001) A nuclear gene encoding mitochondrial delta-pyrroline-5–carboxylate dehydrogenase and its potential role in protection from proline toxicity. *Plant J.* **27**: 345–356.

Deveraux, Q.L., Leo, E., Stennicke, H.R., Welsh, K., Salvesen, G.S. and Reed, J.C. (1999) Cleavage of human inhibitor of apoptosis protein XIAP results in fragments with distinct specificities for caspases. *EMBO J.* **18**: 5242–5251.

Deveraux, Q.L. and Reed, J.C. (1999) IAP family proteins – suppressors of apoptosis. *Genes Dev.* **13**: 239–252.

Deveraux, Q.L., Roy, N., Stennicke, H.R., Van Arsdale, T., Zhou, Q., Srinivasula, S.M., Alnemri, E.S., Salvesen, G.S. and Reed, J.C. (1998) IAPs block apoptotic events induced by caspase-8 and cytochrome c by direct inhibition of distinct caspases. *EMBO J.* **17**: 2215–2223.

Deveraux, Q.L., Takahashi, R., Salvesen, G.S. and Reed, J.C. (1997) X-linked IAP is a direct inhibitor of cell-death proteases. *Nature* **388**: 300–304.

Devireddy, L.R., Teodoro, J.G., Richard, F.A. and Green, M.R. (2001) Induction of apoptosis by a secreted lipocalin that is transcriptionally regulated by IL-3 deprivation. *Science* **293**: 829–834.

Dhein, J., Walczak, H., Baumler, C., Debatin, K.M. and Krammer, P.H. (1995) Autocrine T-cell suicide mediated by APO-1/(Fas/CD95). *Nature* **373**: 438–441.

Diaz, J.L., Oltersdorf, T., Horne, W., McConnell, M., Wilson, G., Weeks, S., Garcia, T. and Fritz, L.C. (1997) A common binding site mediates heterodimerization and homodimerization of Bcl-2 family members. *J. Biol. Chem.* **272**: 11350–11355.

DiBello, P.R., Withers, D.A., Bayer, C.A., Fristrom, J.W. and Guild, G.M. (1991) The *Drosophila* Broad-Complex encodes a family of related proteins containing zinc fingers. *Genetics* **129**: 385–397.

Distelhorst, C.W., Lam, M. and McCormick, T.S. (1996) Bcl-2 inhibits hydrogen peroxide-induced ER Ca^{2+} pool depletion. *Oncogene* **12**: 2051–2055.

Doerfler, P., Forbush, K.A. and Perlmutter, R.M. (2000) Caspase enzyme activity is not essential for apoptosis during thymocyte development. *J. Immunol.* **164**: 4071–4079.

Doi, T.S., Marino, M.W., Takahashi, T., Yoshida, T., Sakakura, T., Old, L.J. and Obata, Y. (1999) Absence of tumor necrosis factor rescues RelA-deficient mice from embryonic lethality. *Proc. Natl Acad. Sci. U. S. A.* **96**: 2994–2999.

Dorner, A., Schulze, K., Rauch, U. and Schultheiss, H.P. (1997) Adenine nucleotide translocator in dilated cardiomyopathy: pathophysiological alterations in expression and function. *Mol. Cell. Biochem.* **174**: 261–269.

Dorstyn, L., Colussi, P., Quinn, L., Richardson, H. and Kumar, S. (1999a) DRONC, an ecdysone-inducible *Drosophila* caspase. *Proc. Natl Acad. Sci. U. S. A.* **96**: 4307–4312.

Dorstyn, L., Read, S., Cakouros, D., Huh, J.R., Hay, B.A. and Kumar, S. (2002) The role of cytochrome c in caspase activation in *Drosophila melanogaster* cells. *J. Cell Biol.* **156**: 1089–1098.

Dorstyn, L., Read, S.H., Quinn, L.M., Richardson, H. and Kumar, S. (1999b) DECAY, a novel *Drosophila* caspase related to mammalian caspase-3 and caspase-7. *J. Biol. Chem.* **274**: 30778–30783.

Doumanis, J., Quinn, L., Richardson, H. and Kumar, S. (2001) STRICA, a novel *Drosophila melanogaster* caspase with an unusual serine/threonine-rich prodomain, interacts with DIAP1 and DIAP2. *Cell Death Differ.* 8: 387–394.

Dramsi, S., Scheid, M. P., Maiti, A., *et al.* (2002) Identification of a novel phosphorylation site, Ser-170, as a regulator of bad pro-apoptotic activity. *J. Biol. Chem.* 277: 6399–6405.

Drexler, H.C.A. (1997) Activation of the cell death program by inhibition of proteasome function. *Proc. Natl Acad. Sci. U. S. A.* 94: 855–860.

Driever, W. and Nüsslein-Volhard, C. (1988) The BICOID protein determines position in the *Drosophila* embryo in a concentration-dependent manner. *Cell* 54: 95–104.

Du, C., Fang, M., Li, Y., Li, L. and Wang, X. (2000) Smac, a mitochondrial protein that promotes cytochrome c-dependent caspase activation by eliminating IAP inhibition. *Cell* 102: 33–42.

Duan, H. and Dixit, V.M. (1997) RAIDD is a new 'death' adaptor molecule. *Nature* 385: 86–89.

Duchen, M.R. (2000) Mitochondria and calcium: from cell signalling to cell death. *J. Physiol.* 529 (Pt 1): 57–68.

Duckett, C. S. and Thompson, C.B. (1997) CD30-dependent degradation of TRAF2: implications for negative regulation of TRAF signaling and the control of cell survival. *Genes Dev.* 11: 2810–2821.

Duckett, C.S., Nava, V.E., Gedrich, R.W., Clem, R.J., Van Dongen, J.L., Gilfillan, M.C., Shiels, H., Hardwick, J.M. and Thompson, C.B. (1996) A conserved family of cellular genes related to the baculovirus iap gene and encoding apoptosis inhibitors. *EMBO J.* 15: 2685–2694.

Duffy, M. J. (1996) Proteases as prognostic markers in cancer. *Clin. Cancer Res.* 2: 613–618.

Duriez, P. J., Wong, F., Dorovini-Zis, K., Shahidi, R. and Karsan, A. (2000) A1 functions at the mitochondria to delay endothelial apoptosis in response to tumor necrosis factor. *J. Biol. Chem.* 275: 18099–18107.

Eagle, H. (1959) Amino acid metabolism in human cell cultures. *Science* 130: 432–436.

Earnshaw, W.C., Martins, L.M. and Kaufmann, S.H. (1999) Mammalian caspases: structure, activation, substrates, and functions during apoptosis. *Annu. Rev. Biochem.* 68: 383–424.

Eberstadt, M., Huang, B., Chen, Z., Meadows, R.P., Ng, S.C., Zheng, L., Lenardo, M.J. and Fesik, S.W. (1998) NMR structure and mutagenesis of the FADD (Mort1) death-effector domain. *Nature* 392: 941–945.

Echtay, K.S., Winkler, E., Frischmuth, K. and Klingenberg, M. (2001) Uncoupling proteins 2 and 3 are highly active H(+) transporters and highly nucleotide sensitive when activated by coenzyme Q (ubiquinone) *Proc. Natl Acad. Sci. U. S. A.* 98: 1416–1421.

Eddy, S.R. (1998) Profile hidden Markov models. *Bioinformatics* 14: 755–763.

Eguchi, Y., Shimizu, S. and Tsujimoto, Y. (1997) Intracellular ATP levels determine cell death fate by apoptosis or necrosis. *Cancer Res.* 57: 1835–1840.

Eischen, C.M., Kottke, T.J., Martins, L.M., Basi, G.S., Tung, J.S., Earnshaw, W.C.,

Leibson, P.J. and Kaufmann, S.H. (1997) Comparison of apoptosis in wild-type and Fas-resistant cells: chemotherapy-induced apoptosis is not dependent on Fas/Fas ligand interactions. *Blood* **90**: 935–943.

Ekegren, T., Grundstrom, E., Lindholm, D. and Aquilonius, S.M. (1999) Upregulation of Bax protein and increased DNA degradation in ALS spinal cord motor neurons. *Acta Neurol. Scand.* **100**: 317–321.

Ekert, P.G., Silke, J. and Vaux, D.L. (1999) Caspase inhibitors. *Cell Death Differ.* **6**: 1081–1086.

el Deiry, W.S. (1998) Regulation of p53 downstream genes. *Semin. Cancer Biol.* **8**: 345–357.

Elbashir, S.M., Harborth, J., Lendeckel, W., Yalcin, A., Weber, K. and Tuschl, T. (2001) Duplexes of 21-nucleotide RNAs mediate RNA interference in cultured mammalian cells. *Nature* **411**: 494–498.

Elder, R.T., Yu, M., Chen, M., Zhu, X., Yanagida, M. and Zhao, Y. (2001) HIV-1 Vpr induces cell cycle G2 arrest in fission yeast (*Schizosaccharomyces pombe*) through a pathway involving regulatory and catalytic subunits of PP2A and acting on both Wee1 and Cdc25. *Virology* **287**: 359–370.

Elliott, K., Ge, K., Du, W. and Prendergast, G.C. (2000) The c-Myc-interacting adaptor protein Bin1 activates a caspase-independent cell death program. *Oncogene* **19**: 4669–4684.

Ellis, H.M. and Horvitz, H.R. (1986) Genetic control of programmed cell death in the nematode *C. elegans*. *Cell* **44**: 817–829.

Ellis, R.E. and Horvitz, H.R. (1991) Two *C. elegans* genes control the programmed deaths of specific cells in the pharynx. *Development* **112**: 591–603.

Ellis, R.E., Jacobson, D.M. and Horvitz, H.R. (1991) Genes required for the engulfment of cell corpses during programmed cell death in *Caenorhabditis elegans*. *Genetics* **129**: 79–94.

Elrod-Erickson, M., Mishra, S. and Schneider, D. (2000) Interactions between the cellular and humoral immune responses in *Drosophila*. *Curr. Biol.* **10**: 781–784.

Enari, M., Sakahira, H., Yokoyama, H., Okawa, K., Iwamatsu, A. and Nagata, S. (1998) A caspase-activated DNase that degrades DNA during apoptosis, and its inhibitor ICAD. *Nature* **391**: 43–50.

Enyedy, I.J., Ling, Y., Nacro, K., *et al.* (2001) Discovery of small-molecule inhibitors of Bcl-2 through structure-based computer screening. *J. Med. Chem.* **44**: 4313–4324.

Erickson, M.R., Galletta, B.J. and Abmayr, S.M. (1997) *Drosophila* myoblast city encodes a conserved protein that is essential for myoblast fusion, dorsal closure, and cytoskeletal organization. *J. Cell Biol.* **138**: 589–603.

Eskes, R., Antonsson, B., Osen-Sand, A., Montessuit, S., Richter, C., Sadoul, R., Mazzei, G., Nichols, A. and Martinou, J.C. (1998) Bax-induced cytochrome C release from mitochondria is independent of the permeability transition pore but highly dependent on Mg^{2+} ions. *J. Cell Biol.* **143**: 217–224.

Eskes, R., Desagher, S., Antonsson, B. and Martinou, J.C. (2000) Bid induces the oligomerization and insertion of Bax into the outer mitochondrial membrane. *Mol. Cell. Biol.* **20**: 929–935.

Espagne, E., Balhadere, P., Begueret, J. and Turcq, B. (1997) Reactivity in vegetative incompatibility of the HET-E protein of the fungus *Podospora anserina* is dependent on GTP-binding activity and a WD40 repeated domain. *Mol. Gen. Genet.* **256**: 620–627.

Evans, E., Kuwana, T., Strum, S., Smith, J., Newmeyer, D. and Kornbluth, S. (1997) Reaper-induced apoptosis in a vertebrate system. *EMBO J.* **16**: 7372–7381.

Fadok, V. A., Xue, D. and Henson, P. (2001) If phosphatidylserine is the death knell, a new phosphatidylserine-specific receptor is the bellringer. *Cell Death Differ.* **8**: 582–587.

Fairbrother, W.J., Gordon, N.C., Humke, E.W., O'Rourke, K.M., Starovasnik, M.A., Yin, J.P. and Dixit, V.M. (2001) The PYRIN domain: a member of the death domain-fold superfamily. *Protein Sci.* **10**: 1911–1918.

Fedor-Chaiken, M., Deschenes, R.J. and Broach, J.R. (1990) SRV2, a gene required for RAS activation of adenylate cyclase in yeast. *Cell* **61**: 329–340.

Feinstein, E., Kimchi, A., Wallach, D., Boldin, M. and Varfolomeev, E. (1995) The death domain: a module shared by proteins with diverse cellular functions letter. *Trends Biochem. Sci.* **20**: 342–344.

Feller, S.M. (2001) Crk family adaptors – signalling complex formation and biological roles. *Oncogene* **20**: 6348–6371.

Ferri, K.F. and Kroemer, G. (2001) Organelle-specific initiation of cell death pathways. *Nat. Cell Biol.* **3**: E255–E263.

Fesik, S.W. (2000) Insights into programmed cell death through structural biology. *Cell* **103**: 273–282.

Fields, S. and Song, O. (1989) A novel genetic system to detect proteinprotein interactions. *Nature* **340**: 245–246.

Finkel, T.H., Tudor-Williams, G., Banda, N.K., Cotton, M.F., Curiel, T., Monks, C., Baba, T.W., Ruprecht, R.M. and Kupfer, A. (1995) Apoptosis occurs predominantly in bystander cells and not in productively infected cells of HIV- and SIV-infected lymph nodes. *Nat. Med.* **1**: 129–134.

Finn, J.T., Weil, M., Archer, F., Siman, R., Srinivasan, A. and Raff, M.C. (2000) Evidence that Wallerian degeneration and localized axon degeneration induced by local neurotrophin deprivation do not involve caspases. *J. Neurosci.* **20**: 1333–1341.

Finucane, D.M., Bossy-Wetzel, E., Waterhouse, N.J., Cotter, T.G. and Green, D.R. (1999) Bax-induced caspase activation and apoptosis via cytochrome c release from mitochondria is inhibitable by Bcl-xL. *J. Biol. Chem.* **274**: 2225–2233.

Fisher, G.H., Rosenberg, F.J., Straus, S.E., Dale, J.K., Middleton, L.A., Lin, A.Y., Strober, W., Lenardo, M.J. and Puck, J.M. (1995) Dominant interfering Fas gene mutations impair apoptosis in a human autoimmune lymphoproliferative syndrome. *Cell* **81**: 935–946.

Fitzgerald, K.A., Palsson-McDermott, E.M., Bowie, A.G., *et al.* (2001) Mal (MyD88–adapter-like) is required for Toll-like receptor-4 signal transduction. *Nature* **413**: 78–83.

Flower, D.R. (1996) The lipocalin protein family: structure and function. *Biochem. J.* **318**: 1–14.

Flybase (1994) The *Drosophila* genetic database. *Nucleic Acids Res.* **22**: 3456–3458.

Foghsgaard, L., Wissing, D., Mauch, D., Lademann, U., Bastholm, L., Boes, M.,

Elling, F., Leist, M. and Jäättelä, M. (2001) Cathepsin B acts as a dominant execution protease in tumor cell apoptosis induced by tumor necrosis factor. *J. Cell Biol.* 153: 999–1009.

Foiani, M., Pellicioli, A., Lopes, M., Lucca, C., Ferrari, M., Liberi, G., Muzi, F.M. and Plevani, P. (2000) DNA damage checkpoints and DNA replication controls in *Saccharomyces cerevisiae*. *Mutat. Res.* 451: 187–196.

Foley, K. and Cooley, L. (1998) Apoptosis in late stage *Drosophila* nurse cells does not require genes within the H99 deficiency. *Development* 125: 1075–1082.

Foy, T.M., Aruffo, A., Bajorath, J., Buhlmann, J.E. and Noelle, R.J. (1996) Immune regulation by CD40 and its ligand GP39. *Annu. Rev. Immunol.* 14: 591–617.

Foy, T.M., Laman, J.D., Ledbetter, J.A., Aruffo, A., Claassen, E. and Noelle, R.J. (1994) gp39–CD40 interactions are essential for germinal center formation and the development of B cell memory. *J. Exp. Med.* 180: 157–163.

Foyouzi-Youssefi, R., Arnaudeau, S., Borner, C., Kelley, W.L., Tschopp, J., Lew, D.P., Demaurex, N. and Krause, K.H. (2000) Bcl-2 decreases the free Ca^{2+} concentration within the endoplasmic reticulum. *Proc. Natl Acad. Sci. U. S. A.* 97: 5723–5728.

Franzoso, G., Carlson, L., Poljak, L., *et al.* (1998) Mice deficient in nuclear factor (NF)-kappa B/p52 present with defects in humoral responses, germinal center reactions, and splenic microarchitecture. *J. Exp. Med.* 187: 147–159.

Fraser, A.G. and Evan, G.I. (1997) Identification of a *Drosophila melanogaster* ICE/CED-3–related protease, drICE. *EMBO J.* 16: 2805–2813.

Fraser, A.G., James, C., Evan, G.I. and Hengartner, M.O. (1999) *Caenorhabditis elegans* inhibitor of apoptosis protein (IAP) homologue BIR-1 plays a conserved role in cytokinesis. *Curr. Biol.* 9: 292–301.

Fraser, A.G., McCarthy, N.J. and Evan, G.I. (1997) drICE is an essential caspase required for apoptotic activity in *Drosophila* cells. *EMBO J.* 16: 6192–6199.

Freeman, M. (1996) Reiterative use of the EGF receptor triggers differentiation of all cell types in the *Drosophila* eye. *Cell* 87: 651–660.

Fröhlich, K.U. and Madeo, F. (2001) Apoptosis in yeast: a new model for aging research. *Exp. Gerontol.* 37: 27–31.

Fröhlich, K.U. and Madeo, F. (2000) Apoptosis in yeast: a monocellular organism exhibits altruistic behaviour. *FEBS Lett.* 473: 6–9.

Fuchs, E.J., McKenna, K.A. and Bedi, A. (1997) p53–dependent DNA damage-induced apoptosis requires Fas/APO-1–independent activation of CPP32beta. *Cancer Res.* 57: 2550–2554.

Fujikawa, D.G. (2000) Confusion between neuronal apoptosis and activation of programmed cell death mechanisms in acute necrotic insults. *Trends Neurosci.* 23: 410–411.

Fujita, N., Nagahashi, A., Nagashima, K., Rokudai, S. and Tsuruo, T. (1998) Acceleration of apoptotic cell death after the cleavage of Bcl-XL protein by caspase-3–like proteases. *Oncogene* 17: 1295–1304.

Furukawa, Y., Iwase, S., Kikuchi, J., Terui, Y., Nakamura, M., Yamada, H., Kano, Y. and Matsuda, M. (2000) Phosphorylation of Bcl-2 protein by CDC2 kinase during G2/M phases and its role in cell cycle regulation. *J. Biol. Chem.* 275: 21661–21667.

Gabig, T.G., Mantel, P.L., Rosli, R. and Crean, C.D. (1994) Requiem: a novel zinc finger gene essential for apoptosis in myeloid cells. *J. Biol. Chem.* **269**: 29515–29519.

Garlid, K.D., Jaburek, M. and Jezek, P. (1998) The mechanism of proton transport mediated by mitochondrial uncoupling proteins. Review. *FEBS Lett.* **438**: 10–14.

Gartner, A., Milstein, S., Ahmed, S., Hodgkin, J. and Hengartner, M.O. (2000) A conserved checkpoint pathway mediates DNA damage-induced apoptosis and cell cycle arrest in *C. elegans*. *Mol. Cell* **5**: 435–443.

Gavrieli, Y., Sherman, Y. and Ben-Sasson, S.A. (1992) Identification of programmed cell death *in situ* via specific labeling of nuclear DNA fragmentation. *J. Cell Biol.* **119**: 493–501.

Georgel, P., Naitza, S., Kappler, C., *et al.* (2001) *Drosophila* immune deficiency (IMD) is a death domain protein that activates antibacterial defense and can promote apoptosis. *Dev. Cell* **1**: 503–514.

Ghosh, C. and Iversen, P.L. (2000) Intracellular delivery strategies for antisense phosphorodiamidate morpholino oligomers. *Antisense Nucleic Acid Drug Dev.* **10**: 263–274.

Gierer, A. (1974) *Hydra* as a model for the development of biological form. *Sci. Am.* **231**: 44–54.

Gingras, M.C., Champagne, C., Roy, M. and Lavoie, J.N. (2002) Cytoplasmic death signal triggered by SRC-mediated phosphorylation of the adenovirus E4orf4 protein. *Mol. Cell. Biol.* **22**: 41–56.

Gogvadze, V., Robertson, J.D., Zhivotovsky, B. and Orrenius, S. (2001) Cytochrome c release occurs via Ca^{2+}-dependent and Ca^{2+}-independent mechanisms that are regulated by Bax. *J. Biol. Chem.* **276**: 19066–19071.

Goldstein, J.C., Waterhouse, N.J., Juin, P., Evan, G.I. and Green, D.R. (2000) The coordinate release of cytochrome c during apoptosis is rapid, complete and kinetically invariant. *Nat. Cell Biol.* **2**: 156–162.

Golstein, P. (1997) Cell death: TRAIL and its receptors. *Curr. Biol.* **7**: R750–R753.

Goltsev, Y.V., Kovalenko, A.V., Arnold, E., Varfolomeev, E.E., Brodianskii, V.M. and Wallach, D. (1997) CASH, a novel caspase homologue with death effector domains. *J. Biol. Chem.* **272**: 19641–19644.

Gonzalez-Garcia, M., Perez-Ballestero, R., Ding, L., Duan, L., Boise, L.H., Thompson, C.B. and Nunez, G. (1994) bcl-XL is the major bcl-x mRNA form expressed during murine development and its product localizes to mitochondria. *Development* **120**: 3033–3042.

Goping, I.S., Gross, A., Lavoie, J.N., Nguyen, M., Jemmerson, R., Roth, K., Korsmeyer, S.J. and Shore, G.C. (1998) Regulated targeting of BAX to mitochondria. *J. Cell Biol.* **143**: 207–215.

Gotow, T., Shibata, M., Kanamori, S., *et al.* (2000) Selective localization of Bcl-2 to the inner mitochondrial and smooth endoplasmic reticulum membranes in mammalian cells. *Cell Death Differ.* **7**: 666–674.

Gougeon, M.L. (1999) Programmed cell death as a mechanism of CD4 and CD8 T cell deletion in AIDS. Molecular control and effect of highly active anti-retroviral therapy. *Ann. N. Y. Acad. Sci.* **887**: 199–212.

Goyal, L. (2001) Cell death inhibition: keeping caspases in check. *Cell* **104**: 805–808.

Goyal, L., McCall, K., Agapite, J., Hartwieg, E. and Steller, H. (2000) Induction of apoptosis by *Drosophila* reaper, hid and grim through inhibition of IAP function. *EMBO J.* **19**: 589–597.

Gratiot-Deans, J., Merino, R., Nunez, G. and Turka, L.A. (1994) Bcl-2 expression during T-cell development: early loss and late return occur at specific stages of commitment to differentiation and survival. *Proc. Natl Acad. Sci. U. S. A.* **91**: 10685–10689.

Green, D.R. and Ferguson, T.A. (2001) The role of Fas ligand in immune privilege. *Nat. Rev. Mol. Cell. Biol.* **2**: 917–924.

Green, D.R. (2000) Apoptotic pathways: paper wraps stone blunts scissors. *Cell* **102**: 1–4.

Green, D.R. (1998) Apoptotic pathways: the roads to ruin. *Cell* **94**: 695–698.

Green, D.R. and Reed, J.C. (1998) Mitochondria and apoptosis. *Science* **281**: 1309–1312.

Greenhalf, W., Stephan, C. and Chaudhuri, B. (1996) Role of mitochondria and C-terminal membrane anchor of Bcl-2 in Bax induced growth arrest and mortality in *Saccharomyces cerevisiae. FEBS Lett.* **380**: 169–175.

Grether, M.E., Abrams, J.M., Agapite, J., White, K. and Steller, H. (1995) The head involution defective gene of *Drosophila melanogaster* functions in programmed cell death. *Genes Dev.* **9**: 1694–1708.

Gribskov, M., McLachlan, A.D. and Eisenberg, D. (1987) Profile analysis: detection of distantly related proteins. *Proc. Natl Acad. Sci. U. S. A.* **84**: 4355–4358.

Griffith, T.S. and Lynch, D.H. (1998) TRAIL: a molecule with multiple receptors and control mechanisms. *Curr. Opin. Immunol.* **10**: 559–563.

Griffiths, G.J., Dubrez, L., Morgan, C.P., Jones, N.A., Whitehouse, J., Corfe, B.M., Dive, C. and Hickman, J.A. (1999) Cell damage-induced conformational changes of the pro-apoptotic protein Bak in vivo precede the onset of apoptosis. *J. Cell Biol.* **144**: 903–914.

Grillot, D.A., Merino, R., Pena, J.C., Fanslow, W.C., Finkelman, F.D., Thompson, C.B. and Nunez, G. (1996) bcl-x exhibits regulated expression during B cell development and activation and modulates lymphocyte survival in transgenic mice. *J. Exp. Med.* **183**: 381–391.

Grimm, S. and Baeuerle, P.A. (1993) The inducible transcription factor NF-kappa B: structure-function relationship of its protein subunits. *Biochem. J.* **290**: 297–308.

Grimm, S. and Leder, P. (1997) An apoptosis-inducing isoform of neu differentiation factor (NDF) identified using a novel screen for dominant, apoptosis-inducing genes. *J. Exp. Med.* **185**: 1137–1142.

Grimm, S., Weinstein, E., Krane, I. M. and Leder, P. (1998) Neu differentiation factor (NDF), a dominant oncogene, causes apoptosis in vitro and in vivo. *J. Exp. Med.* **188**: 1535–1539.

Grinberg, M., Sarig, R., Zaltsman, Y., Frumkin, D., Grammatikakis, N., Reuveny, E. and Gross, A. (2002) tBID homooligomerizes in the mitochondrial membrane to induce apoptosis. *J. Biol. Chem.* **277**: 12237–12245.

Gross, A., Jockel, J., Wei, M.C. and Korsmeyer, S.J. (1998) Enforced dimerization of BAX results in its translocation, mitochondrial dysfunction and apoptosis. *EMBO J.* **17**: 3878–3885.

Gross, A., McDonnell, J.M. and Korsmeyer, S.J. (1999a) BCL-2 family members and the mitochondria in apoptosis. *Genes Dev.* **13**: 1899–1911.

Gross, A., Pilcher, K., Blachly-Dyson, E., Basso, E., Jockel, J., Bassik, M.C., Korsmeyer, S.J. and Forte, M. (2000) Biochemical and genetic analysis of the mitochondrial response of yeast to BAX and BCL-X(L) *Mol. Cell. Biol.* **20**: 3125–3136.

Gross, A., Yin, X.M., Wang, K., Wei, M.C., Jockel, J., Milliman, C., Erdjument-Bromage, H., Tempst, P. and Korsmeyer, S.J. (1999b) Caspase cleaved BID targets mitochondria and is required for cytochrome c release, while BCL-XL prevents this release but not tumor necrosis factor-R1/Fas death. *J. Biol. Chem.* **274**: 1156–1163.

Grutter, M.G. (2000) Caspases: key players in programmed cell death. *Curr. Opin. Struct. Biol.* **10**: 649–655.

Gudkov, A.V., Kazarov, A.R., Thimmapaya, R., Axenovich, S.A., Mazo, I.A. and Roninson, I.B. (1994) Cloning mammalian genes by expression selection of genetic suppressor elements: association of kinesin with drug resistance and cell immortalization. *Proc. Natl Acad. Sci. U. S. A.* **91**: 3744–3748.

Guicciardi, M.E., Deussing, J., Miyoshi, H., Bronk, S.F., Svingen, P.A., Peters, C., Kaufmann, S.H. and Gores, G.J. (2000) Cathepsin B contributes to TNF-alpha-mediated hepatocyte apoptosis by promoting mitochondrial release of cytochrome c. *J. Clin. Invest.* **106**: 1127–1137.

Guicciardi, M.E., Miyoshi, H., Bronk, S.F. and Gores, G.J. (2001) Cathepsin B knockout mice are resistant to tumor necrosis factor-alpha-mediated hepatocyte apoptosis and liver injury: implications for therapeutic applications. *Am. J. Pathol.* **159**: 2045–2054.

Gumienny, T.L., Brugnera, E., Tosello-Trampont, A.C. *et al.* (2001) CED-12/ELMO, a novel member of the CrkII/Dock180/Rac pathway, is required for phagocytosis and cell migration. *Cell* **107**: 27–41.

Gumienny, T.L., Lambie, E., Hartwieg, E., Horvitz, H.R. and Hengartner, M.O. (1999) Genetic control of programmed cell death in the *Caenorhabditis elegans* hermaphrodite germline. *Development* **126**: 1011–1022.

Guo, Q., Furukawa, K., Sopher, B.L., Pham, D.G., Xie, J., Robinson, N., Martin, G.M. and Mattson, M.P. (1996) Alzheimer's PS-1 mutation perturbs calcium homeostasis and sensitizes PC12 cells to death induced by amyloid beta-peptide. *Neuroreport* **8**: 379–383.

Guo, Y., Srinivasula, S.M., Druilhe, A., Fernandes-Alnemri, T. and Alnemri, E.S. (2002) Caspase-2 induces apoptosis by releasing proapoptotic proteins from mitochondria. *J. Biol. Chem.* **277**: 13430–13437.

Ha, H.C. and Snyder, S.H. (1999) Poly(ADP-ribose) polymerase is a mediator of necrotic cell death by ATP depletion. *Proc. Natl Acad. Sci. U. S. A.* **96**: 13978–13982.

Hacki, J., Egger, L., Monney, L., Conus, S., Rosse, T., Fellay, I. and Borner, C. (2000) Apoptotic crosstalk between the endoplasmic reticulum and mitochondria controlled by Bcl-2. *Oncogene* **19**: 2286–2295.

Haining, W., Carboy-Newcomb, C., Wei, C. and Steller, H. (1999) The proapoptotic function of *Drosophila* Hid is conserved in mammalian cells. *Proc. Natl Acad. Sci. U. S. A.* **96**: 4936–4941.

Hajnoczky, G., Csordas, G., Madesh, M. and Pacher, P. (2000a) Control of apoptosis by IP(3) and ryanodine receptor driven calcium signals. *Cell Calcium* **28**: 349–363.

Hajnoczky, G., Csordas, G., Madesh, M. and Pacher, P. (2000b) The machinery of local Ca^{2+} signalling between sarco-endoplasmic reticulum and mitochondria. *J. Physiol.* **529** (Pt 1): 69–81.

Hakem, R., Hakem, A., Duncan, G.S., *et al.* (1998) Differential requirement for caspase-9 in apoptotic pathways in vivo. *Cell* **94**: 339–352.

Halestrap, A.P., Connern, C.P., Griffiths, E.J. and Kerr, P.M. (1997) Cyclosporin A binding to mitochondrial cyclophilin inhibits the permeability transition pore and protects hearts from ischaemia/reperfusion injury. *Mol. Cell. Biochem.* **174**: 167–172.

Hall, D.H., Winfrey, V.P., Blaeuer, G., Hoffman, L.H., Furuta, T., Rose, K.L., Hobert, O. and Greenstein, D. (1999) Ultrastructural features of the adult hermaphrodite gonad of *Caenorhabditis elegans*: relations between the germ line and soma. *Dev. Biol.* **212**: 101–123.

Hamasaki, A., Sendo, F., Nakayama, K., Ishida, N., Negishi, I. and Hatakeyama, S. (1998) Accelerated neutrophil apoptosis in mice lacking A1-a, a subtype of the bcl-2-related A1 gene. *J. Exp. Med.* **188**: 1985–1992.

Han, D.K., Chaudhary, P.M., Wright, M.E., Friedman, C., Trask, B.J., Riedel, R.T., Baskin, D.G., Schwartz, S.M. and Hood, L. (1997) MRIT, a novel death-effector domain-containing protein, interacts with caspases and BclXL and initiates cell death. *Proc. Natl Acad. Sci. U. S. A.* **94**: 11333–11338.

Hanada, M., Aime-Sempe, C., Sato, T. and Reed, J.C. (1995) Structure-function analysis of Bcl-2: protein identification of conserved domains important for homodimerization with Bcl-2 and heterodimerization with Bax. *J. Biol. Chem.* **270**: 11962–11969.

Hansson, O., Petersen, A., Leist, M., Nicotera, P., Castilho, R.F. and Brundin, P. (1999) Transgenic mice expressing a Huntington's disease mutation are resistant to quinolinic acid-induced striatal excitotoxicity. *Proc. Natl Acad. Sci. U. S. A.* **96**: 8727–8732.

Harada, H., Andersen, J.S., Mann, M., Terada, N. and Korsmeyer, S.J. (2001) p70S6 kinase signals cell survival as well as growth, inactivating the pro-apoptotic molecule BAD. *Proc. Natl Acad. Sci. U. S. A.* **98**: 9666–9670.

Harada, H., Becknell, B., Wilm, M., Mann, M., Huang, L.J., Taylor, S.S., Scott, J.D. and Korsmeyer, S.J. (1999) Phosphorylation and inactivation of BAD by mitochondria-anchored protein kinase A. *Mol. Cell* **3**: 413–422.

Harding, H.P., Novoa, I., Zhang, Y., Zeng, H., Wek, R., Schapira, M. and Ron, D. (2000a) Regulated translation initiation controls stress-induced gene expression in mammalian cells. *Mol. Cell* **6**: 1099–1108.

Harding, H.P., Zhang, Y., Bertolotti, A., Zeng, H. and Ron, D. (2000b) Perk is essential for translational regulation and cell survival during the unfolded protein response. *Mol. Cell* **5**: 897–904.

Harding, H.P., Zhang, Y. and Ron, D. (1999) Protein translation and folding are coupled by an endoplasmic-reticulum-resident kinase. *Nature* **397**: 271–274.

Hardy, J. (1997) Amyloid, the presenilins and Alzheimer's disease. *Trends Neurosci.* **20**: 154–159.

Harris, M.H., Vander Heiden, M.G., Kron, S.J. and Thompson, C.B. (2000) Role of oxidative phosphorylation in Bax toxicity. *Mol. Cell. Biol.* **20**: 3590–3596.

Hartman, P.S. and Herman, R.K. (1982) Radiation-sensitive mutants of *Caenorhabditis elegans*. *Genetics* **102**: 159–178.

Harvey, K.J., Lukovic, D. and Ucker, D.S. (2000) Caspase-dependent Cdk activity is a requisite effector of apoptotic death events. *J. Cell Biol.* **148**: 59–72.

Harvey, N., Trapani, J.A., Fernandes-Alnemri, T., Litwack, G., Alnemri, E.S. and Kumar, S. (1996) Processing of the Nedd2 precursor by ICElike proteases and granzyme B. *Genes Cells* **1**: 673–685.

Harvey, N.L., Butt, A. and Kumar, S. (1997) Functional activation of Nedd2/ICH-1 (caspase-2) is an early process in apoptosis. *J. Biol. Chem.* **272**: 13134–13139.

Harvey, N.L., Daish, T., Mills, K., Dorstyn, L., Quinn, L.M., Read, S.H., Richardson, H. and Kumar, S. (2001) Characterization of the *Drosophila* caspase, DAMM. *J. Biol. Chem.* **276**: 25342–25350.

Hasegawa, H., Kiyokawa, E., Tanaka, S., Nagashima, K., Gotoh, N., Shibuya, M., Kurata, T. and Matsuda, M. (1996) DOCK180, a major CRK-binding protein, alters cell morphology upon translocation to the cell membrane. *Mol. Cell. Biol.* **16**: 1770–1776.

Hashimoto, Y., Niikura, T., Tajima, H., *et al.* (2001) A rescue factor abolishing neuronal cell death by a wide spectrum of familial Alzheimer's disease genes and Abeta. *Proc. Natl Acad. Sci. U. S. A.* **98**: 6336–6341.

Hassane, D.C., Lee, R.B., Mendenhall, M.D. and Pickett, C.L. (2001) Cytolethal distending toxin demonstrates genotoxic activity in a yeast model. *Infect. Immun.* **69**: 5752–5759.

Hawkins, C., Wang, S. and Hay, B. (1999) A cloning method to identify caspases and their regulators in yeast: identification of *Drosophila* IAP1 as an inhibitor of the *Drosophila* caspase DCP-1. *Proc. Natl Acad. Sci. U. S. A.* **96**: 2885–2890.

Hawkins, C.J., Yoo, S.J., Peterson, E.P., Wang, S.L., Vernooy, S.Y. and Hay, B.A. (2000) The *Drosophila* caspase DRONC cleaves following glutamate or aspartate and is regulated by DIAP1, HID, and GRIM. *J. Biol. Chem.* **275**: 27084–27093.

Hay, B.A. (2000) Understanding IAP function and regulation: a view from *Drosophila*. *Cell Death Differ.* **7**: 1045–1056.

Hay, B.A., Wassarman, D.A. and Rubin, G.M. (1995) *Drosophila* homologs of baculovirus inhibitor of apoptosis proteins function to block cell death. *Cell* **83**: 1253–1262.

He, H., Lam, M., McCormick, T.S. and Distelhorst, C.W. (1997) Maintenance of calcium homeostasis in the endoplasmic reticulum by Bcl-2. *J. Cell Biol.* **138**: 1219–1228.

Heath, M.C. (2000) Hypersensitive response-related death. *Plant Mol. Biol.* **44**: 321–334.

Hedge, R., Srinivasula, S.M., Wassell, R., *et al.* (2002) Identification of Omi/HtrA2 as a mitochondrial apoptotic serine protease that disrupts IAP-caspase interaction. *J. Biol. Chem.* **277**: 432–438.

Hedgecock, E.M., Sulston, J.E. and Thomson, J.N. (1983) Mutations affecting programmed cell deaths in the nematode *Caenorhabditis elegans*. *Science* **220**: 1277–1279.

Hegde, R., Srinivasula, S.M., Zhang, Z., *et al.* (2002) Identification of Omi/HtrA2 as a mitochondrial apoptotic serine protease that disrupts inhibitor of apoptosis protein–caspase interaction. *J. Biol. Chem.* **277**: 432–438.

Heibein, J.A., Goping, I.S., Barry, M., Pinkoski, M.J., Shore, G.C., Green, D.R. and Bleackley, R.C. (2000) Granzyme B-mediated cytochrome c release is regulated by the Bcl-2 family members bid and Bax. *J. Exp. Med.* **192**: 1391–1402.

Heinrich, M., Wickel, M., Schneider-Brachert, W., *et al.* (1999) Cathepsin D targeted by acid sphingomyelinase-derived ceramide. *EMBO J.* **18**: 5252–5263 [erratum: **19**:15].

Helmerhorst, E.J., Troxler, R.F. and Oppenheim, F. (2001) The human salivary peptide histatin 5 exerts its antifungal activity through the formation of reactive oxygen species. *Proc. Natl Acad. Sci. U. S. A.* **98**: 14637–14642.

Hengartner, M.O. (2000) The biochemistry of apoptosis. *Nature* **407**: 770–776.

Hengartner, M.O., Ellis, R.E. and Horvitz, H.R. (1992) *Caenorhabditis elegans* gene ced-9 protects cells from programmed cell death. *Nature* **356**: 494–499.

Hengartner, M.O. and Horvitz, H.R. (1994) Programmed cell death in *Caenorhabditis elegans*. *Curr. Opin. Genet. Dev.* **4**: 581–586.

Hengartner, M.O. and Horvitz, H.R. (1994) *C. elegans* cell survival gene ced-9 encodes a functional homolog of the mammalian proto-oncogene bcl-2. *Cell* **76**: 665–676.

Hentze, H., Schwoebel, F., Lund, S., Kehl, M., Ertel, W., Wendel, A., Jäättelä, M. and Leist, M. (2001) In vivo and in vitro evidence for extracellular caspase activity released from apoptotic cells. *Biochem. Biophys. Res. Commun.* **283**: 1111–1117.

Herrera, P.L., Harlan, D. M. and Vassalli, P. (2000) A mouse CD8 T cell-mediated acute autoimmune diabetes independent of the perforin and Fas cytotoxic pathways: possible role of membrane TNF. *Proc. Natl Acad. Sci. U. S. A.* **97**: 279–284.

Heusel, J.W., Wesselschmidt, R.L., Shresta, S., Russell, J.H. and Ley, T.J. (1994) Cytotoxic lymphocytes require granzyme B for the rapid induction of DNA fragmentation and apoptosis in allogeneic target cells. *Cell* **76**: 977–987.

Heussler, V.T., Kuenzi, P. and Rottenberg, S. (2001) Inhibition of apoptosis by intracellular protozoan parasites. *Int. J. Parasitol.* **31**: 1166–1176.

Hildeman, D.A., Mitchell, T., Teague, T.K., Henson, P., Day, B.J., Kappler, J. and Marrack, P.C. (1999) Reactive oxygen species regulate activation-induced T cell apoptosis. *Immunity* **10**: 735–744.

Hinds, M.G., Norton, R.S., Vaux, D.L. and Day, C.L. (1999) Solution structure of a baculoviral inhibitor of apoptosis (IAP) repeat. *Nat. Struct. Biol.* **6**: 648–651.

Hirotani, M., Zhang, Y., Fujita, N., Naito, M. and Tsuruo, T. (1999) NH2-terminal BH4 domain of Bcl-2 is functional for heterodimerization with Bax and inhibition of apoptosis. *J. Biol. Chem.* **274**: 20415–20420.

Hirsch, T., Marchetti, P., Susin, S.A., Dallaporta, B., Zamzami, N., Marzo, I., Geuskens, M. and Kroemer, G. (1997) The apoptosis-necrosis paradox. Apoptogenic proteases activated after mitochondrial permeability transition determine the mode of cell death. *Oncogene* **15**: 1573–1581.

Hirt, U.A., Gantner, F. and Leist, M. (2000) Phagocytosis of nonapoptotic cells dying by caspase-independent mechanisms. *J. Immunol.* **164**: 6520–6529.

Hobmayer, B., Rentzsch, F., Kuhn, K., Happel, C.M., von Laue, C.C., Snyder, P., Rothbacher, U. and Holstein, T.W. (2000) WNT signalling molecules act in axis formation in the diploblastic metazoan *Hydra. Nature* **407**: 186–189.

Hockenbery, D.M., Oltvai, Z.N., Yin, X.M., Milliman, C.L. and Korsmeyer, S.J. (1993) Bcl-2 functions in an antioxidant pathway to prevent apoptosis. *Cell* **75**: 241–251.

Hoeppner, D.J., Hengartner, M.O. and Schnabel, R. (2001) Engulfment genes cooperate with ced-3 to promote cell death in *Caenorhabditis elegans. Nature* **412**: 202–206.

Hoffman, H.M., Mueller, J.L., Broide, D.H., Wanderer, A.A. and Kolodner, R.D. (2001) Mutation of a new gene encoding a putative pyrin-like protein causes familial cold autoinflammatory syndrome and Muckle-Wells syndrome. *Nat. Genet.* **29**: 301–305.

Hoffmann, B., Stockl, A., Schlame, M., Beyer, K. and Klingenberg, M. (1994) The reconstituted ADP/ATP carrier activity has an absolute requirement for cardiolipin as shown in cysteine mutants. *J. Biol. Chem.* **269**: 1940–1944.

Hoffmann, J. and Reichhart, J.M. (1997) *Drosophila* immunity. *Trends Cell Biol.* **7**: 309–316.

Hofmann, K. (2000) Sensitive protein comparisons with profiles and Hidden Markov Models. *Briefings in Bioinformatics* **1**: 167–178.

Hofmann, K. (1999) The modular nature of apoptotic signaling proteins. *Cell Mol. Life Sci.* **55**: 1113–1128.

Hofmann, K. (1998) Protein classification and functional assignment. The modular nature of apoptotic signaling proteins. *Trends Guide Bioinformatics Suppl S:* **55**: 1113–1128.

Hofmann, K., Bucher, P., Falquet, L. and Bairoch, A. (1999) The PROSITE database, its status in 1999. *Nucleic Acids Res.* **27**: 215–219.

Hofmann, K., Bucher, P. and Tschopp, J. (1997) The CARD domain: a new apoptotic signalling motif. *Trends Biochem. Sci.* **22**: 155–156.

Hofmann, K. and Tschopp, J. (1995) The death domain motif found in Fas (Apo-1) and TNF receptor is present in proteins involved in apoptosis and axonal guidance. *FEBS Lett.* **371**: 321–323.

Holcik, M., Sonenberg, N. and Korneluk, R.G. (2000) Internal ribosome initiation of translation and the control of cell death. *Trends Genet.* **16**: 469–473.

Holinger, E.P., Chittenden, T. and Lutz, R.J. (1999) Bak BH3 peptides antagonize Bcl-xL function and induce apoptosis through cytochrome c-independent activation of caspases. *J. Biol. Chem.* **274**: 13298–13304.

Holler, N., Zaru, R., Micheau, O., Thome, M., Attinger, A., Valitutti, S., Bodmer, J.L., Schneider, P., Seed, B. and Tschopp, J. (2000) Fas triggers an alternative, caspase-8-independent cell death pathway using the kinase RIP as effector molecule. *Nat. Immunol.* 1: 489–495.

Holmstrom, T.H., Schmitz, I., Soderstrom, T.S., Poukkula, M., Johnson, V.L., Chow, S.C., Krammer, P.H. and Eriksson, J.E. (2000) MAPK/ERK signaling in activated T cells inhibits CD95/Fas-mediated apoptosis downstream of DISC assembly. *EMBO J.* 19: 5418–5428.

Honegger, T., Zürrer, D. and Tardent, P. (1989) Oogenesis in *Hydra carnea*: a new model based on light and electron microscopic analyses of oocyte and nurse cell differentiation. *Tissue Cell* 21: 381–393.

Hong, S.Y., Yoon, W.H., Park, J.H., Kang, S.G., Ahn, J.H. and Lee, T.H. (2000) Involvement of two NF-kappa B binding elements in tumor necrosis factor alpha-, CD40-, and Epstein-Barr virus latent membrane protein 1-mediated induction of the cellular inhibitor of apoptosis protein 2 gene. *J. Biol. Chem.* 275: 18022–18028.

Horng, T. and Medzhitov, R. (2001) *Drosophila* MyD88 is an adapter in the Toll signaling pathway. *Proc. Natl Acad. Sci. U. S. A.* 98: 12654–12658.

Hsu, H., Shu, H.B., Pan, M.G. and Goeddel, D.V. (1996) TRADD-TRAF2 and TRADD-FADD interactions define two distinct TNF receptor 1 signal transduction pathways. *Cell* 84: 299–308.

Hsu, H., Xiong, J. and Goeddel, D.V. (1995) The TNF receptor 1-associated protein TRADD signals cell death and NF-kappa B activation. *Cell* 81: 495–504.

Hsu, S.Y., Kaipia, A., McGee, E., Lomeli, M. and Hsueh, A.J. (1997) Bok is a pro-apoptotic Bcl-2 protein with restricted expression in reproductive tissues and heterodimerizes with selective anti-apoptotic Bcl-2 family members. *Proc. Natl Acad. Sci. U. S. A.* 94: 12401–12406.

Hsu, Y.T., Wolter, K.G. and Youle, R.J. (1997) Cytosol-to-membrane redistribution of Bax and Bcl-X(L) during apoptosis. *Proc. Natl Acad. Sci. U. S. A.* 94: 3668–3672.

Hsu, Y.T. and Youle, R.J. (1998) Bax in murine thymus is a soluble monomeric protein that displays differential detergent-induced conformations. *J. Biol. Chem.* 273: 10777–10783.

Hu, S., Vincenz, C., Buller, M. and Dixit, V.M. (1997) A novel family of viral death effector domain-containing molecules that inhibit both CD-95- and tumor necrosis factor receptor-1-induced apoptosis. *J. Biol. Chem.* 272: 9621–9624.

Hu, S. and Yang, X. (2000) dFADD, a novel death domain-containing adapter protein for the *Drosophila* caspase DREDD. *J. Biol. Chem.* 275: 30761–30764.

Huang, B., Eberstadt, M., Olejniczak, E.T., Meadows, R.P. and Fesik, S.W. (1996) NMR structure and mutagenesis of the Fas (APO-1/CD95) death domain. *Nature* 384: 638–641.

Huang, D.C., O'Reilly, L.A., Strasser, A., Cory, S., Chou, J.J., Li, H., Salvesen, G.S., Yuan, J. and Wagner, G. (1997) The anti-apoptosis function of Bcl-2 can be genetically separated from its inhibitory effect on cell cycle entry. *EMBO J.* 16: 4628–4638.

Huang, D.C.S. and Strasser, A. (2000) BH3-only proteins – essential initiators of apoptotic cell death. *Cell* 103: 839–842.

Huang, Y., Park, Y.C., Rich, R.L., Segal, D., Myszka, D.G. and Wu, H. (2001) Structural basis of caspase inhibition by XIAP: differential roles of the linker versus the BIR domain. *Cell* 104: 781–790.

Huang, Z. (2000) Bcl-2 family proteins as targets for anticancer drug design. *Oncogene* 19: 6627–6631.

Hunter, J.J. and Parslow, T.G. (1996) A peptide sequence from Bax that converts Bcl-2 into an activator of apoptosis. *J. Biol. Chem.* 271: 8521–8524.

Hymowitz, S.G., Christinger, H.W., Fuh, G., Ultsch, M., O'Connell, M., Kelley, R.F., Ashkenazi, A. and de Vos, A.M. (1999) Triggering cell death: the crystal structure of Apo2L/TRAIL in a complex with death receptor 5. *Mol. Cell* 4: 563–571.

Ichijo, H., Nishida, E., Irie, K., *et al.* (1997) Induction of apoptosis by ASK1, a mammalian MAPKKK that activates SAPK/JNK and p38 signaling pathways. *Science* 275: 90–94.

Ichikawa, K., Liu, W., Zhao, L., *et al.* (2001) Tumoricidal activity of a novel anti-human DR5 monoclonal antibody without hepatocyte cytotoxicity. *Nat. Med.* 7: 954–960.

Ichimiya, M., Chang, S.H., Liu, H., Berezesky, I.K., Trump, B.F. and Amstad, P.A. (1998) Effect of Bcl-2 on oxidant-induced cell death and intracellular Ca^{2+} mobilization. *Am. J. Physiol.* 275: 832–839.

Igaki, T., Kanuka, H., Inohara, N., Sawamoto, K., Nunez, G., Okano, H. and Miura, M. (2000) Drob-1, a *Drosophila* member of the Bcl-2/CED-9 family that promotes cell death. *Proc. Natl Acad. Sci. U. S. A.* 97: 662–667.

Ikemoto, H., Tani, E., Ozaki, I., Kitagawa, H. and Arita, N. (2000) Calphostin C-mediated translocation and integration of Bax into mitochondria induces cytochrome c release before mitochondrial dysfunction. *Cell Death Differ.* 7: 511–520.

Imaizumi, K., Miyoshi, K., Katayama, T., Yoneda, T., Taniguchi, M., Kudo, T. and Tohyama, M. (2001) The unfolded protein response and Alzheimer's disease. *Biochim. Biophys. Acta* 1536: 85–96.

Inbal, B., Cohen, O., Polak-Charcon, S., Kopolovic, J., Vadai, E., Eisenbach, L. and Kimchi, A. (1997) DAP kinase links the control of apoptosis to metastasis. *Nature* 390: 180–184.

Ink, B., Zörnig, M., Baum, B., Hajibagheri, N., James, C., Chittenden, T. and Evan, G. (1997) Human bak induces cell death in *Schizosaccharomyces pombe* with morphological changes similar to those with apoptosis in mammalian cells. *Mol. Cell. Biol.* 17: 2468–2474.

Inohara, N., Koseki, T., Chen, S., Wu, X. and Nunez, G. (1998) CIDE, a novel family of cell death activators with homology to the 45 kDa subunit of the DNA fragmentation factor. *EMBO J.* 17: 2526–2533.

Inohara, N., Koseki, T., Hu, Y., Chen, S. and Nunez, G. (1997) CLARP, a death effector domain-containing protein interacts with caspase-8 and regulates apoptosis. *Proc. Natl Acad. Sci. U. S. A.* 94: 10717–10722.

Irmler, M., Thome, M., Hahne, M., *et al.* (1997) Inhibition of death receptor signals by cellular FLIP. *Nature* 388: 190–195.

Ishida, Y., Agata, Y., Shibahara, K. and Honjo, T. (1992) Induced expression of PD-1, a novel member of the immunoglobulin gene superfamily, upon programmed cell death. *EMBO J.* **11**: 3887–3895.

Ito, T., Deng, X., Carr, B. and May, W.S. (1997) Bcl-2 phosphorylation required for anti-apoptosis function. *J. Biol. Chem.* **272**: 11671–11673.

Ito, T., Fujieda, S., Tsuzuki, H., Sunaga, H., Fan, G., Sugimoto, C., Fukuda, M. and Saito, H. (1999) Decreased expression of Bax is correlated with poor prognosis in oral and oropharyngeal carcinoma. *Cancer Lett.* **140**: 81–91.

Ito, Y., Mishra, N. C., Yoshida, K., Kharbanda, S., Saxena, S. and Kufe, D. (2001a) Mitochondrial targeting of JNK/SAPK in the phorbol ester response of myeloid leukemia cells. *Cell Death Differ.* **8**: 794–800.

Ito, Y., Pandey, P., Mishra, N., Kumar, S., Narula, N., Kharbanda, S., Saxena, S. and Kufe, D. (2001b) Targeting of the c-Abl tyrosine kinase to mitochondria in endoplasmic reticulum stress-induced apoptosis. *Mol. Cell. Biol.* **21**: 6233–6242.

Itoh, N. and Nagata, S. (1993) A novel protein domain required for apoptosis. Mutational analysis of human Fas antigen. *J. Biol. Chem.* **268**: 10932–10937.

Jäättelä, M. (1999) Escaping cell death: survival proteins in cancer. *Exp. Cell Res.* **248**: 30–43.

Jäättelä, M., Benedict, M., Tewari, M., Shayman, J.A. and Dixit, V.M. (1995) Bcl-x and Bcl-2 inhibit TNF and Fas-induced apoptosis and activation of phospholipase A2 in breast carcinoma cells. *Oncogene* **10**: 2297–2305.

Jäättelä, M., Wissing, D., Kokholm, K., Kallunki, T. and Egeblad, M. (1998) Hsp70 exerts its anti-apoptotic function downstream of caspase-3-like proteases. *EMBO J.* **17**: 6124–6134.

Jacobson, M.D., Burne, J.F. and Raff, M.C. (1994) Programmed cell death and Bcl-2 protection in the absence of a nucleus. *EMBO J.* **13**: 1899–1910.

Jacobson, M.D., Weil, M. and Raff, M.C. (1997) Programmed cell death in animal development. *Cell* **88**: 347–354.

James, C., Gschmeissner, S., Fraser, A. and Evan, G.I. (1997) CED-4 induces chromatin condensation in *Schizosaccharomyces pombe* and is inhibited by direct physical association with CED-9. *Curr. Biol.* **7**: 246–252.

Javelaud, D. (2001) NF-kappa B activation results in rapid inactivation of JNK in TNF alpha-treated Ewing sarcoma cells: a mechanism for the anti-apoptotic effect of NF-kappa B. *Oncogene* **20**: 4365–4372.

Jayaraman, T. and Marks, A.R. (1997) T cells deficient in inositol 1,4,5-trisphosphate receptor are resistant to apoptosis. *Mol. Cell. Biol.* **17**: 3005–3012.

Jazwinski, S.M. (1999) Molecular mechanisms of yeast longevity. *Trends Microbiol.* **7**: 247–252.

Jellinger, K.A. and Stadelmann, C.H. (2000) The enigma of cell death in neurodegenerative disorders. *J. Neural Transm.* **8** (Suppl 60): 21–36.

Jeong, E.J., Bang, S., Lee, T.H., Park, Y.I., Sim, W.S. and Kim, K.S. (1999) The solution structure of FADD death domain. Structural basis of death domain interactions of Fas and FADD. *J. Biol. Chem.* **274**: 16337–16342.

Jeremias, I., Herr, I., Boehler, T. and Debatin, K.M. (1998) TRAIL/Apo-2–ligand-induced apoptosis in human T cells. *Eur. J. Immunol.* **28**: 143–152.

Jiang, C.A., Baehrecke, E.H. and Thummel, C.S. (1997) Steroid regulated programmed cell death during *Drosophila* metamorphosis. *Development* **124**: 4673–4683.

Jiang, C.A., Lamblin, A.F.J., Steller, H. and Thummel, C.S. (2000) A steroid-triggered transcriptional hierarchy controls salivary gland cell death during *Drosophila* metamorphosis. *Mol. Cell* **5**: 445–455.

Jiang, Y., Woronicz, J.D., Liu, W. and Goeddel, D.V. (1999) Prevention of constitutive TNF receptor 1 signaling by silencer of death domains. *Science* **283**: 543–546.

Jo, M., Kim, T.H., Seol, D.W., Esplen, J.E., Dorko, K., Billiar, T.R. and Strom, S.C. (2000) Apoptosis induced in normal human hepatocytes by tumor necrosis factor-related apoptosis- inducing ligand. *Nat. Med.* **6**: 564–567.

Johnson, D.E. (2000) Noncaspase proteases in apoptosis. *Leukemia* **14**: 1695–1703.

Johnson, F.B., Sinclair, D.A. and Guarente, L. (1999) Molecular biology of aging. *Cell* **96**: 291–302.

Johnson, M.D., Kinoshita, Y., Xiang, H., Ghatan, S. and Morrison, R.S. (1999) Contribution of p53- dependent caspase activation to neuronal cell death declines with neuronal maturation. *J. Neurosci.* **19**: 2996–3006.

Jones, E.Y., Stuart, D.I. and Walker, N.P. (1989) Structure of tumour necrosis factor. *Nature* **338**: 225–228.

Jones, G., Jones, D., Zhou, L., Steller, H. and Chu, Y. (2000) Deterin, a new inhibitor of apoptosis from *Drosophila melanogaster*. *J. Biol. Chem.* **275**: 22157–22165.

Joza, N., Susin, S.A., Daugas, E., *et al.* (2001) Essential role of the mitochondrial apoptosis-inducing factor in programmed cell death. *Nature* **410**: 549–554.

Ju, S.T., Panka, D.J., Cui, H., Ettinger, R., el Khatib, M., Sherr, D.H., Stanger, B.Z. and Marshak-Rothstein, A. (1995) Fas(CD95)/FasL interactions required for programmed cell death after T-cell activation. *Nature* **373**: 444–448.

Jurgensmeier, J.M., Xie, Z., Deveraux, Q., Ellerby, L., Bredesen, D. and Reed, J.C. (1998) Bax directly induces release of cytochrome c from isolated mitochondria. *Proc. Natl Acad. Sci. U. S. A.* **95**: 4997–5002.

Kagedal, K., Zhao, M., Svensson, I. and Brunk, U.T. (2001) Sphingosine-induced apoptosis is dependent on lysosomal proteases. *Biochem. J.* **359**: 335–343.

Kaiser, W.J., Vucic, D. and Miller, L.K. (1998) The *Drosophila* inhibitor of apoptosis D-IAP1 suppresses cell death induced by the caspase drICE. *FEBS Letts.* **440**: 243–248.

Kang, J.J., Schaber, M.D., Srinivasula, S., Alnemri, E.S., Litwack, G., Hall, D.J. and Bjornsti, M.A. (1999) Cascades of mammalian caspase activation in the yeast *Saccharomyces cerevisiae*. *J. Biol. Chem.* **274**: 3189–3198.

Kannan, K., Kaminski, N., Rechavi, G., Jakob-Hirsch, J., Amariglio, N. and Givol, D. (2001) DNA microarray analysis of genes involved in p53 mediated apoptosis: activation of Apaf-1. *Oncogene* **20**: 3449–3455.

Kanuka, H., Sawamoto, K., Inohara, N., Matsuno, K., Okano, H. and Miura, M. (1999) Control of the cell death pathway by Dapaf-1, a *Drosophila* Apaf-1/CED-4- related caspase activator. *Mol. Cell* **4**: 757–769.

Karim, F.D. and Rubin, G.M. (1998) Ectopic expression of activated Ras1 induces hyperplastic growth and increased cell death in *Drosophila* imaginal tissues. *Development* **125**: 1–9.

Karin, M. and Lin, A. (2002) NF-kappaB at the crossroads of life and death. *Nat. Immunol.* **3**: 221–227.

Karin, M. and Ben Neriah, Y. (2000) Phosphorylation meets ubiquitination: the control of NF-kappa B activity. *Annu. Rev. Immunol.* **18**: 621–663.

Karpusas, M., Hsu, Y.M., Wang, J.H., Thompson, J., Lederman, S., Chess, L. and Thomas, D. (1995) A crystal structure of an extracellular fragment of human CD40 ligand. *Structure* **3**: 1426.

Karre, K., Ljunggren, H.G., Piontek, G. and Kiessling, R. (1986) Selective rejection of H-2-deficient lymphoma variants suggests alternative immune defence strategy. *Nature* **319**: 675–678.

Kashii, Y., Giorda, R., Herberman, R.B., Whiteside, T.L. and Vujanovic, N.L. (1999) Constitutive expression and role of the TNF family ligands in apoptotic killing of tumor cells by human NK cells. *J. Immunol.* **163**: 5358–5366.

Kataoka, T., Schroter, M., Hahne, M., Schneider, P., Irmler, M., Thome, M., Froelich, C.J. and Tschopp, J. (1998) FLIP prevents apoptosis induced by death receptors but not by perforin/granzyme B, chemotherapeutic drugs, and gamma irradiation. *J. Immunol.* **161**: 3936–3942.

Katz, E., Deehan, M.R., Seatter, S., Lord, C., Sturrock, R.D. and Harnett, M.M. (2001) B cell receptor-stimulated mitochondrial phospholipase A$_2$ activation and resultant disruption of mitochondrial membrane potential correlate with the induction of apoptosis in WEHI-231 B cells. *J. Immunol.* **166**: 137–147.

Kaufman, R.J. (1999) Stress signaling from the lumen of the endoplasmic reticulum: coordination of gene transcriptional and translational controls. *Genes Dev.* **13**: 1211–1233.

Kaufmann, S.H. and Earnshaw, W.C. (2000) Induction of apoptosis by cancer chemotherapy. *Exp. Cell Res.* **256**: 42–49.

Kaufmann, S.H. and Hengartner, M.O. (2001) Programmed cell death: alive and well in the new millennium. *Trends Cell Biol.* **11**: 526–534.

Kawahara, A., Ohsawa, Y., Matsumura, H., Uchiyama, Y. and Nagata, S. (1998) Caspase-independent cell killing by Fas-associated protein with death domain. *J. Cell Biol.* **143**: 1353–1360.

Kawai, T., Matsumoto, M., Takeda, K., Sanjo, H. and Akira, S. (1998) ZIP kinase, a novel serine/threonine kinase which mediates apoptosis. *Mol. Cell. Biol.* **18**: 1642–1651.

Kawai-Yamada, M., Jin, L., Yoshinaga, K., Hirata, A. and Uchimiya, H. (2001) Mammalian Bax-induced plant cell death can be down-regulated by overexpression of *Arabidopsis* Bax Inhibitor-1 (AtBI-1) *Proc. Natl Acad. Sci. U. S. A.* **98**: 12295–12300.

Kayagaki, N., Yamaguchi, N., Nakayama, M., Eto, H., Okumura, K. and Yagita, H. (1999) Type I interferons (IFNs) regulate tumor necrosis factor-related apoptosis-inducing ligand (TRAIL) expression on human T cells: A novel mechanism for the antitumor effects of type I IFNs. *J. Exp. Med.* **189**: 1451–1460.

Kelekar, A., Chang, B.S., Harlan, J.E., Fesik, S.W. and Thompson, C.B. (1997) Bad is a BH3 domain-containing protein that forms an inactivating dimer with Bcl-XL. *Mol. Cell. Biol.* **17**: 7040–7046.

Kelekar, A. and Thompson, C.B. (1998) Bcl-2–family proteins: the role of the BH3 domain in apoptosis. *Trends Cell Biol.* **8**: 324–330.

Kelliher, M.A., Grimm, S., Ishida, Y., Kuo, F., Stanger, B.Z. and Leder, P. (1998) The death domain kinase RIP mediates the TNF-induced NF-kappaB signal. *Immunity* **8**: 297–303.

Kerr, J.F.R., Winterford, C.M. and Harmon, B.V. (1994) Apoptosis. Its significance in cancer and cancer therapy. *Cancer* **73**: 2013–2026.

Kerr, J.F.R., Wyllie, A.H. and Currie, A.R. (1972) Apoptosis: a basic biological phenomenon with wide-ranging implications in tissue kinetics. *Br. J. Cancer.* **26**: 239–257.

Khaled, A.R., Kim, K., Hofmeister, R., Muegge, K., Durum, S.K., Chou, J.J., Li, H., Salvesen, G.S., Yuan, J. and Wagner, G. (1999) Withdrawal of IL-7 induces Bax translocation from cytosol to mitochondria through a rise in intracellular pH. *Proc. Natl Acad. Sci. U. S. A.* **96**: 14476–14481.

Khare, S.D., Sarosi, I., Xia, X.Z., *et al.* (2000) Severe B cell hyperplasia and autoimmune disease in TALL-1 transgenic mice. *Proc. Natl Acad. Sci. U. S. A.* **97**: 3370–3375.

Khoshnan, A., Tindell, C., Laux, I., Bae, D., Bennett, B. and Nel, A.E. (2000) The NF-kappa B cascade is important in Bcl-xL expression and for the anti-apoptotic effects of the CD28 receptor in primary human CD4+ lymphocytes. *J. Immunol.* **165**: 1743–1754.

Khwaja, A. and Tatton, L. (1999) Resistance to the cytotoxic effects of tumor necrosis factor alpha can be overcome by inhibition of a FADD/caspase-dependent signaling pathway. *J. Biol. Chem.* **274**: 36817–36823.

Kim, K., Fisher, M.J., Xu, S.Q. and el Deiry, W.S. (2000) Molecular determinants of response to TRAIL in killing of normal and cancer cells. *Clin. Cancer Res.* **6**: 335–346.

Kim, K.M., Adachi, T., Nielsen, P.J., Terashima, M., Lamers, M.C., Kohler, G. and Reth, M. (1994) Two new proteins preferentially associated with membrane immunoglobulin D. *EMBO J.* **13**: 3793–3800.

Kim, K.M., Giedt, C.D., Basañez, G., *et al.* (2001) Biophysical characterization of recombinant human Bcl-2 and its interactions with an inhibitory ligand, antimycin A. *Biochemistry* **40**: 4911–4922.

Kim, T.H., Zhao, Y., Barber, M.J., Kuharsky, D.K. and Yin, X.M. (2000) Bid-induced cytochrome c release is mediated by a pathway independent of mitochondrial permeability transition pore and Bax. *J. Biol. Chem.* **275**: 39474–39481.

Kimble, J. and Hirsh, D. (1979) The postembryonic cell lineages of the hermaphrodite and male gonads in *Caenorhabditis elegans*. *Dev. Biol.* **70**: 396–417.

Kinoshita, M., Tomimoto, H., Kinoshita, A., Kumar, S. and Noda, M. (1997) Upregulation of the Nedd2 gene encoding an ICE/Ced-3–like cysteine protease in the gerbil brain after transient global ischemia. *J. Cereb. Blood Flow Metab.* **17**: 507–514.

Kirchman, P.A., Kim, S., Lai, C.Y. and Jazwinski, S.M. (1999) Interorganelle signaling is a determinant of longevity in *Saccharomyces cerevisiae*. *Genetics* **152**: 179–190.

Kirsch, D. G. and Kastan, M.B. (1998) Tumor-suppressor p53: implications for tumor development and prognosis. *J. Clin. Oncol.* **16**: 3158–3168.

Kirsch, D.G., Doseff, A., Chau, B.N., Lim, D. S., de Souza-Pinto, N.C., Hansford, R., Kastan, M.B., Lazebnik, Y.A. and Hardwick, J.M. (1999) Caspase-3-dependent cleavage of Bcl-2 promotes release of cytochrome c. *J. Biol. Chem.* **274**: 21155–21161.

Kischkel, F.C., Hellbardt, S., Behrmann, I., Germer, M., Pawlita, M., Krammer, P.H. and Peter, M.E. (1995) Cytotoxicity-dependent APO-1 (Fas/CD95)-associated proteins form a death-inducing signaling complex (DISC) with the receptor. *EMBO J.* **14**: 5579–5588.

Kischkel, F.C., Lawrence, D.A., Chuntharapai, A., Schow, P., Kim, K.J. and Ashkenazi, A. (2000) Apo2L/TRAIL-dependent recruitment of endogenous FADD and caspase-8 to death receptors 4 and 5. *Immunity* **12**: 611–620.

Kischkel, F.C., Lawrence, D.A., Tinel, A., *et al.* (2001) Death receptor recruitment of endogenous caspase-10 and apoptosis initiation in the absence of caspase-8. *J. Biol. Chem.* **276**: 46639–46646.

Kishikawa, K., Chalfant, C.E., Perry, D.K., Bielawska, A. and Hannun, Y.A. (1999) Phosphatidic acid is a potent and selective inhibitor of protein phosphatase 1 and an inhibitor of ceramide-mediated responses. *J. Biol. Chem.* **274**: 21335–21341.

Kishimoto, H., Surh, C.D. and Sprent, J. (1998) A role for Fas in negative selection of thymocytes in vivo. *J. Exp. Med.* **187**: 1427–1438.

Kitanaka, C. and Kuchino, Y. (1999) Caspase-independent programmed cell death with necrotic morphology. *Cell Death Differ.* **6**: 508–515.

Kitson, J., Raven, T., Jiang, Y.P., Goeddel, D.V., Giles, K.M., Pun, K.T., Grinham, C.J., Brown, R. and Farrow, S.N. (1996) A death-domain-containing receptor that mediates apoptosis. *Nature* **384**: 372–375.

Kluck, R.M., Bossy-Wetzel, E., Green, D.R. and Newmeyer, D.D. (1997) The release of cytochrome c from mitochondria: a primary site for Bcl-2 regulation of apoptosis. *Science* **275**: 1132–1136.

Kluck, R.M., Esposti, M.D., Perkins, G., *et al.* (1999) The pro-apoptotic proteins, Bid and Bax, cause a limited permeabilization of the mitochondrial outer membrane that is enhanced by cytosol. *J. Cell Biol.* **147**: 809–822.

Knudson, C.M. and Korsmeyer, S.J. (1997) Bcl-2 and Bax function independently to regulate cell death. *Nat. Genet.* **16**: 358–363.

Ko, Y.G., Kang, Y.S., Park, H., Seol, W., Kim, J., Kim, T., Park, H.S., Choi, E.J. and Kim, S. (2001) Apoptosis signal-regulating kinase 1 controls the proapoptotic function of death-associated protein (Daxx) in the cytoplasm. *J. Biol. Chem.* **276**: 39103–39106.

Kobrinsky, E.M. and Kirchberger, M.A. (2001) Evidence for a role of the sarcoplasmic/endoplasmic reticulum Ca(2+)-ATPase in thapsigargin and Bcl-2 induced changes in *Xenopus laevis* oocyte maturation. *Oncogene* **20**: 933–941.

Kockx, M.M. and Herman, A.G. (2000) Apoptosis in atherosclerosis: beneficial or detrimental? *Cardiovasc. Res.* **45**: 736–746.

Kogel, D., Prehn, J.H. and Scheidtmann, K.H. (2001) The DAP kinase family of pro-apoptotic proteins: novel players in the apoptotic game. *Bioessays* 23: 352–358.

Kohler, C., Gahm, A., Noma, T., Nakazawa, A., Orrenius, S. and Zhivotovsky, B. (1999) Release of adenylate kinase 2 from the mitochondrial intermembrane space during apoptosis. *FEBS Lett.* 447: 10–12.

Komatsu, K., Hopkins, K.M., Lieberman, H.B. and Wang, H. (2000) *Schizosaccharomyces pombe* Rad9 contains a BH3–like region and interacts with the anti-apoptotic protein Bcl-2. *FEBS Lett.* 481: 122–126.

Kondo, T., Wakayama, T., Naiki, T., Matsumoto, K. and Sugimoto, K. (2001) Recruitment of Mec1 and Ddc1 checkpoint proteins to double-strand breaks through distinct mechanisms. *Science* 294: 867–870.

Koonin, E.V. and Aravind, L. (2000) The NACHT family – a new group of predicted NTPases implicated in apoptosis and MHC transcription activation. *Trends Biochem. Sci.* 25: 223–224.

Koopman, G., Reutelingsperger, C.P., Kuijten, G.A., Keehnen, R.M., Pals, S.T. and van Oers, M.H. (1994) Annexin V for flow cytometric detection of phosphatidylserine expression on B cells undergoing apoptosis. *Blood* 84: 1415–1420.

Kornitzer, D., Sharf, R. and Kleinberger, T. (2001) Adenovirus E4orf4 protein induces PP2A-dependent growth arrest in *Saccharomyces cerevisiae* and interacts with the anaphase-promoting complex/cyclosome. *J. Cell Biol.* 154: 331–344.

Korsmeyer, S.J., Shutter, J.R., Veis, D.J., Merry, D.E. and Oltvai, Z.N. (1993) Bcl-2/Bax: a rheostat that regulates an anti-oxidant pathway and cell death. *Semin. Cancer Biol.* 4: 327–332.

Korsmeyer, S.J., Wei, M.C., Saito, M., Weiler, S., Oh, K.J. and Schlesinger, P.H. (2000) Pro-apoptotic cascade activates BID, which oligomerizes BAK or BAX into pores that result in the release of cytochrome c. *Cell Death Differ.* 7: 1166–1173.

Kothakota, S., Azuma, T., Reinhard, C., *et al.* (1997) Caspase-3 generated fragment of gelsolin: effector of morphological change in apoptosis. *Science* 278: 294–298.

Krajewski, S., Blomqvist, C., Franssila, K., Krajewska, M., Wasenius, V.M., Niskanen, E. and Nordling, S. and Reed, J. C. (1995a) Reduced expression of proapoptotic gene BAX is associated with poor response rates to combination chemotherapy and shorter survival in women with metastatic breast adenocarcinoma. *Cancer Res.* 55: 4471–4478.

Krajewski, S., Mai, J.K., Krajewska, M., Sikorska, M., Mossakowski, M.J. and Reed, J.C. (1995b) Upregulation of bax protein levels in neurons following cerebral ischemia. *J. Neurosci.* 15: 6364–6376.

Krajewski, S., Tanaka, S., Takayama, S., Schibler, M.J., Fenton, W. and Reed, J.C. (1993) Investigation of the subcellular distribution of the bcl-2 oncoprotein: residence in the nuclear envelope, endoplasmic reticulum, and outer mitochondrial membranes. *Cancer Res.* 53: 4701–4714.

Krammer, P.H. (2000) CD95's deadly mission in the immune system. *Nature* 407: 789–795.

Krane, I.M. and Leder, P. (1996) NDF/heregulin induces persistence of terminal end buds and adenocarcinomas in the mammary glands of transgenic mice. *Oncogene* 12: 1781–1788.

Kreuz, S., Siegmund, D., Scheurich, P. and Wajant, H. (2001) NF-kappaB inducers upregulate cFLIP, a cycloheximide-sensitive inhibitor of death receptor signaling. *Mol. Cell. Biol.* 21: 3964–3973.

Kroemer, G. (1999) Mitochondrial control of apoptosis: an overview. *Biochem. Soc. Symp.* 66: 1–15.

Krogh, A., Brown, M., Mian, I.S., Sjolander, K. and Haussler, D. (1994) Hidden Markov models in computational biology. Applications to protein modeling. *J. Mol. Biol.* 235: 1501–1531.

Kruman, I.I. and Mattson, M.P. (1999) Pivotal role of mitochondrial calcium uptake in neural cell apoptosis and necrosis. *J. Neurochem.* 72: 529–540.

Kudla, G., Montessuit, S., Eskes, R., Berrier, C., Martinou, J.C., Ghazi, A. and Antonsson, B. (2000) The destabilization of lipid membranes induced by the C-terminal fragment of caspase 8-cleaved bid is inhibited by the N-terminal fragment. *J. Biol. Chem.* 275: 22713–22718.

Kuhlmann, T., Lucchinetti, C., Zettl, U.K., Bitsch, A., Lassmann, H. and Bruck, W. (1999) Bcl-2-expressing oligodendrocytes in multiple sclerosis lesions. *Glia* 28: 34–39.

Kuida, K., Zheng, T.S., Na, S., Kuan, C., Yang, D., Karasuyama, H., Rakic, P. and Flavell, R.A. (1996) Decreased apoptosis in the brain and premature lethality in CPP32-deficient mice. *Nature* 384: 368–372.

Kuida, K., Haydar, T.F., Kuan, C.Y., Gu, Y., Taya, C., Karasuyama, H., Su, M.S.S., Rakic, P. and Flavell, R.A. (1998) Reduced apoptosis and cytochrome c-mediated caspase activation in mice lacking caspase 9. *Cell* 94: 325–337.

Kumar, S. (1999) Mechanisms mediating caspase activation in cell death. *Cell Death Differ.* 6: 1060–1066.

Kumar, S. and Colussi, P.A. (1999) Prodomains–adaptors–oligomerization: the pursuit of caspase activation in apoptosis. *Trends Biochem. Sci.* 24: 1–4.

Kumar, S. and Doumanis, J. (2000) The fly caspases. *Cell Death Differ.* 7: 1039–1044.

Kumar, S., Kinoshita, M. and Noda, M. (1997) Origin, expression and possible functions of the two alternatively spliced forms of the mouse Nedd2 mRNA. *Cell Death Differ.* 4: 378–387.

Kumar, S., Kinoshita, M., Noda, M., Copeland, N.G. and Jenkins, N.A. (1994) Induction of apoptosis by the mouse Nedd2 gene, which encodes a protein similar to the product of the *Caenorhabditis elegans* cell death gene ced-3 and the mammalian IL-1-converting enzyme. *Genes Dev.* 8: 1613–1626.

Kumar, S. and Lavin, M.F. (1996) The ICE family of cysteine proteases as effectors of cell death. *Cell Death Differ.* 3: 255–267.

Kunstle, G., Hentze, H., Germann, P.G., Tiegs, G., Meergans, T. and Wendel, A. (1999) Concanavalin A hepatotoxicity in mice: tumor necrosis factor-mediated organ failure independent of caspase-3-like protease activation. *Hepatology* 30: 1241–1251.

Kuo, T.H., Kim, H.R., Zhu, L., Yu, Y., Lin, H.M. and Tsang, W. (1998) Modulation of endoplasmic reticulum calcium pump by Bcl-2. *Oncogene* 17: 1903–1910.

Kurada, P. and White, K. (1998) Ras promotes cell survival in *Drosophila* by down-regulating hid expression. *Cell* **95**: 319–329.

Kuznetsov, S., Lyanguzowa, M. and Bosch, C.G. (2001) Role of epithelial cells and programmed cell death in *Hydra* spermatogenesis. *Zoology* **104**: 25–31.

Lacana, E. and D'Adamio, L. (1999) Regulation of Fas ligand expression and cell death by apoptosis-linked gene 4. *Nat. Med.* **5**: 542–547.

Lacana, E., Ganjei, J.K., Vito, P. and D'Adamio, L. (1997) Dissociation of apoptosis and activation of IL-1beta-converting enzyme/Ced-3 proteases by ALG-2 and the truncated Alzheimer's gene ALG-3. *J. Immunol.* **158**: 5129–5135.

Lagresle, C., Mondiere, P., Bella, C., Krammer, P.H. and Defrance, T. (1996) Concurrent engagement of CD40 and the antigen receptor protects naive and memory human B cells from APO-1/Fas-mediated apoptosis. *J. Exp. Med.* **183**: 1377–1388.

Lakowski, B. and Hekimi, S. (1996) Determination of life span in *Caenorhabditis elegans* by four clock genes. *Science* **272**: 1010–1013.

Lam, K.P. (1998) Rapid elimination of mature autoreactive B cells demonstrated by Cre-induced change in B cell antigen receptor specificity in vivo. *Proc. Natl Acad. Sci. U. S. A.* **95**: 13171–13175.

Lam, M., Dubyak, G., Chen, L., Nunez, G., Miesfeld, R.L. and Distelhorst, C.W. (1994) Evidence that BCL-2 represses apoptosis by regulating endoplasmic reticulum-associated Ca^{2+} fluxes. *Proc. Natl Acad. Sci. U. S. A.* **91**: 6569–6573.

Lambert, G., Becker, B., Schreiber, R., Boucherot, A., Reth, M. and Kunzelmann, K. (2001) Control of cystic fibrosis transmembrane conductance regulator expression by BAP31. *J. Biol. Chem.* **276**: 20340–20345.

Lamkanfi, M., Declercq, W., Kalai, M., Saelens, X. and Vandenabeele, P. (2002) Alice in caspase land. A phylogenetic analysis of caspases from worm to man. *Cell Death Differ.* **9**: 358–361.

Landschulz, W.H., Johnson, P.F. and McKnight, S.L. (1988) The leucine zipper: a hypothetical structure common to a new class of DNA binding proteins. *Science* **240**: 1759–1764.

Lankiewicz, S., Marc Luetjens, C., Truc Bui, N., Krohn, A.J., Poppe, M., Cole, G.M., Saido, T.C. and Prehn, J.H. (2000) Activation of calpain I converts excitotoxic neuron death into a caspase-independent cell death. *J. Biol. Chem.* **275**: 17064–17071.

Larisch, S., Yi, Y., Lotan, R., *et al.* (2000) A novel mitochondrial septin-like protein, ARTS, mediates apoptosis dependent on its P-loop motif. *Nat. Cell Biol.* **2**: 915–921.

Laster, S.M., Wood, J.G. and Gooding, L.R. (1988) Tumor necrosis factor can induce both apoptic and necrotic forms of cell lysis. *J. Immunol.* **141**: 2629–2634.

Latterich, M., Fröhlich, K.U. and Schekman, R. (1995) Membrane fusion and the cell cycle: Cdc48p participates in the fusion of ER membranes. *Cell* **82**: 885–893.

Laun, P., Pichova, A., Madeo, F., Fuchs, J., Ellinger, A., Kohlwein, S., Dawes, I., Fröhlich, K.U. and Breitenbach, M. (2001) Aged mother cells of *Saccharomyces cerevisiae* show markers of oxidative stress and apoptosis. *Mol. Microbiol.* **39**: 1166–1173.

Lavoie, J.N., Champagne, C., Gingras, M.C. and Robert, A. (2000) Adenovirus E4 open reading frame 4-induced apoptosis involves dysregulation of Src family kinases. *J. Cell Biol.* **150**: 1037–1056.

Lavoie, J.N., Nguyen, M., Marcellus, R.C., Branton, P.E. and Shore, G.C. (1998) E4orf4, a novel adenovirus death factor that induces p53-independent apoptosis by a pathway that is not inhibited by zVAD-fmk. *J. Cell Biol.* **140**: 637–645.

Lawrence, D., Shahrokh, Z., Marsters, S., *et al.* (2001) Differential hepatocyte toxicity of recombinant Apo2L/TRAIL versions. *Nat. Med.* **7**: 383–385.

LeBlanc, H., Lawrence, D., Varfolomeev, E., *et al.* (2002) Tumor-cell resistance to death receptor-induced apoptosis through mutational inactivation of the proapoptotic Bcl-2 homolog Bax. *Nat. Med.* **8**: 274–281.

Lee, C.Y., Wendel, D.P., Reid, P., Lam, G., Thummel, C.S. and Baehrecke, E.H. (2000) E93 directs steroid-triggered programmed cell death in *Drosophila*. *Mol. Cell* **6**: 433–443.

Lee, L.C., Hunter, J.J., Mujeeb, A., Turck, C. and Parslow, T.G. (1996) Evidence for alpha-helical conformation of an essential N-terminal region in the human Bcl2 protein. *J. Biol. Chem.* **271**: 23284–23288.

Lee, R., Chen, J., Matthews, C.P., McDougall, J.K. and Neiman, P.E. (2001) Characterization of NR13-related human cell death regulator, Boo/Diva, in normal and cancer tissues. *Biochim. Biophys. Acta* **1520**: 187–194.

Lee, S.T., Hoeflich, K.P., Wasfy, G.W., Woodgett, J.R., Leber, B., Andrews, D.W., Hedley, D.W. and Penn, L.Z. (1999) Bcl-2 targeted to the endoplasmic reticulum can inhibit apoptosis induced by Myc but not etoposide in Rat-1 fibroblasts. *Oncogene* **18**: 3520–3528.

Lee, S.Y., Reichlin, A., Santana, A., Sokol, K.A., Nussenzweig, M.C. and Choi, Y. (1997) TRAF2 is essential for JNK but not NF-kappaB activation and regulates lymphocyte proliferation and survival. *Immunity* **7**: 703–713.

Leist, M. and Jäättelä, M. (2001a) Four deaths and a funeral: from caspases to alternative mechanisms. *Nat. Rev. Mol. Cell. Biol.* **2**: 589–598.

Leist, M. and Jäättelä, M. (2001b) Triggering of apoptosis by cathepsins. *Cell Death Differ.* **8**: 324–326.

Leist, M., Single, B., Castoldi, A.F., Kuhnle, S. and Nicotera, P. (1997) Intracellular adenosine triphosphate (ATP) concentration: a switch in the decision between apoptosis and necrosis. *J. Exp. Med.* **185**: 1481–1486.

Leist, M., Single, B., Naumann, H., Fava, E., Simon, B., Kuhnle, S. and Nicotera, P. (1999) Inhibition of mitochondrial ATP generation by nitric oxide switches apoptosis to necrosis. *Exp. Cell Res.* **249**: 396–403.

Leist, M., Volbracht, C., Fava, E. and Nicotera, P. (1998) 1-Methyl-4-phenylpyridinium induces autocrine excitotoxicity, protease activation, and neuronal apoptosis. *Mol. Pharmacol.* **54**: 789–801.

Leithauser, F., Dhein, J., Mechtersheimer, G., *et al.* (1993) Constitutive and induced expression of APO-1, a new member of the nerve growth factor/tumor necrosis factor receptor superfamily, in normal and neoplastic cells. *Lab. Invest.* **69**: 415–429.

Leo, E., Deveraux, Q.L., Buchholtz, C., Welsh, K., Matsuzawa, S., Stennicke, H.R.,

Salvesen, G.S. and Reed, J.C. (2001) TRAF1 is a substrate of caspases activated during tumor necrosis factor receptor-alpha-induced apoptosis. *J. Biol. Chem.* **276**: 8087–8093.

Leulier, F., Rodriguez, A., Khush, R.S., Abrams, J.M. and Lemaitre, B. (2000) The *Drosophila* caspase Dredd is required to resist gram-negative bacterial infection. *EMBO Rep.* **1**: 353–358.

Levine, A., Belenghi, B., Damari-Weisler, H. and Granot, D. (2001) Vesicle associated membrane protein of *Arabidopsis* suppresses Bax-induced apoptosis in yeast downstream of oxidative burst. *J. Biol. Chem.* **276**: 46284–46289.

Levkau, B., Scatena, M., Giachelli, C.M., Ross, R. and Raines, E.W. (1999) Apoptosis overrides survival signals through a caspase-mediated dominant-negative NF-kappa B loop. *Nat. Cell Biol.* **1**: 227–233.

Levy-Strumpf, N. and Kimchi, A. (1998) Death associated proteins (DAPs): from gene identification to the analysis of their apoptotic and tumor suppressive functions. *Oncogene* **17**: 3331–3340.

Li, C.J., Friedman, D.J., Wang, C., Metelev, V. and Pardee, A.B. (1995) Induction of apoptosis in uninfected lymphocytes by HIV-1 Tat protein. *Science* **268**: 429–431.

Li, H., Kolluri, S.K., Gu, J., *et al.* (2000) Cytochrome c release and apoptosis induced by mitochondrial targeting of nuclear orphan receptor TR3. *Science* **289**: 1159–1164.

Li, H., Zhu, H., Xu, C.J., *et al.* (1998) Cleavage of BID by caspase 8 mediates the mitochondrial damage in the Fas pathway of apoptosis. *Cell* **94**: 491–501.

Li, J.H., Rosen, D., Ronen, D., Behrens, C.K., Krammer, P.H., Clark, W.R. and Berke, G. (1998) The regulation of CD95 ligand expression and function in CTL. *J. Immunol.* **161**: 3943–3949.

Li, L.Y., Luo, X. and Wang, X. (2001) Endonuclease G is an apoptotic DNase when released from mitochondria. *Nature* **412**: 95–99.

Li, P., Nijhawan, D., Budihardjo, I., Srinivasula, S.M., Ahmad, M., Alnemri, E.S. and Wang, X. (1997) Cytochrome c and dATP-dependent formation of Apaf1/caspase-9 complex initiates the apoptotic protease cascade. *Cell* **91**: 479–489.

Li, Q., Pazdera, T. and Minden, J. (1999a) *Drosophila* embryonic pattern repair: how embryos respond to cyclin E-induced ectopic division. *Development* **126**: 2299–2307.

Li, Q., Van Antwerp, D., Mercurio, F., Lee, K.F. and Verma, I.M. (1999b) Severe liver degeneration in mice lacking the IkappaB kinase 2 gene. *Science* **284**: 321–325.

Li, Q.X., Evin, G., Small, D.H., Multhaup, G., Beyreuther, K. and Masters, C.L. (1995) Proteolytic processing of Alzheimer's disease beta A4 amyloid precursor protein in human platelets. *J. Biol. Chem.* **270**: 14140–14147.

Li, Z.W., Chu, W., Hu, Y., Delhase, M., Deerinck, T., Ellisman, M., Johnson, R. and Karin, M. (1999) The IKKbeta subunit of IkappaB kinase (IKK) is essential for nuclear factor kappaB activation and prevention of apoptosis. *J. Exp. Med.* **189**: 1839–1845.

Liang, H. and Fesik, S.W. (1997) Three-dimensional structures of proteins involved in programmed cell death. *J. Mol. Biol.* **274**: 291–302.

Liepinsh, E., Ilag, L.L., Otting, G. and Ibanez, C.F. (1997) NMR structure of the death domain of the p75 neurotrophin receptor. *EMBO J.* **16**: 4999–5005.

Ligr, M., Madeo, F., Fröhlich, E., Hilt, W., Fröhlich, K.U. and Wolf, D.H. (1998) Mammalian Bax triggers apoptotic changes in yeast. *FEBS Lett.* **438**: 61–65.

Ligr, M., Velten, I., Fröhlich, E., Madeo, F., Ledig, M., Fröhlich, K.U., Wolf, D.H. and Hilt, W. (2001) The proteasomal substrate Stm1 participates in apoptosis-like cell death in yeast. *Mol. Biol. Cell* **12**: 2422–2432.

Lim, C.S., Mian, I.S., Dernburg, A.F. and Campisi, J. (2001) *C. elegans* clk-2, a gene that limits life span, encodes a telomere length regulator similar to yeast telomere binding protein Tel2p. *Curr. Biol.* **11**: 1706–1710.

Lin, Y., Devin, A., Rodriguez, Y. and Liu, Z.G. (1999) Cleavage of the death domain kinase RIP by caspase-8 prompts TNF-induced apoptosis. *Genes Dev.* **13**: 2514–2526.

Lind, E.F., Wayne, J., Wang, Q.Z., *et al.* (1999) Bcl-2–induced changes in E2F regulatory complexes reveal the potential for integrated cell cycle and cell death functions. *J. Immunol.* **162**: 5374–5379.

Lindenboim, L., Yuan, J. and Stein, R. (2000) Bcl-xS and Bax induce different apoptotic pathways in PC12 cells. *Oncogene* **19**: 1783–1793.

Lindsten, T., Ross, A.J., King, A., *et al.* (2000) The combined functions of proapoptotic Bcl-2 family members bak and bax are essential for normal development of multiple tissues. *Mol. Cell* **6**: 1389–1399.

Linette, G.P., Li, Y., Roth, K., Korsmeyer, S.J., Chou, J.J., Li, H., Salvesen, G.S., Yuan, J. and Wagner, G. (1996) Cross talk between cell death and cell cycle progression: BCL-2 regulates NFAT-mediated activation. *Proc. Natl Acad. Sci. U. S. A.* **93**: 9545–9552.

Lisi, S., Mazzon, I. and White, K. (2000) Diverse domains of THREAD/DIAP1 are required to inhibit apoptosis induced by REAPER and HID in *Drosophila*. *Genetics* **154**: 669–678.

Liston, P., Roy, N., Tamai, K., *et al.* (1996) Suppression of apoptosis in mammalian cells by NAIP and a related family of IAP genes. *Nature* **379**: 349–353.

Liu, Q.A. and Hengartner, M.O. (1999a) The molecular mechanism of programmed cell death in *C. elegans. Ann. N. Y. Acad. Sci.* **887**: 92–104.

Liu, Q.A. and Hengartner, M.O. (1999b) Human CED-6 encodes a functional homologue of the *Caenorhabditis elegans* engulfment protein CED-6. *Curr. Biol.* **9**: 1347–1350.

Liu, Q.A. and Hengartner, M.O. (1998) Candidate adaptor protein CED-6 promotes the engulfment of apoptotic cells in *C. elegans. Cell* **93**: 961–972.

Liu, X., Kim, C.N., Yang, J., Jemmerson, R. and Wang, X. (1996) Induction of apoptotic program in cell-free extracts: requirement for dATP and cytochrome c. *Cell* **86**: 147–157.

Liu, X., Li, P., Widlak, P., Zou, H., Luo, X., Garrard, T.W. and Wang, X. (1998) The 40-kDa subunit of DNA fragmentation factor induces DNA fragmentation and chromatin condensation during apoptosis. *Proc. Natl Acad. Sci. U. S. A.* **95**: 8461–8466.

Liu, X., Zou, H., Slaughter, C. and Wang, X. (1997) DFF, a heterodimeric protein that functions downstream of caspase-3 to trigger DNA fragmentation during apoptosis. *Cell* **89**: 175–184.

Liu, Z., Sun, C., Olejniczak, E.T., Meadows, R.P., Betz, S.F., Oost, T., Herrmann, J., Wu, J.C. and Fesik, S.W. (2000) Structural basis for binding of Smac/DIABLO to the XIAP BIR3 domain. *Nature* **408**: 1004–1008.

Liu, Z.G., Hsu, H., Goeddel, D.V. and Karin, M. (1996) Dissection of TNF receptor 1 effector functions: JNK activation is not linked to apoptosis while. *Cell* **87**: 565–576.

Lockshin, R.A. (1969) Programmed cell death. Activation of lysis by a mechanism involving the synthesis of protein. *J. Insect Physiol.* **15**: 1505–1516.

Lockshin, R.A. and Zakeri, Z. (2001) Programmed cell death and apoptosis: origins of the theory. *Nat. Rev. Mol. Cell. Biol.* **2**: 545–550.

Locksley, R.M., Killeen, N. and Lenardo, M.J. (2001) The TNF and TNF receptor superfamilies: integrating mammalian biology. *Cell* **104**: 487–501.

Lombardi, L., Frigerio, S., Collini, P. and Pilotti, S. (1997) Immunocytochemical and immunoelectron microscopical analysis of BCL2 expression in thyroid oxyphilic tumors. *Ultrastruct. Pathol.* **21**: 33–39.

Longo, V.D., Ellerby, L.M., Bredesen, D.E., Valentine, J.S. and Gralla, E.B. (1997) Human bcl-2 reverses survival defects in yeast lacking superoxide dismutase and delays death of wild type yeast. *J. Cell Biol.* **137**: 1581–1588.

Los, M., Stroh, C., Janicke, R.U., Engels, I.H. and Schulze-Osthoff, K. (2001) Caspases: more than just killers? *Trends Immunol.* **22**: 31–34.

Los, M., Wesselborg, S. and Schulze-Osthoff, K. (1999) The role of caspases in development, immunity, and apoptotic signal transduction: lessons from knockout mice. *Immunity* **10**: 629–639.

Losonczi, J.A., Olejniczak, E.T., Betz, S.F., Harlan, J.E., Mack, J. and Fesik, S.W. (2000) NMR studies of the anti-apoptotic protein Bcl-xL in micelles. *Biochemistry (Mosc.)* **39**: 11024–11033.

Lovering, R., Hanson, I.M., Borden, K.L., *et al.* (1993) Identification and preliminary characterization of a protein motif related to the zinc finger. *Proc. Natl Acad. Sci. U. S. A.* **90**: 2112–2116.

Lowary, P.T. and Widom, J. (1989) Higher-order structure of *Saccharomyces cerevisiae* chromatin. *Proc. Natl Acad. Sci. U. S. A.* **86**: 8266–8270.

Lu, E. and Wolfe, J. (2001) Lysosomal enzymes in the macronucleus of *Tetrahymena* during its apoptosis-like degradation. *Cell Death Differ.* **8**: 289–297.

Luciani, M.F. and Chimini, G. (1996) The ATP binding cassette transporter, ABC1, is required for the engulfment of corpses generated by apoptotic cell death. *EMBO J.* **15**: 226–235.

Ludovico, P., Sousa, M.J., Silva, M.T., Leao, C. and Corte-Real, M. (2001) *Saccharomyces cerevisiae* commits to a programmed cell death process in response to acetic acid. *Microbiology* **147**: 2409–2415.

Luetjens, C.M., Bui, N.T., Sengpiel, B., Munstermann, G., Poppe, M., Krohn, A.J., Bauerbach, E., Krieglstein, J. and Prehn, J.H. (2000) Delayed mitochondrial dysfunction in excitotoxic neuron death: cytochrome c release and a secondary increase in superoxide production. *J. Neurosci.* **20**: 5715–5723.

Luo, X., Budihardjo, I., Zou, H., *et al.* (1998) Bid, a Bcl2 interacting protein, mediates cytochrome c release from mitochondria in response to activation of cell surface death receptors. *Cell* **94**: 481–490.

Luschen, S., Ussat, S., Scherer, G., Kabelitz, D. and Adam-Klages, S. (2000) Sensitization to death receptor cytotoxicity by inhibition of FADD/caspase signaling: requirement of cell cycle progression. *J. Biol. Chem.* **275**: 24670–24678.

Lutter, M., Fang, M., Luo, X., Nishijima, M., Xie, X. and Wang, X. (2000) Cardiolipin provides specificity for targeting of tBid to mitochondria. *Nat. Cell Biol.* **2**: 754–761.

Lyon, C.J., Evans, C.J., Bill, B.R., Otsuka, A.J. and Aguilera, R.J. (2000) The *C. elegans* apoptotic nuclease NUC-1 is related in sequence and activity to mammalian DNase II. *Gene* **252**: 147–154.

Ma, Y. and Hendershot, L.M. (2001) The unfolding tale of the unfolded protein response. *Cell* **107**: 827–830.

MacCorkle, R.A., Freeman, K.W. and Spencer, D.M. (1998) Synthetic activation of caspases: artificial death switches. *Proc. Natl Acad. Sci. U. S. A.* **95**: 3655–3660.

MacDonald, G., Shi, L., Vande, V.C., Lieberman, J. and Greenberg, A.H. (1999) Mitochondria-dependent and -independent regulation of granzyme B-induced apoptosis. *J. Exp. Med.* **189**: 131–144.

MacFarlane, M., Ahmad, M., Srinivasula, S.M., Fernandes-Alnemri, T., Cohen, G.M. and Alnemri, E.S. (1997) Identification and molecular cloning of two novel receptors for the cytotoxic ligand TRAIL. *J. Biol. Chem.* **272**: 25417–25420.

Mackay, F., Woodcock, S.A., Lawton, P., Ambrose, C., Baetscher, M., Schneider, P., Tschopp, J. and Browning, J.L. (1999) Mice transgenic for BAFF develop lymphocytic disorders along with autoimmune manifestations. *J. Exp. Med.* **190**: 1697–1710.

Madeo, F., Fröhlich, E. and Fröhlich, K.U. (1997) A yeast mutant showing diagnostic markers of early and late apoptosis. *J. Cell Biol.* **139**: 729–734.

Madeo, F., Fröhlich, E., Ligr, M., Grey, M., Sigrist, S.J., Wolf, D.H. and Fröhlich, K.U. (1999) Oxygen stress: a regulator of apoptosis in yeast. *J. Cell Biol.* **145**: 757–767.

Madrid, L.V., Wang, C.Y., Guttridge, D.C., Schottelius, A.J., Baldwin, A.S., Jr. and Mayo, M.W. (2000) Akt suppresses apoptosis by stimulating the transactivation potential of the RelA/p65 subunit of. *Mol. Cell. Biol.* **20**: 1626–1638.

Mahajan, N.P., Linder, K., Berry, G., Gordon, G.W., Heim, R. and Herman, B. (1998) Bcl-2 and Bax interactions in mitochondria probed with green fluorescent protein and fluorescence resonance energy transfer. *Nat. Biotechnol.* **16**: 547–552.

Makin, G.W., Corfe, B.M., Griffiths, G.J., Thistlethwaite, A., Hickman, J.A. and Dive, C. (2001) Damage-induced Bax N-terminal change, translocation to mitochondria and formation of Bax dimers/complexes occur regardless of cell fate. *EMBO J.* **20**: 6306–6315.

Malinin, N.L., Boldin, M.P., Kovalenko, A.V. and Wallach, D. (1997) MAP3K-related kinase involved in NF-kappaB induction by TNF, CD95 and IL-1. *Nature* **385**: 540–544.

Manon, S., Chaudhuri, B. and Guerin, M. (1997) Release of cytochrome c and decrease of cytochrome c oxidase in Bax-expressing yeast cells, and prevention of these effects by coexpression of Bcl-xL. *FEBS Lett.* **415**: 29–32.

Manon, S., Priault, M. and Camougrand, N. (2001) Mitochondrial AAAtype protease Yme1p is involved in Bax effects on cytochrome c oxidase. *Biochem. Biophys. Res. Commun.* **289**: 1314–1319.

Marsters, S.A., Pitti, R.A., Sheridan, J.P. and Ashkenazi, A. (1999) Control of apoptosis signaling by Apo2 ligand. *Recent Prog. Horm. Res.* **54**: 225–234.

Marsters, S.A., Sheridan, J.P., Donahue, C.J., Pitti, R.M., Gray, C.L., Goddard, A.D., Bauer, K.D. and Ashkenazi, A. (1996) Apo-3, a new member of the tumor necrosis factor receptor family, contains a death domain and activates apoptosis and NF-kappa B. *Curr. Biol.* **6**: 1669–1676.

Marsters, S.A., Sheridan, J.P., Pitti, R.M., Brush, J., Goddard, A. and Ashkenazi, A. (1998) Identification of a ligand for the death-domain-containing receptor Apo3. *Curr. Biol.* **8**: 525–528.

Marsters, S.A., Sheridan, J.P., Pitti, R.M., *et al.* (1997) A novel receptor for Apo2L/TRAIL contains a truncated death domain. *Curr. Biol.* **7**: 1003–1006.

Martin, D.A., Siegel, R.M., Zheng, L. and Lenardo, M.J. (1998) Membrane oligomerization and cleavage activates the caspase-8 (FLICE/MACHalpha1) death signal. *J. Biol. Chem.* **273**: 4345–4349.

Martin, L.M., Iaccarino, I., Tenev, T., *et al.* (2002) The serine protease Omi/HtrA2 regulates apoptosis by binding XUIAP through a Reaper-like motif. *J. Biol. Chem.* **277**: 439–444.

Martinez-Lorenzo, M.J., Alava, M.A., Gamen, S., Kim, K.J., Chuntharapai, A., Pineiro, A., Naval, J. and Anel, A. (1998) Involvement of APO2 ligand/TRAIL in activation-induced death of Jurkat and human peripheral blood T cells. *Eur. J. Immunol.* **28**: 2714–2725.

Martinon, F., Hofmann, K. and Tschopp, J. (2001) The pyrin domain: a possible member of the death domain-fold family implicated in apoptosis and inflammation. *Curr. Biol.* **11**: R118–120.

Martinou, I., Desagher, S., Eskes, R., Antonsson, B., Andre, E., Fakan, S. and Martinou, J.C. (1999) The release of cytochrome c from mitochondria during apoptosis of NGF-deprived sympathetic neurons is a reversible event. *J. Cell Biol.* **144**: 883–889.

Martinou, J.C. and Green, D.R. (2001) Breaking the mitochondrial barrier. *Nat. Rev. Mol. Cell. Biol.* **2**: 63–67.

Martins, L.M., Iaccarino, I., Tenev, T., *et al.* (2002) The serine protease Omi/HtrA2 regulates apoptosis by binding XIAP through a reaper-like motif. *J. Biol. Chem.* **277**: 439–444.

Maruyama, K., Usami, M., Kametani, F., Tomita, T., Iwatsubo, T., Saido, T.C., Mori, H. and Ishiura, S. (2000) Molecular interactions between presenilin and calpain: inhibition of m-calpain protease activity by presenilin-1, 2 and cleavage of presenilin-1 by m , mu calpain. *Int. J. Mol. Med.* **5**: 269–273.

Marzo, I., Brenner, C., Zamzami, N., *et al.* (1998a) Bax and adenine nucleotide translocator cooperate in the mitochondrial control of apoptosis. *Science* **281**: 2027–2031.

Marzo, I., Brenner, C., Zamzami, N., *et al.* (1998b) The permeability transition pore complex: a target for apoptosis regulation by caspases and bcl-2-related proteins. *J. Exp. Med.* **187**: 1261–1271.

Maser, R.S., Antoku, K., Scully, W.J., Cho, R.L. and Johnson, D.E. (2000) Analysis of the role of conserved cysteine residues in the Bcl-2 oncoprotein. *Biochem. Biophys. Res. Commun.* **277**: 171–178.

Masters, J.R. (2000) Human cancer cell lines: fact and fantasy. *Nat. Rev. Mol. Cell. Biol.* **1**: 233–236.

Mateo, V., Lagneaux, L., Bron, D., Biron, G., Armant, M., Delespesse, G. and Sarfati, M. (1999) CD47 ligation induces caspase-independent cell death in chronic lymphocytic leukemia. *Nat. Med.* **5**: 1277–1284.

Mathai, J.P., Germain, M., Marcellus, R.C. and Shore, G.C. (2002) Induction and endoplasmic reticulum localization of proapoptotic BIK/NBK in response to apoptotic signaling by E1A and p53. *Oncogene* **21**: 2534–2544.

Mathiasen, I.S., Lademann, U. and Jäättelä, M. (1999) Apoptosis induced by vitamin D compounds in breast cancer cells is inhibited by Bcl-2 but does not involve known caspases or p53. *Cancer Res.* **59**: 4848–4856.

Matsuda, M., Mayer, B.J. and Hanafusa, H. (1991) Identification of domains of the v-crk oncogene product sufficient for association with phosphotyrosine-containing proteins. *Mol. Cell. Biol.* **11**: 1607–1613.

Matsumura, H., Shimizu, Y., Ohsawa, Y., Kawahara, A., Uchiyama, Y. and Nagata, S. (2000) Necrotic death pathway in fas receptor signaling. *J. Cell Biol.* **151**: 1247–1256.

Matsuyama, S., Llopis, J., Deveraux, Q.L., Tsien, R.Y. and Reed, J.C. (2000) Changes in intramitochondrial and cytosolic pH: early events that modulate caspase activation during apoptosis. *Nat. Cell Biol.* **2**: 318–325.

Matsuyama, S., Xu, Q., Velours, J. and Reed, J.C. (1998) The mitochondrial F0F1-ATPase proton pump is required for function of the proapoptotic protein Bax in yeast and mammalian cells. *Mol. Cell* **1**: 327–336.

Mattson, M.P. (2000) Apoptosis in neurodegenerative disorders. *Nat. Rev. Mol. Cell. Biol.* **1**: 120–129.

Mattson, M.P., Chan, S.L. and Camandola, S. (2001) Presenilin mutations and calcium signaling defects in the nervous and immune systems. *Bioessays* **23**: 733–744.

Mattson, M.P., Duan, W., Chan, S.L. and Camandola, S. (1999) Par-4: an emerging pivotal player in neuronal apoptosis and neurodegenerative disorders. *J. Mol. Neurosci.* **13**: 17–30.

May, R.C. and Machesky, L.M. (2001) Phagocytosis and the actin cytoskeleton. *J. Cell Sci.* **114**: 1061–1077.

Mazel, S., Burtrum, D., Petrie, H.T., Chou, J.J., Li, H., Salvesen, G.S., Yuan, J. and Wagner, G. (1996) Regulation of cell division cycle progression by bcl-2 expression: a potential mechanism for inhibition of programmed cell death. *J. Exp. Med.* **183**: 2219–2226.

McCall, K. and Steller, H. (1998) Requirement for DCP-1 caspase during *Drosophila* oogenesis. *Science* **279**: 230–234.

McCarthy, J.V. and Dixit, V.M. (1998) Apoptosis induced by *Drosophila* reaper and grim in a human system. Attenuation by 10 inhibitor of apoptosis proteins (cIAPs) *J. Biol. Chem.* **273**: 24009–24015.

McCarthy, N.J., Whyte, M.K.B., Gilbert, C.S. and Evan, G.I. (1997) Inhibition of Ced-3/ICE-related proteases does not prevent cell death induced by oncogenes, DNA damage, or the Bcl-2 homologue Bak. *J. Cell Biol.* **136**: 215–227.

McCormick, T.S., McColl, K.S. and Distelhorst, C.W. (1997) Mouse lymphoma cells destined to undergo apoptosis in response to thapsigargin treatment fail to generate a calcium-mediated grp78/grp94 stress response. *J. Biol. Chem.* **272**: 6087–6092.

McCullough, K.D., Martindale, J.L., Klotz, L.O., Aw, T.Y. and Holbrook, N.J. (2001) Gadd153 sensitizes cells to endoplasmic reticulum stress by down-regulating Bcl2 and perturbing the cellular redox state. *Mol. Cell. Biol.* **21**: 1249–1259.

McDonnell, J. M., Fushman, D., Milliman, C. L., Korsmeyer, S. J. and Cowburn, D. (1999) Solution structure of the proapoptotic molecule BID: a structural basis for apoptotic agonists and antagonists. *Cell.* **96**: 625–634.

McGinnis, K.M., Gnegy, M.E. and Wang, K.K. (1999) Endogenous bax translocation in SH-SY5Y human neuroblastoma cells and cerebellar granule neurons undergoing apoptosis. *J. Neurochem.* **72**: 1899–1906.

Medema, J.P., Scaffidi, C., Kischkel, F.C., Shevchenko, A., Mann, M., Krammer, P.H. and Peter, M. E. (1997) FLICE is activated by association with the CD95 death-inducing signaling complex (DISC) *EMBO J.* **16**: 2794–2804.

Meier, P., Finch, A. and Evan, G. (2000) Apoptosis in development. *Nature* **407**: 796–801.

Meier, P., Silke, J., Leevers, S.J. and Evan, G.I. (2000) The *Drosophila* caspase DRONC is regulated by DIAP1. *EMBO J.* **19**: 598–611.

Meijerink, J.P., Mensink, E.J., Wang, K., Sedlak, T.W., Sloetjes, A.W., de Witte, T., Waksman, G. and Korsmeyer, S.J. (1998) Hematopoietic malignancies demonstrate loss-of-function mutations of BAX. *Blood* **91**: 2991–2997.

Melo, J.A., Cohen, J. and Toczyski, D.P. (2001) Two checkpoint complexes are independently recruited to sites of DNA damage in vivo. *Genes Dev.* **15**: 2809–2821.

Merino, R., Ding, L., Veis, D.J., Korsmeyer, S.J. and Nunez, G. (1994) Developmental regulation of the Bcl-2 protein and susceptibility to cell death in B lymphocytes. *EMBO J.* **13**: 683–691.

Mesaeli, N., Nakamura, K., Zvaritch, E., Dickie, P., Dziak, E., Krause, K.H., Opas, M., MacLennan, D.H. and Michalak, M. (1999) Calreticulin is essential for cardiac development. *J. Cell Biol.* **144**: 857–868.

Metkar, S.S., Wang, B., Aguilar-Santelises, M., Raja, S.M., Uhlin-Hansen, L., Podack, E., Trapani, J.A. and Froelich, C.J. (2002) Cytotoxic cell granule-mediated apoptosis: perforin delivers granzyme B-serglycin complexes into target cells without plasma membrane pore formation. *Immunity* **16**: 417–428.

Metzstein, M.M., Hengartner, M.O., Tsung, N., Ellis, R.E. and Horvitz, H.R. (1996) Transcriptional regulator of programmed cell death encoded by *Caenorhabditis elegans* gene ces-2. *Nature* **382**: 545–547.

Metzstein, M.M. and Horvitz, H.R. (1999) The *C. elegans* cell death specification gene ces-1 encodes a snail family zinc finger protein. *Mol. Cell* **4**: 309–319.

Micheau, O., Lens, S., Gaide, O., Alevizopoulos, K. and Tschopp, J. (2001) NF-kappaB signals induce the expression of c-FLIP. *Mol. Cell. Biol.* **21**: 5299–5305.

Middleton, G., Cox, S.W., Korsmeyer, S. and Davies, A.M. (2000) Differences in bcl-2– and bax-independent function in regulating apoptosis in sensory neuron populations. *Eur. J. Neurosci.* **12**: 819–827.

Middleton, G., Nunez, G. and Davies, A.M. (1996) Bax promotes neuronal survival and antagonises the survival effects of neurotrophic factors. *Development* **122**: 695–701.

Mignery, G.A., Sudhof, T.C., Takei, K. and De Camilli, P. (1989) Putative receptor for inositol 1,4,5-trisphosphate similar to ryanodine receptor. *Nature* **342**: 192–195.

Mikhailov, V., Mikhailova, M., Pulkrabek, D.J., Dong, Z., Venkatachalam, M.A. and Saikumar, P. (2001) Bcl-2 prevents Bax oligomerization in the mitochondrial outer membrane. *J. Biol. Chem.* **276**: 18361–18374.

Milán, M., Campuzano, S. and García-Bellido, A. (1997) Developmental parameters of cell death in the wing disc of *Drosophila*. *Proc. Natl Acad. Sci. U. S. A.* **94**: 5691–5696.

Miller, D.T. and Cagan, R.L. (1998) Local induction of patterning and programmed cell death in the developing *Drosophila* retina. *Development* **125**: 2327–2335.

Miller, M.A., Technau, U., Smith, K.M. and Steele, R.E. (2000) Oocyte development in *Hydra* involves selection from competent precursor cells. *Dev. Biol.* **224**: 326–338.

Miller, T.M., Moulder, K.L., Knudson, C.M., Creedon, D.J., Deshmukh, M., Korsmeyer, S.J. and Johnson, E.M. (1997) Bax deletion further orders the cell death pathway in cerebellar granule cells and suggests a caspase-independent pathway to cell death. *J. Cell Biol.* **139**: 205–217.

Minn, A.J., Kettlun, C.S., Liang, H., Kelekar, A., Vander Heiden, M.G., Chang, B.S., Fesik, S.W., Fill, M. and Thompson, C.B. (1999) Bcl-xL regulates apoptosis by heterodimerization-dependent and -independent mechanisms. *EMBO J.* **18**: 632–643.

Minn, A.J., Velez, P., Schendel, S.L., Liang, H., Muchmore, S.W., Fesik, S.W., Fill, M. and Thompson, C.B. (1997) Bcl-x(L) forms an ion channel in synthetic lipid membranes. *Nature* **385**: 353–357.

Miramar, M.D., Costantini, P., Ravagnan, L., *et al.* (2001) NADH oxidase activity of mitochondrial apoptosis-inducing factor. *J. Biol. Chem.* **276**: 16391–16398.

Mitsuhara, I., Malik, K.A., Miura, M. and Ohashi, Y. (1999) Animal cell-death suppressors Bcl-xL and Ced-9 inhibit cell death in tobacco plants. *Curr. Biol.* **9**: 775–778.

Mittl, P.R., Di Marco, S., Krebs, J.F., Bai, X., Karanewsky, D.S., Priestle, J.P., Tomaselli, K.J. and Grutter, M.G. (1997) Structure of recombinant human CPP32 in complex with the tetrapeptide acetyl-Asp-Val-Ala-Asp fluoromethyl ketone. *J. Biol. Chem.* **272**: 6539–6547.

Miura, M., Zhu, H., Rotello, R., Hartwieg, E.A. and Yuan, J. (1993) Induction of apoptosis in fibroblasts by IL-1 beta-converting enzyme, a mammalian homolog of the *C. elegans* cell death gene ced-3. *Cell* 75: 653–660.

Miyashita, T. and Reed, J.C. (1995) Tumor suppressor p53 is a direct transcriptional activator of the human bax gene. *Cell* 80: 293–299.

Miyazaki, T. and Reed, J.C. (2001) A GTP-binding adapter protein couples TRAIL receptors to apoptosis-inducing proteins. *Nat. Immunol.* 2: 493–500.

Monaghan, P., Robertson, D., Amos, T.A., Dyer, M.J., Mason, D.Y. and Greaves, M.F. (1992) Ultrastructural localization of bcl-2 protein. *J. Histochem. Cytochem.* 40: 1819–1825.

Mongkolsapaya, J., Cowper, A.E., Xu, X.N., Morris, G., McMichael, A.J., Bell, J.I. and Screaton, G. R. (1998) Lymphocyte inhibitor of TRAIL (TNF-related apoptosis-inducing ligand): a new receptor protecting lymphocytes from the death ligand TRAIL. *J. Immunol.* 160: 3–6.

Monney, L., Otter, I., Olivier, R., Ozer, H.L., Haas, A.L., Omura, S. and Borner, C. (1998) Defects in the ubiquitin pathway induce caspase-independent apoptosis blocked by Bcl-2. *J. Biol. Chem.* 273: 6121–6131.

Montessuit, S., Mazzei, G., Magnenat, E. and Antonsson, B. (1999) Expression and purification of full-length human Bax alpha. *Protein Expr. Purif.* 15: 202–206.

Moon, H., Baek, D., Lee, B., Prasad, D. T., Lee, S. Y., Cho, M. J., Lim, C. O., Choi, M. S., Bahk, J., Kim, M. O., *et al.* (2002) Soybean Ascorbate Peroxidase Suppresses Bax-Induced Apoptosis in Yeast by Inhibiting Oxygen Radical Generation. *Biochem Biophys Rex Commun.* 290: 457–462.

Moore, P.A., Belvedere, O., Orr, A., *et al.* (1999) BLyS: member of the tumor necrosis factor family and B lymphocyte stimulator. *Science* 285: 260–263.

Moreira, M.E., Del Portillo, H.A., Milder, R.V., Balanco, J.M. and Barcinski, M.A. (1996) Heat shock induction of apoptosis in promastigotes of the unicellular organism *Leishmania amazonensis. J. Cell. Physiol.* 167: 305–313.

Motoyama, N., Wang, F., Roth, K.A., Sawa, H., Nakayama, K., Negishi, I., Senju, S., Zhang, Q. and Fujii, S. (1995) Massive cell death of immature hematopoietic cells and neurons in Bcl-x- deficient mice. *Science* 267: 1506–1510.

Motoyama, S., Kitamura, M., Saito, S., Minamiya, Y., Suzuki, H., Saito, R., Terada, K., Ogawa, J. and Inaba, H. (1998) Bcl-2 is located predominantly in the inner membrane and crista of mitochondria in rat liver. *Biochem. Biophys. Res. Commun.* 249: 628–636.

Mpoke, S. and Wolfe, J. (1996) DNA digestion and chromatin condensation during nuclear death in *Tetrahymena. Exp. Cell Res.* 225: 357–365.

Mpoke, S. S. and Wolfe, J. (1997) Differential staining of apoptotic nuclei in living cells: application to macronuclear elimination in *Tetrahymena. J. Histochem. Cytochem.* 45: 675–683.

Muchmore, S.W., Sattler, M., Liang, H., *et al.* (1996) X-ray and NMR structure of human Bcl-xL, an inhibitor of programmed cell death. *Nature* 381: 335–341.

Mukhopadhyay, A., Ni, J., Zhai, Y., Yu, G.L. and Aggarwal, B.B. (1999) Identification and characterization of a novel cytokine, THANK, a TNF homologue that activates apoptosis, nuclear factor-kappaB, and c-Jun NH2–terminal kinase. *J. Biol. Chem.* 274: 15978–15981.

Muller, M., Wilder, S., Bannasch, D., *et al.* (1998) p53 activates the CD95 (APO-

1/Fas) gene in response to DNA damage by anticancer drugs. *J. Exp. Med.* **188**: 2033–2045.

Murakami, H. and Nurse, P. (2000) DNA replication and damage checkpoints and meiotic cell cycle controls in the fission and budding yeasts. *Biochem. J.* **349**: 1–12.

Muriel, M.P., Lambeng, N., Darios, F., Michel, P.P., Hirsch, E. C., Agid, Y. and Ruberg, M. (2000) Mitochondrial free calcium levels (Rhod-2 fluorescence) and ultrastructural alterations in neuronally differentiated PC12 cells during ceramidedependent cell death. *J. Comp. Neurol.* **426**: 297–315.

Murphy, A.N., Bredesen, D.E., Cortopassi, G., Wang, E. and Fiskum, G. (1996) Bcl-2 potentiates the maximal calcium uptake capacity of neural cell mitochondria. *Proc. Natl Acad. Sci. U. S. A.* **93**: 9893–9898.

Murphy, K.M., Streips, U.N. and Lock, R.B. (2000) Bcl-2 inhibits a Fas-induced conformational change in the Bax N terminus and Bax mitochondrial translocation. *J. Biol. Chem.* **275**: 17225–17228.

Murphy, R.C., Schneider, E. and Kinnally, K.W. (2001) Overexpression of Bcl-2 suppresses the calcium activation of a mitochondrial megachannel. *FEBS Lett.* **497**: 73–76.

Muzio, M., Chinnaiyan, A.M., Kischkel, F.C., *et al.* (1996) FLICE, a novel FADD-homologous ICE/CED-3–like protease, is recruited to the CD95 (Fas/APO-1) death-inducing signaling complex. *Cell* **85**: 817–827.

Muzio, M., Stockwell, B.R., Stennicke, H.R., Salvesen, G.S. and Dixit, V.M. (1998) An induced proximity model for caspase-8 activation. *J. Biol. Chem.* **273**: 2926–2930.

Nagata, S. (1998) Human autoimmune lymphoproliferative syndrome, a defect in the apoptosis-inducing Fas receptor: a lesson from the mouse model. *J. Hum. Genet.* **43**: 2–8.

Nagata, S. (1997) Apoptosis by death factor. *Cell* **88**: 355–365.

Naismith, J.H. and Sprang, S.R. (1998) Modularity in the TNF-receptor family. *Trends Biochem. Sci.* **23**: 74–79.

Nakagawa, T. and Yuan, J. (2000) Cross-talk between two cysteine protease families. Activation of caspase-12 by calpain in apoptosis. *J. Cell Biol.* **150**: 887–894.

Nakagawa, T., Zhu, H., Morishima, N., Li, E., Xu, J., Yankner, B.A. and Yuan, J. (2000) Caspase-12 mediates endoplasmic-reticulum-specific apoptosis and cytotoxicity by amyloid-beta. *Nature* **403**: 98–103.

Nakamura, K., Bossy-Wetzel, E., Burns, K., *et al.* (2000) Changes in endoplasmic reticulum luminal environment affect cell sensitivity to apoptosis. *J. Cell Biol.* **150**: 731–740.

Nakano, K. and Vousden, K.H. (2001) PUMA, a novel proapoptotic gene, is induced by p53. *Mol. Cell* **7**: 683–694.

Namba, R., Pazdera, T., Cerrone, R. and Minden, J. (1997) *Drosophila* embryonic pattern repair: how embryos respond to bicoid dosage alteration. *Development* **124**: 1393–1403.

Narasimhan, M.L., Damsz, B., Coca, M.A., Ibeas, J.I., Yun, D.J., Pardo, J.M., Hasegawa, P.M. and Bressan, R.A. (2001) A plant defense response effector induces microbial apoptosis. *Mol. Cell* **8**: 921–930.

Narita, M., Shimizu, S., Ito, T., Chittenden, T., Lutz, R.J., Matsuda, H. and Tsujimoto, Y. (1998) Bax interacts with the permeability transition pore to induce permeability transition and cytochrome c release in isolated mitochondria. *Proc. Natl Acad. Sci. U. S. A.* **95**: 14681–14686.

Nassif, C., Daniel, A., Lengyel, J. and Hartenstein, V. (1998) The role of morphogenetic cell death during *Drosophila* embryonic head development. *Dev. Biol.* **197**: 170–186.

Nechushtan, A., Smith, C.L., Hsu, Y.T., Youle, R.J., Chou, J.J., Li, H., Salvesen, G.S., Yuan, J. and Wagner, G. (1999) Conformation of the Bax C-terminus regulates subcellular location and cell death. *EMBO J.* **18**: 2330–2341.

Nechushtan, A., Smith, C.L., Lamensdorf, I., Yoon, S.H. and Youle, R.J. (2001) Bax and Bak coalesce into novel mitochondria-associated clusters during apoptosis. *J. Cell Biol.* **153**: 1265–1276.

Nestelbacher, R., Laun, P., Vondrakova, D., Pichova, A., Schüller, C. and Breitenbach, M. (2000) The influence of oxygen toxicity on yeast mother cell-specific aging. *Exp. Gerontol.* **35**: 63–70.

Neudecker, F. and Grimm, S. (2000) High-throughput method for isolating plasmid DNA with reduced lipopolysaccharide content. *Biotechniques* **28**: 107–109.

Newmeyer, D.D., Farschon, D.M. and Reed, J.C. (1994) Cell-free apoptosis in *Xenopus* egg extracts: inhibition by Bcl-2 and requirement for an organelle fraction enriched in mitochondria. *Cell* **79**: 353–364.

Newton, K. and Strasser, A. (2000) Ionizing radiation and chemotherapeutic drugs induce apoptosis in lymphocytes in the absence of Fas or FADD/MORT1 signaling. Implications for cancer therapy. *J. Exp. Med.* **191**: 195–200.

Newton, K., Harris, A.W., Bath, M.L., Smith, K.G. and Strasser, A. (1998) A dominant interfering mutant of FADD/MORT1 enhances deletion of autoreactive thymocytes and inhibits proliferation of mature T lymphocytes. *EMBO J.* **17**: 706–718.

Newton, K., Harris, A.W. and Strasser, A. (2000) FADD/MORT1 regulates the pre-TCR checkpoint and can function as a tumour suppressor. *EMBO J.* **19**: 931–941.

Ng, F.W., Nguyen, M., Kwan, T., Branton, P.E., Nicholson, D.W., Cromlish, J.A. and Shore, G.C. (1997) p28 Bap31, a Bcl-2/Bcl-XL- and procaspase-8–associated protein in the endoplasmic reticulum. *J. Cell Biol.* **139**: 327–338.

Ng, F.W. and Shore, G.C. (1998) Bcl-XL cooperatively associates with the Bap31 complex in the endoplasmic reticulum, dependent on procaspase-8 and Ced-4 adaptor. *J. Biol. Chem.* **273**: 3140–3143.

Nguyen, M., Breckenridge, D.G., Ducret, A. and Shore, G.C. (2000) Caspase-resistant BAP31 inhibits fas-mediated apoptotic membrane fragmentation and release of cytochrome c from mitochondria. *Mol. Cell. Biol.* **20**: 6731–6740.

Nguyen, M., Millar, D.G., Yong, V.W., Korsmeyer, S.J. and Shore, G.C. (1993) Targeting of Bcl-2 to the mitochondrial outer membrane by a COOH-terminal signal anchor sequence. *J. Biol. Chem.* **268**: 25265–25268.

Nicholson, D.W. (1999) Caspase structure, proteolytic substrates, and function during apoptotic cell death. *Cell Death Differ.* **6**: 1028–1042.

Nickels, J.T. and Broach, J.R. (1996) A ceramide-activated protein phosphatase mediates ceramide-induced G1 arrest of *Saccharomyces cerevisiae*. *Genes Dev.* **10**: 382–394.

Nicotera, P., Leist, M. and Ferrando-May, E. (1998) Intracellular ATP, a switch in the decision between apoptosis and necrosis. *Toxicol. Lett.* **102–103**: 139–142.

Nicotera, P., Leist, M. and Manzo, L. (1999) Neuronal cell death: a demise with different shapes. *Trends Pharmacol. Sci.* **20**: 46–51.

Nishitoh, H., Saitoh, M., Mochida, Y., Takeda, K., Nakano, H., Rothe, M., Miyazono, K. and Ichijo, H. (1998) ASK1 is essential for JNK/SAPK activation by TRAF2. *Mol. Cell* **2**: 389–395.

Niwa, M. and Walter, P. (2000) Pausing to decide. *Proc. Natl Acad. Sci. U. S. A.* **97**: 12396–12397.

Nomura, M., Shimizu, S., Ito, T., Narita, M., Matsuda, H. and Tsujimoto, Y. (1999) Apoptotic cytosol facilitates Bax translocation to mitochondria that involves cytosolic factor regulated by Bcl-2. *Cancer Res.* **59**: 5542–5548.

Nordstrom, W., Chen, P., Steller, H. and Abrams, J.M. (1996) Activation of the reaper gene during ectopic cell killing in *Drosophila*. *Dev. Biol.* **180**: 213–226.

Nutt, L.K., Pataer, A., Pahler, J., Fang, B., Roth, J., McConkey, D.J. and Swisher, S.G. (2002) Bax and Bak promote apoptosis by modulating endoplasmic reticular and mitochondrial Ca^{2+} stores. *J. Biol. Chem.* **277**: 9219–9225.

Nylandsted, J., Brand, K. and Jäättelä, M. (2000a) Heat shock protein 70 is required for the survival of cancer cells. *Ann. N. Y. Acad. Sci.* **926**: 122–125.

Nylandsted, J., Rohde, M., Brand, K., Bastholm, L., Elling, F. and Jäättelä, M. (2000b) Selective depletion of heat shock protein 70 (Hsp70) activates a tumor-specific death program that is independent of caspases and bypasses Bcl-2. *Proc. Natl Acad. Sci. U. S. A.* **97**: 7871–7876.

O'Brien, I.E.W., Reutelingsperger, C.P.M. and Holdaway, K.M. (1997) Annexin-V and TUNEL use in monitoring the progression of apoptosis in plants. *Cytometry* **29**: 28–33.

O'Connor, L., Strasser, A., O'Reilly, L.A., Hausmann, G., Adams, J.M., Cory, S. and Huang, D.C. (1998) Bim: a novel member of the Bcl-2 family that promotes apoptosis. *EMBO J.* **17**: 384–395.

Oda, E., Ohki, R., Murasawa, H., Nemoto, J., Shibue, T., Yamashita, T., Tokino, T., Taniguchi, T. and Tanaka, N. (2000) Noxa, a BH3-only member of the Bcl-2 family and candidate mediator of p53-induced apoptosis. *Science* **288**: 1053–1058.

Ojala, P.M., Yamamoto, K., Castanos-Velez, E., *et al.* (2000) The apoptotic v-cyclin-CDK6 complex phosphorylates and inactivates Bcl-2. *Nat. Cell. Biol.* **2**: 819–825.

Olie, R.A., Durrieu, F., Cornillon, S., Loughran, G., Gross, J., Earnshaw, W.C. and Golstein, P. (1998) Apparent caspase independence of programmed cell death in *Dictyostelium*. *Curr. Biol.* **8**: 955–958.

Ollmann, M., Young, L.M., Di Como, C.J., *et al.* (2000) *Drosophila* p53 is a structural and functional homolog of the tumor suppressor p53. *Cell* **101**: 91–101.

Oltvai, Z.N. and Korsmeyer, S.J. (1994) Checkpoints of dueling dimers foil death wishes. *Cell* **79**: 189–192.

Oltvai, Z.N., Milliman, C.L. and Korsmeyer, S.J. (1993) Bcl-2 heterodimerizes in vivo with a conserved homolog, Bax, that accelerates programmed cell death. *Cell* 74: 609–619.

O'Neill, L.A. (2000) The interleukin-1 receptor/Toll-like receptor superfamily: signal transduction during inflammation and host defense. *Sci STKE.* 2000: RE1.

O'Neill, L.A. and Greene, C. (1998) Signal transduction pathways activated by the IL-1 receptor family: ancient signaling machinery in mammals, insects, and plants. *J. Leukoc. Biol.* 63: 650–657.

Oppenheim, R.W., Flavell, R.A., Vinsant, S., Prevette, D., Kuan, C.Y. and Rakic, P. (2001) Programmed cell death of developing mammalian neurons after genetic deletion of caspases. *J. Neurosci.* 21: 4752–4760.

O'Reilly, L.A., Ekert, P., Harvey, N., *et al.* (2002) Caspase-2 is not required for thymocyte or neuronal apoptosis even though cleavage of caspase-2 is mediated by Apaf-1 and caspase-9. *Cell Death Differ.* 9: in press.

O'Reilly, L.A., Huang, D.C., Strasser, A., *et al.* (1996) The cell death inhibitor Bcl-2 and its homologues influence control of cell cycle entry. *EMBO J.* 15: 6979–6990.

O'Reilly, L.A., Print, C., Hausmann, G., Moriishi, K., Cory, S., Huang, D.C. and Strasser, A. (2001) Tissue expression and subcellular localization of the pro-survival molecule Bcl-w. *Cell Death Differ.* 8: 486–494.

Ottilie, S., Diaz, J.L., Horne, W., Chang, J., Wang, Y., Wilson, G., Chang, S., Weeks, S., Fritz, L.C. and Oltersdorf, T. (1997) Dimerization properties of human BAD: identification of a BH-3 domain and analysis of its binding to mutant BCL-2 and BCL-XL proteins. *J. Biol. Chem.* 272: 30866–30872.

Owen-Schaub, L.B., van Golen, K.L., Hill, L.L. and Price, J.E. (1998) Fas and Fas ligand interactions suppress melanoma lung metastasis. *J. Exp. Med.* 188: 1717–1723.

Pacher, P. and Hajnoczky, G. (2001) Propagation of the apoptotic signal by mitochondrial waves. *EMBO J.* 20: 4107–4121.

Pan, G., Bauer, J.H., Haridas, V., *et al.* (1998a) Identification and functional characterization of DR6, a novel death domain-containing TNF receptor. *FEBS Lett.* 431: 351–356.

Pan, G., Ni, J., Wei, Y.F., Yu, G., Gentz, R. and Dixit, V.M. (1997a) An antagonist decoy receptor and a death domain-containing receptor for TRAIL. *Science* 277: 815–818.

Pan, G., Ni, J., Yu, G., Wei, Y.F. and Dixit, V.M. (1998b) TRUNDD, a new member of the TRAIL receptor family that antagonizes TRAIL signalling. *FEBS Lett.* 424: 41–45.

Pan, G., O'Rourke, K., Chinnaiyan, A.M., Gentz, R., Ebner, R., Ni, J. and Dixit, V.M. (1997b) The receptor for the cytotoxic ligand TRAIL. *Science* 276: 111–113.

Pan, L., Kawai, M., Yu, L.H., Kim, K.M., Hirata, A., Umeda, M. and Uchimiya, H. (2001) The *Arabidopsis thaliana* ethylene-responsive element binding protein (AtEBP) can function as a dominant suppressor of Bax-induced cell death of yeast. *FEBS Lett.* 508: 375–378.

Pan, Z., Bhat, M.B., Nieminen, A.L. and Ma, J. (2001) Synergistic movements of Ca(2+) and Bax in cells undergoing apoptosis. *J. Biol. Chem.* 276: 32257–32263.

Park, J.R., Bernstein, I.D. and Hockenbery, D.M. (1995) Primitive human hematopoietic precursors express Bcl-x but not Bcl-2. *Blood* 86: 868–876.

Parker, M.W. and Pattus, F. (1993) Rendering a membrane protein soluble in water: a common packing motif in bacterial protein toxins. *Trends Biochem. Sci.* 18: 391–395.

Parrish, J., Li, L., Klotz, K., Ledwich, D., Wang, X. and Xue, D. (2001) Mitochondrial endonuclease G is important for apoptosis in *C. elegans. Nature* 412: 90–94.

Parrish, J., Metters, H., Chen, L. and Xue, D. (2000) Demonstration of the in vivo interaction of key cell death regulators by structure-based design of second-site suppressors. *Proc. Natl Acad. Sci. U. S. A.* 97: 11916–11921.

Pastorino, J.G., Chen, S.T., Tafani, M., Snyder, J.W. and Farber, J.L. (1998) The overexpression of Bax produces cell death upon induction of the mitochondrial permeability transition. *J. Biol. Chem.* 273: 7770–7775.

Pastorino, J.G., Tafani, M., Rothman, R.J., Marcinkeviciute, A., Hoek, J.B., Farber, J.L. and Marcineviciute, A. (1999) Functional consequences of the sustained or transient activation by Bax of the mitochondrial permeability transition pore. *J. Biol. Chem.* 274: 31734–31739.

Patil, C. and Walter, P. (2001) Intracellular signaling from the endoplasmic reticulum to the nucleus: the unfolded protein response in yeast and mammals. *Curr. Opin. Cell Biol.* 13: 349–355.

Pavlov, E.V., Priault, M., Pietkiewicz, D., Cheng, E.H., Antonsson, B., Manon, S., Korsmeyer, S.J., Mannella, C.A. and Kinnally, K.W. (2001) A novel, high conductance channel of mitochondria linked to apoptosis in mammalian cells and Bax expression in yeast. *J. Cell Biol.* 155: 725–731.

Pawlowski, J., Kraft, A.S., O'Reilly, L.A., *et al.* (2000) Bax-induced apoptotic cell death. *Proc. Natl Acad. Sci. U. S. A.* 97: 529–531.

Pawlowski, K., Pio, F., Chu, Z., Reed, J.C. and Godzik, A. (2001) PAAD – a new protein domain associated with apoptosis, cancer and autoimmune diseases. *Trends Biochem. Sci.* 26: 85–87.

Pazdera, T., Janardhan, P. and Minden, J. (1998) Patterned epidermal cell death in wild-type and segment polarity mutant *Drosophila* embryos. *Development* 125: 3427–3436.

Pennacchio, L.A., Bouley, D.M., Higgins, K.M., Scott, M.P., Noebels, J.L. and Myers, R.M. (1998) Progressive ataxia, myoclonic epilepsy and cerebellar apoptosis in cystatin B-deficient mice. *Nat. Genet.* 20: 251–258.

Peterson, C., Carney, G.E., Taylor, B.J. and White, K. (2002) reaper is required for neuroblast apoptosis during *Drosophila* development. *Development* 129: 1467–1476.

Petros, A.M., Medek, A., Nettesheim, D.G., Kim, D.H., Yoon, H.S., Swift, K., Matayoshi, E.D., Oltersdorf, T. and Fesik, S.W. (2001) Solution structure of the antiapoptotic protein bcl-2. *Proc. Natl Acad. Sci. U. S. A.* 98: 3012–3017.

Pettersen, R.D., Bernard, G., Olafsen, M.K., Pourtein, M. and Lie, S.O. (2001) CD99 signals caspase-independent T cell death. *J. Immunol.* 166: 4931–4942.

Pietenpol, J.A., Papadopoulos, N., Markowitz, S., *et al.* (1994) Paradoxical inhibition of solid tumor cell growth by bcl2. *Cancer Res.* 54: 3714–3717.

Pinton, P., Ferrari, D., Magalhaes, P., Schulze-Osthoff, K., Di Virgilio, F., Pozzan, T. and Rizzuto, R. (2000) Reduced loading of intracellular Ca(2+) stores and downregulation of capacitative Ca(2+) influx in Bcl-2-overexpressing cells. *J. Cell Biol.* 148: 857–862.

Pinton, P., Ferrari, D., Rapizzi, E., Di Virgilio, F.D., Pozzan, T. and Rizzuto, R. (2001) The Ca^{2+} concentration of the endoplasmic reticulum is a key determinant of ceramide-induced apoptosis: significance for the molecular mechanism of Bcl-2 action. *EMBO J.* 20: 2690–2701.

Pitti, R.M., Marsters, S.A., Lawrence, D.A., *et al.* (1998) Genomic amplification of a decoy receptor for Fas ligand in lung and colon cancer. *Nature* 396: 699–703.

Pitti, R.M., Marsters, S.A., Ruppert, S., Donahue, C.J., Moore, A. and Ashkenazi, A. (1996) Induction of apoptosis by Apo-2 ligand, a new member of the tumor necrosis factor cytokine family. *J. Biol. Chem.* 271: 12687–12690.

Polyak, K., Xia, Y., Zweier, J.L., Kinzler, K.W. and Vogelstein, B. (1997) A model for p53-induced apoptosis. *Nature* 389: 300–305.

Pozzan, T., Rizzuto, R., Volpe, P. and Meldolesi, J. (1994) Molecular and cellular physiology of intracellular calcium stores. *Physiol. Rev.* 74: 595–636.

Priault, M., Chaudhuri, B., Clow, A., Camougrand, N. and Manon, S. (1999) Investigation of bax-induced release of cytochrome c from yeast mitochondria: permeability of mitochondrial membranes, role of VDAC and ATP requirement. *Eur. J. Biochem.* 260: 684–691.

Print, C.G. and Loveland, K.L. (2000) Germ cell suicide: new insights into apoptosis during spermatogenesis. *Bioessays* 22: 423–430.

Pronk, G.J., Ramer, K., Amiri, P. and Williams, L.T. (1996) Requirement of an ICE-like protease for induction of apoptosis and ceramide generation by REAPER. *Science* 271: 808–810.

Putcha, G.V., Deshmukh, M. and Johnson, E.M. Jr (1999) BAX translocation is a critical event in neuronal apoptosis: regulation by neuroprotectants, BCL-2, and caspases. *J. Neurosci.* 19: 7476–7485.

Puthalakath, H., Huang, D.C., O'Reilly, L.A., *et al.* (1999) The proapoptotic activity of the Bcl-2 family member Bim is regulated by interaction with the dynein motor complex. *Mol. Cell* 3: 287–296.

Puthalakath, H., Villunger, A., O'Reilly, L.A., *et al.* (2001) Bmf: a proapoptotic BH3-only protein regulated by interaction with the myosin V actin motor complex, activated by anoikis. *Science* 293: 1829–1832.

Qin, H., Srinivasula, S.M., Wu, G., Fernandes-Alnemri, T., Alnemri, E.S. and Shi, Y. (1999) Structural basis of procaspase-9 recruitment by the apoptotic protease-activating factor 1. *Nature* 399: 549–557.

Querfurth, H.W. and Selkoe, D.J. (1994) Calcium ionophore increases amyloid beta peptide production by cultured cells. *Biochemistry (Mosc.)* 33: 4550–4561.

Quignon, F., De Bels, F., Koken, M., Feuntcun, J., Ameisen, J.C. and de The, H. (1998) PML induces a novel caspase-independent death process. *Nat. Genet.* 20: 259–265.

Quillet-Mary, A., Jaffrezou, J.P., Mansat, V., Bordier, C., Naval, J. and Laurent, G.

(1997) Implication of mitochondrial hydrogen peroxide generation in ceramide-induced apoptosis. *J. Biol. Chem.* **272**: 21388–21395.

Quinn, L.M., Dorstyn, L., Mills, K., Colussi, P.A., Chen, P., Coombe, M., Abrams, J.M., Kumar, S. and Richardson, H. (2000) An essential role for the caspase dronc in developmentally programmed cell death in *Drosophila*. *J. Biol. Chem.* **275**: 40416–40424.

Rabinovich, E., Kerem, A., Fröhlich, K.U., Diamant, N. and Bar-Nun, S. (2002) AAA-ATPase p97/Cdc48p, a cytosolic chaperone required for endoplasmic reticulum-associated protein degradation. *Mol. Cell. Biol.* **22**: 626–634.

Raff, M.C. (1992) Social controls on cell survival and cell death. *Nature* **356**: 397–400.

Ramesh, V., Sharma, V.K., Sheu, S.S. and Franzini-Armstrong, C. (1998) Structural proximity of mitochondria to calcium release units in rat ventricular myocardium may suggest a role in Ca^{2+} sequestration. *Ann. N. Y. Acad. Sci.* **853**: 341–344.

Rampino, N., Yamamoto, H., Ionov, Y., Li, Y., Sawai, H., Reed, J.C. and Perucho, M. (1997) Somatic frameshift mutations in the BAX gene in colon cancers of the microsatellite mutator phenotype. *Science* **275**: 967–969.

Rao, L., Perez, D. and White, E. (1996) Lamin proteolysis facilitates nuclear events during apoptosis. *J. Cell Biol.* **135**: 1441–1455.

Rao, R.V., Hermel, E., Castro-Obregon, S., del Rio, G., Ellerby, L.M., Ellerby, H.M. and Bredesen, D.E. (2001) Coupling endoplasmic reticulum stress to the cell death program. Mechanism of caspase activation. *J. Biol. Chem.* **276**: 33869–33874.

Raoul, C., Henderson, C.E. and Pettmann, B. (1999) Programmed cell death of embryonic motoneurons triggered through the Fas death receptor. *J. Cell Biol.* **147**: 1049–1062.

Rasper, D.M., Vaillancourt, J.P., Hadano, S., *et al.* (1998) Cell death attenuation by Usurpin, a mammalian DED-caspase homologue that precludes caspase-8 recruitment and activation by the CD-95 (Fas, APO-1) receptor complex. *Cell Death Differ.* **5**: 271–288.

Raveh, T., Droguett, G., Horwitz, M.S., DePinho, R.A. and Kimchi, A. (2001) DAP kinase activates a p19ARF/p53-mediated apoptotic checkpoint to suppress oncogenic transformation. *Nat. Cell. Biol.* **3**: 1–7.

Ravi, R. and Bedi, A. (2002) Requirement of BAX for TRAIL/Apo2L-induced apoptosis of colorectal cancers: synergism with sulindac-mediated inhibition of Bcl-x(L). *Cancer Res.* **62**: 1583–1587.

Ravi, R., Bedi, A. and Fuchs, E.J. (1998a) CD95 (Fas)-induced caspase-mediated proteolysis of NF-kappaB. *Cancer Res.* **58**: 882–886.

Ravi, R., Bedi, G.C., Engstrom, L.W., Zeng, Q., Mookerjee, B., Gelinas, C., Fuchs, E.J. and Bedi, A. (2001) Regulation of death receptor expression and TRAIL/Apo2L-induced apoptosis by NF-kappaB. *Nat. Cell. Biol.* **3**: 409–416.

Ravi, R., Mookerjee, B., van Hensbergen, Y., Bedi, G.C., Giordano, A., el Deiry, W.S., Fuchs, E.J. and Bedi, A. (1998b) p53-mediated repression of nuclear factor-kappaB RelA via the transcriptional integrator p300. *Cancer Res.* **58**: 4531–4536.

Reddien, P.W., Cameron, S. and Horvitz, H.R. (2001) Phagocytosis promotes programmed cell death in *C. elegans*. *Nature* **412**: 198–202.

Reddien, P.W. and Horvitz, H.R. (2000) CED-2/CrkII and CED-10/Rac control phagocytosis and cell migration in *Caenorhabditis elegans*. *Nat. Cell. Biol.* **2**: 131–136.

Reimertz, C., Kogel, D., Lankiewicz, S., Poppe, M. and Prehn, J.H. (2001) Ca(2+)-induced inhibition of apoptosis in human SH-SY5Y neuroblastoma cells: degradation of apoptotic protease activating factor-1 (APAF-1) *J. Neurochem.* **78**: 1256–1266.

Reinheckel, T., Deussing, J., Roth, W. and Peters, C. (2001) Towards specific functions of lysosomal cysteine peptidases: phenotypes of mice deficient for cathepsin B or cathepsin L. *Biol. Chem.* **382**: 735–741.

Renatus, M., Stennicke, H.R., Scott, F.L., Liddington, R.C. and Salvesen, G.S. (2001) Dimer formation drives the activation of the cell death protease caspase 9. *Proc. Natl Acad. Sci. U. S. A.* **98**: 14250–14255.

Renshaw, B.R., Fanslow, W.C., Iii, A.R.J., Campbell, K.A., Liggitt, D., Wright, B., Davison, B.L. and Maliszewski, C.R. (1994) Humoral immune responses in CD40 ligand-deficient mice. *J. Exp. Med.* **180**: 1889–1900.

Reuther, J.Y. and Baldwin, A.S., Jr. (1999) Apoptosis promotes a caspase-induced amino-terminal truncation of IkappaBalpha that functions as a stable inhibitor of NF-kappaB. *J. Biol. Chem.* **274**: 20664–20670.

Rhind, N. and Russell, P. (2000) Checkpoints: it takes more than time to heal some wounds. *Curr. Biol.* **10**: R908–R911.

Rich, T., Allen, R.L. and Wyllie, A.H. (2000) Defying death after DNA damage. *Nature* **407**: 777–783.

Riddiford, L.M. (1993) Hormones and *Drosophila* development. In: *The Development of* Drosophila melanogaster, vol. II (eds M. Bate and A. Martinez Arias). Cold Spring Harbor Laboratory Press, Cold Spring Harbor, NY, pp. 899–939.

Riedl, S.J., Fuentes-Prior, P., Renatus, M., Kairies, N., Krapp, S., Huber, R., Salvesen, G.S. and Bode, W. (2001a) Structural basis for the activation of human procaspase-7. *Proc. Natl Acad. Sci. U. S. A.* **98**: 14790–14795.

Riedl, S.J., Renatus, M., Schwarzenbacher, R., Zhou, Q., Sun, C., Fesik, S.W., Liddington, R.C. and Salvesen, G.S. (2001b) Structural basis for the inhibition of caspase-3 by XIAP. *Cell* **104**: 791–800.

Rieux-Laucat, F., Le Deist, F., Hivroz, C., Roberts, I.A., Debatin, K.M., Fischer, A. and de Villartay, J.P. (1995) Mutations in Fas associated with human lymphoproliferative syndrome and autoimmunity. *Science* **268**: 1347–1349.

Rinaldo, C., Ederle, S., Rocco, V. and La Volpe, A. (1998) The *Caenorhabditis elegans* RAD51 homolog is transcribed into two alternative mRNAs potentially encoding proteins of different sizes. *Mol. Gen. Genet.* **260**: 289–294.

Rizzuto, R., Bernardi, P. and Pozzan, T. (2000) Mitochondria as all-round players of the calcium game. *J. Physiol.* **529** (Pt 1): 37–47.

Rizzuto, R., Pinton, P., Carrington, W., Fay, F.S., Fogarty, K.E., Lifshitz, L.M., Tuft, R.A. and Pozzan, T. (1998) Close contacts with the endoplasmic reticulum as determinants of mitochondrial Ca^{2+} responses. *Science* **280**: 1763–1766.

Roach, H.I. and Clarke, N.M. (2000) Physiological cell death of chondrocytes in vivo is not confined to apoptosis. New observations on the mammalian growth plate. *J. Bone Joint Surg. Br.* **82**: 601–613.

Roberg, K. (2001) Relocalization of cathepsin D and cytochrome c early in apoptosis revealed by immunoelectron microscopy. *Lab. Invest.* **81**: 149–158.

Roberts, L.R., Adjei, P.N. and Gores, G.J. (1999) Cathepsins as effector proteases in hepatocyte apoptosis. *Cell Biochem. Biophys.* **30**: 71–88.

Roberts, L.R., Kurosawa, H., Bronk, S.F., Fesmier, P.J., Agellon, L.B., Leung, W.Y., Mao, F. and Gores, G.J. (1997) Cathepsin B contributes to bile salt-induced apoptosis of rat hepatocytes. *Gastroenterology* **113**: 1714–1726.

Robinow, S., Draizen, T.A. and Truman, J.W. (1997) Genes that induce apoptosis: transcriptional regulation in identified, doomed neurons of the *Drosophila* CNS. *Dev. Biol.* **190**: 206–213.

Robinow, S., Talbot, W.S., Hogness, D.S. and Truman, J.W. (1993) Programmed cell death in the *Drosophila* CNS is ecdysone-regulated and coupled with a specific receptor isoform. *Development* **119**: 1251–1259.

Rodriguez, A., Chen, P., Oliver, H. and Abrams, J. M. (2002) Unrestrained caspase dependent cell death caused by loss of Diap1 function requires the *Drosophila* Apaf-1 homolog, Dark. *EMBO J.* **21**: 2189–2197.

Rodriguez, A., Oliver, H., Zou, H., Chen, P., Wang, X. and Abrams, J.M. (1999) Dark is a *Drosophila* homologue of Apaf-1/CED-4 and functions in an evolutionarily conserved death pathway. *Nat. Cell. Biol.* **1**: 272–279.

Rodriguez, J. and Lazebnik, Y. (1999) Caspase-9 and APAF-1 form an active holoenzyme. *Genes Dev.* **13**: 3179–3184.

Roeder, G.S. (1997) Meiotic chromosomes: it takes two to tango. *Genes Dev.* **11**: 2600–2621.

Rosenfeld, M.E., Prichard, L., Shiojiri, N. and Fausto, N. (2000) Prevention of hepatic apoptosis and embryonic lethality in RelA/TNFR-1 double knockout mice. *Am. J. Pathol.* **156**: 997–1007.

Rosse, T., Olivier, R., Monney, L., Rager, M., Conus, S., Fellay, I., Jansen, B. and Borner, C. (1998) Bcl-2 prolongs cell survival after Bax-induced release of cytochrome c. *Nature* **391**: 496–499.

Roth, W., Isenmann, S., Naumann, U., Kugler, S., Bahr, M., Dichgans, J., Ashkenazi, A. and Weller, M. (1999) Locoregional Apo2L/TRAIL eradicates intracranial human malignant glioma xenografts in athymic mice in the absence of neurotoxicity. *Biochem. Biophys. Res. Commun.* **265**: 479–483.

Rothe, M., Pan, M.G., Henzel, W.J., Ayres, T.M. and Goeddel, D.V. (1995) The TNFR2–TRAF signaling complex contains two novel proteins related to baculoviral inhibitor of apoptosis proteins. *Cell* **83**: 1243–1252.

Rotonda, J., Nicholson, D.W., Fazil, K.M., *et al.* (1996) The three-dimensional structure of apopain/CPP32, a key mediator of apoptosis. *Nat. Struct. Biol.* **3**: 619–625.

Roucou, X., Prescott, M., Devenish, R.J. and Nagley, P. (2000) A cytochrome c-GFP fusion is not released from mitochondria into the cytoplasm upon expression of Bax in yeast cells. *FEBS Lett.* **471**: 235–239.

Roucou, X., Rostovtseva, T., Montessuit, S., Martinou, J.C. and Antonsson, B.

(2002) Bid induces cytochrome c impermeable Bax channels in liposomes. *Biochem. J.* **363**: 547–552.

Roy, M. and Sapolsky, R. (1999) Neuronal apoptosis in acute necrotic insults: why is this subject such a mess? *Trends Neurosci.* **22**: 419–422.

Roy, N., Deveraux, Q.L., Takahashi, R., Salvesen, G.S. and Reed, J.C. (1997) The c-IAP-1 and c-IAP-2 proteins are direct inhibitors of specific caspases. *EMBO J.* **16**: 6914–6925.

Roy, S. and Nicholson, D.W. (2000) Cross-talk in cell death signaling. *J. Exp. Med.* **192**: F21–F25.

Ruck, A., Dolder, M., Wallimann, T. and Brdiczka, D. (1998) Reconstituted adenine nucleotide translocase forms a channel for small molecules comparable to the mitochondrial permeability transition pore. *FEBS Lett.* **426**: 97–101.

Rudel, T. and Bokoch, G.M. (1997) Membrane and morphological changes in apoptotic cells regulated by caspase-mediated activation of PAK2. *Science* **276**: 1571–1574.

Rudner, J., Lepple-Wienhues, A., Budach, W., Berschauer, J., Friedrich, B., Wesselborg, S., Schulze-Osthoff, K. and Belka, C. (2001) Wild-type, mitochondrial and ER-restricted Bcl-2 inhibit DNA damage-induced apoptosis but do not affect death receptor-induced apoptosis. *J. Cell Sci.* **114**: 4161–4172.

Rudolph, D., Yeh, W.C., Wakeham, A., Rudolph, B., Nallainathan, D., Potter, J., Elia, A.J. and Mak, T.W. (2000) Severe liver degeneration and lack of NF-kappaB activation in NEMO/IKKgamma-deficient mice. *Genes Dev.* **14**: 854–862.

Ruiz-Vela, A., Gonzalez de Buitrago, G. and Martinez, A.C. (1999) Implication of calpain in caspase activation during B cell clonal deletion. *EMBO J.* **18**: 4988–4998.

Ruiz-Vela, A., Serrano, F., Gonzalez, M.A., Abad, J.L., Bernad, A., Maki, M. and Martinez, A.C. (2001) Transplanted long-term cultured pre-BI cells expressing calpastatin are resistant to B cell receptor-induced apoptosis. *J. Exp. Med.* **194**: 247–254.

Sagot, Y., Dubois-Dauphin, M., Tan, S.A., de Bilbao, F., Aebischer, P., Martinou, J.C. and Kato, A.C. (1995) Bcl-2 overexpression prevents motoneuron cell body loss but not axonal degeneration in a mouse model of a neurodegenerative disease. *J. Neurosci.* **15**: 7727–7733.

Saikumar, P., Dong, Z., Patel, Y., Hall, K., Hopfer, U., Weinberg, J.M. and Venkatachalam, M.A. (1998) Role of hypoxia-induced Bax translocation and cytochrome c release in reoxygenation injury. *Oncogene* **17**: 3401–3415.

Saito, M., Korsmeyer, S.J. and Schlesinger, P.H. (2000) BAX-dependent transport of cytochrome c reconstituted in pure liposomes. *Nat. Cell Biol.* **2**: 553–555.

Sakahira, H., Enari, M. and Nagata, S. (1998) Cleavage of CAD inhibitor in CAD activation and DNA degradation during apoptosis. *Nature* **391**: 96–99.

Salvesen, G.S. and Dixit, V.M. (1999) Caspase activation: the induced-proximity model. *Proc. Natl Acad. Sci. U. S. A.* **96**: 10964–10967.

Salvesen, G.S. and Dixit, V.M. (1997) Caspases: intracellular signaling by proteolysis. *Cell* **91**: 443–446.

Samali, A., Zhivotovsky, B., Jones, D., Nagata, S. and Orrenius, S. (1999)

Apoptosis: cell death defined by caspase activation. *Cell Death Differ.* **6**: 495–496.

Sato, N., Hotta, K., Waguri, S., *et al.* (1994) Neuronal differentiation of PC12 cells as a result of prevention of cell death by bcl-2. *J. Neurobiol.* **25**: 1227–1234.

Satoh, T., Ross, C.A., Villa, A., Supattapone, S., Pozzan, T., Snyder, S.H. and Meldolesi, J. (1990) The inositol 1,4,5-trisphosphate receptor in cerebellar Purkinje cells: quantitative immunogold labeling reveals concentration in an ER subcompartment. *J. Cell Biol.* **111**: 615–624.

Sattler, M., Liang, H., Nettesheim, D., *et al.* (1997) Structure of Bcl-xL-Bak peptide complex: recognition between regulators of apoptosis. *Science* **275**: 983–986.

Savill, J. and Fadok, V. (2000) Corpse clearance defines the meaning of cell death. *Nature* **407**: 784–788.

Sawamoto, A., Taguchi, A., Hirota, Y., Yamada, C., Jin, M. and Okano, H. (1998) Argos induces programmed cell death in the developing *Drosophila* eye by inhibition of the ras pathway. *Cell Death Differ.* **5**: 262–270.

Scaffidi, C., Fulda, S., Srinivasan, A., Friesen, C., Li, F., Tomaselli, K.J., Debatin, K.M., Krammer, P.H. and Peter, M.E. (1998) Two CD95 (APO-1/Fas) signaling pathways. *EMBO J.* **17**: 1675–1687.

Scarlett, J.L. and Murphy, M.P. (1997) Release of apoptogenic proteins from the mitochondrial intermembrane space during the mitochondrial permeability transition. *FEBS Lett.* **418**: 282–286.

Scatena, C.D., Stewart, Z.A., Mays, D., *et al.* (1998) Mitotic phosphorylation of Bcl-2 during normal cell cycle progression and Taxol-induced growth arrest. *J. Biol. Chem.* **273**: 30777–30784.

Scheidereit, C. (1998) Signal transduction. Docking IkappaB kinases. *Nature* **395**: 225–226.

Schendel, S.L., Azimov, R., Pawlowski, K., Godzik, A., Kagan, B.L. and Reed, J.C. (1999) Ion channel activity of the BH3 only Bcl-2 family member, BID. *J. Biol. Chem.* **274**: 21932–21936.

Schendel, S.L., Xie, Z., Montal, M.O., Matsuyama, S., Montal, M. and Reed, J.C. (1997) Channel formation by antiapoptotic protein Bcl-2. *Proc. Natl Acad. Sci. U. S. A.* **94**: 5113–5118.

Schierle, G.S., Leist, M., Martinou, J.C., Widner, H., Nicotera, P. and Brundin, P. (1999) Differential effects of Bcl-2 overexpression on fibre outgrowth and survival of embryonic dopaminergic neurons in intracerebral transplants. *Eur. J. Neurosci.* **11**: 3073–3081.

Schlegel, R.A., Callahan, M.K. and Williamson, P. (2000) The central role of phosphatidylserine in the phagocytosis of apoptotic thymocytes. *Ann. N. Y. Acad. Sci.* **926**: 217–225.

Schlesinger, P.H., Gross, A., Yin, X.M., Yamamoto, K., Saito, M., Waksman, G. and Korsmeyer, S.J. (1997) Comparison of the ion channel characteristics of proapoptotic BAX and antiapoptotic BCL-2. *Proc. Natl Acad. Sci. U. S. A.* **94**: 11357–11362.

Schmich, J., Trepel, S. and Leitz, T. (1998) The role of GLWamides in metamorphosis of *Hydractinia echinata*. *Dev. Genes Evol.* **208**: 267–273.

Schmitt, C.A. and Lowe, S.W. (1999) Apoptosis and therapy. *J. Pathol.* **187**: 127–137.

Schmitz, I., Kirchhoff, S. and Krammer, P.H. (2000) Regulation of death receptor-mediated apoptosis pathways. *Int. J. Biochem. Cell Biol.* **32**: 1123–1136.

Schmitz, I., Walczak, H., Krammer, P.H. and Peter, M.E. (1999) Differences between CD95 type I and II cells detected with the CD95 ligand. *Cell Death Differ.* **6**: 821–822.

Schnare, M., Barton, G.M., Holt, A.C., Takeda, K., Akira, S. and Medzhitov, R. (2001) Toll-like receptors control activation of adaptive immune responses. *Nat. Immunol.* **2**: 947–950.

Schneider, P., Bodmer, J.L., Thome, M., Hofmann, K., Holler, N. and Tschopp, J. (1997) Characterization of two receptors for TRAIL. *FEBS Lett.* **416**: 329–334.

Scholz, H., Sadlowski, E., Klaes, A. and Klambt, C. (1997) Control of midline glia development in the embryonic *Drosophila* CNS. *Mech. Dev.* **64**: 137–151.

Schotte, P., Van Criekinge, W., Van de Craen, M., *et al.* (1998) Cathepsin B-mediated activation of the proinflammatory caspase-11. *Biochem. Biophys. Res. Commun.* **251**: 379–387.

Schousboe, P., Wheatley, D.N. and Rasmussen, L. (1998) Autocrine/paracrine activator of cell proliferation: purification of a 4–6 kD compound with growth-factor-like effects in *Tetrahymena thermophila*. *Cell. Physiol. Biochem.* **8**: 130–137.

Schuler, M. and Green, D.R. (2001) Mechanisms of p53-dependent apoptosis. *Biochem. Soc. Trans.* **29**: 684–688.

Schultz, J., Milpetz, F., Bork, P. and Ponting, C.P. (1998) SMART, a simple modular architecture research tool: identification of signaling domains. *Proc. Natl Acad. Sci. U. S. A.* **95**: 5857–5864.

Schulz, J.B., Weller, M. and Klockgether, T. (1996) Potassium deprivation-induced apoptosis of cerebellar granule neurons: a sequential requirement for new mRNA and protein synthesis, ICE-like protease activity, and reactive oxygen species. *J. Neurosci.* **16**: 4696–4706.

Schulze-Osthoff, K., Bakker, A.C., Vanhaesebroeck, B., Beyaert, R., Jacob, W.A. and Fiers, W. (1992) Cytotoxic activity of tumor necrosis factor is mediated by early damage of mitochondrial functions. Evidence for the involvement of mitochondrial radical generation. *J. Biol. Chem.* **267**: 5317–5323.

Schulze-Osthoff, K., Ferrari, D., Los, M., Wesselborg, S. and Peter, M.E. (1998) Apoptosis signaling by death receptors. *Eur. J. Biochem.* **254**: 439–459.

Schulze-Osthoff, K., Walczak, H., Droge, W. and Krammer, P.H. (1994) Cell nucleus and DNA fragmentation are not required for apoptosis. *J. Cell Biol.* **127**: 15–20.

Schumacher, B., Hofmann, K., Boulton, S. and Gartner, A. (2001) The *C. elegans* homolog of the p53 tumor suppressor is required for DNA damage-induced apoptosis. *Curr. Biol.* **11**: 1722–1727.

Schurmann, A., Mooney, A.F., Sanders, L.C., *et al.* (2000) p21-activated kinase 1 phosphorylates the death agonist Bad and protects cells from apoptosis. *Mol. Cell. Biol.* **20**: 453–461.

Schwenzer, R., Siemienski, K., Liptay, S., Schubert, G., Peters, N., Scheurich, P.,

Schmid, R.M. and Wajant, H. (1999) The human tumor necrosis factor (TNF) receptor-associated factor 1 gene (TRAF1) is up-regulated by cytokines of the TNF ligand family and modulates TNF-induced activation of NF-kappaB and c-Jun N-terminal kinase. *J. Biol. Chem.* **274**: 19368–19374.

Scorrano, L., Ashiya, M., Buttle, K., Weiler, S., Oakes, S.A., Mannella, C.A. and Korsmeyer, S.J. (2002) A distinct pathway remodels mitochondrial cristae and mobilizes cytochrome c during apoptosis. *Dev. Cell* **2**: 55–67.

Scott, D.W., Grdina, T. and Shi, Y. (1996) T cells commit suicide, but B cells are murdered! *J. Immunol.* **156**: 2352–2356.

Screaton, G.R., Mongkolsapaya, J., Xu, X.N., Cowper, A.E., McMichael, A.J. and Bell, J.I. (1997a) TRICK2, a new alternatively spliced receptor that transduces the cytotoxic signal from TRAIL. *Curr. Biol.* **7**: 693–696.

Screaton, G.R., Xu, X.N., Olsen, A.L., Cowper, A.E., Tan, R., McMichael, A.J. and Bell, J.I. (1997b) LARD: a new lymphoid-specific death domain containing receptor regulated by alternative pre-mRNA splicing. *Proc. Natl Acad. Sci. U. S. A.* **94**: 4615–4619.

Sebbagh, M., Renvoize, C., Hamelin, J., Riche, N., Bertoglio, J. and Breard, J. (2001) Caspase-3-mediated cleavage of ROCK I induces MLC phosphorylation and apoptotic membrane blebbing. *Nat. Cell. Biol.* **3**: 346–352.

Segraves, W.A. and Hogness, D.S. (1990) The E75 ecdysone-inducible gene responsible for the 75B early puff in *Drosophila* encodes two new members of the steroid receptor superfamily. *Genes Dev.* **4**: 204–219.

Seipp, S., Schmich, J. and Leitz, T. (2001) Apoptosis – a death-inducing mechanism tightly linked with morphogenesis in *Hydractina echinata* (Cnidaria, Hydrozoa). *Development* **128**: 4891–4898.

Sekiguchi, T., Nakashima, T., Hayashida, T., Kuraoka, A., Hashimoto, S., Tsuchida, N., Shibata, Y., Hunter, T. and Nishimoto, T. (1995) Apoptosis is induced in BHK cells by the tsBN462/13 mutation in the CCG1/TAFII250 subunit of the TFIID basal transcription factor. *Exp. Cell Res.* **218**: 490–498.

Sekito, T., Thornton, J. and Butow, R.A. (2000) Mitochondria-to-nuclear signaling is regulated by the subcellular localization of the transcription factors Rtg1p and Rtg3p. *Mol. Biol. Cell* **11**: 2103–2115.

Sells, S.F., Wood, D.P., Jr., Joshi-Barve, S.S., Muthukumar, S., Jacob, R.J., Crist, S.A., Humphreys, S. and Rangnekar, V.M. (1994) Commonality of the gene programs induced by effectors of apoptosis in androgen-dependent and -independent prostate cells. *Cell Growth Differ.* **5**: 457–466.

Senftleben, U., Cao, Y., Xiao, G., *et al.* (2001a) Activation by IKKalpha of a second, evolutionary conserved, NF-kappa B signaling pathway. *Science* **293**: 1495–1499.

Senftleben, U., Li, Z.W., Baud, V. and Karin, M. (2001b) IKKbeta is essential for protecting T cells from TNFalpha-induced apoptosis. *Immunity* **14**: 217–230.

Sennvik, K., Benedikz, E., Fastbom, J., Sundstrom, E., Winblad, B. and Ankarcrona, M. (2001) Calcium ionophore A23187 specifically decreases the secretion of beta-secretase cleaved amyloid precursor protein during apoptosis in primary rat cortical cultures. *J. Neurosci. Res.* **63**: 429–437.

Seydoux, G. and Schedl, T. (2001) The germline in *C. elegans*: origins, proliferation, and silencing. *Int. Rev. Cytol.* 203: 139–185.

Sha, W. C., Liou, H.C., Tuomanen, E.I. and Baltimore, D. (1995) Targeted disruption of the p50 subunit of NF-kappa B leads to multifocal defects in immune responses. *Cell* 80: 321–330.

Shaham, S. (1998) Identification of multiple *Caenorhabditis elegans* caspases and their potential roles in proteolytic cascades. *J. Biol. Chem.* 273: 35109–35117.

Shaham, S. and Horvitz, H.R. (1996) Developing *Caenorhabditis elegans* neurons may contain both cell-death protective and killer activities. *Genes Dev.* 10: 578–591.

Shapiro, L. and Scherer, P.E. (1998) The crystal structure of a complement-1q family protein suggests an evolutionary link to tumor necrosis factor. *Curr. Biol.* 8: 335–338.

Sharma, V.K., Ramesh, V., Franzini-Armstrong, C. and Sheu, S.S. (2000) Transport of Ca^{2+} from sarcoplasmic reticulum to mitochondria in rat ventricular myocytes. *J. Bioenerg. Biomembr.* 32: 97–104.

Sharp, P.A. (1999) RNAi and double-strand RNA. *Genes Dev.* 13: 139–141.

Shearwin-Whyatt, L.M., Harvey, N.L. and Kumar, S. (2000) Subcellular localization and CARD-dependent oligomerization of the death adaptor RAIDD. *Cell Death Differ.* 7: 155–165.

Shen, X., Ellis, R.E., Lee, K., *et al.* (2001) Complementary signaling pathways regulate the unfolded protein response and are required for *C. elegans* development. *Cell* 107: 893–903.

Sheridan, J.P., Marsters, S.A., Pitti, R.M., *et al.* (1997) Control of TRAIL-induced apoptosis by a family of signaling and decoy receptors. *Science* 277: 818–821.

Shi, L., Mai, S., Israels, S., Browne, K., Trapani, J.A. and Greenberg, A.H. (1997) Granzyme B (GraB) autonomously crosses the cell membrane and perforin initiates apoptosis and GraB nuclear localization. *J. Exp. Med.* 185: 855–866.

Shi, Y. (2002) Mechanisms of caspase activation and inhibition during apoptosis. *Mol. Cell* 9: 459–470.

Shi, Y. (2001) A structural view of mitochondria-mediated apoptosis. *Nat. Struc. Biol.* 8: 394–401.

Shibasaki, F., Kondo, E., Akagi, T., McKeon, F., Chou, J.J., Li, H., Salvesen, G.S., Yuan, J. and Wagner, G. (1997) Suppression of signalling through transcription factor NF-AT by interactions between calcineurin and Bcl-2. *Nature* 386: 728–731.

Shimizu, S., Eguchi, Y., Kamiike, W., Funahashi, Y., Mignon, A., Lacronique, V., Matsuda, H. and Tsujimoto, Y. (1998) Bcl-2 prevents apoptotic mitochondrial dysfunction by regulating proton flux. *Proc. Natl Acad. Sci. U. S. A.* 95: 1455–1459.

Shimizu, S., Ide, T., Yanagida, T. and Tsujimoto, Y. (2000a) Electrophysiological study of a novel large pore formed by Bax and the voltage-dependent anion channel that is permeable to cytochrome c. *J. Biol. Chem.* 275: 12321–12325.

Shimizu, S., Konishi, A., Kodama, T. and Tsujimoto, Y. (2000b) BH4 domain of antiapoptotic Bcl-2 family members closes voltage-dependent anion channel and inhibits apoptotic mitochondrial changes and cell death. *Proc. Natl Acad. Sci. U. S. A.* **97**: 3100–3105.

Shimizu, S., Matsuoka, Y., Shinohara, Y., Yoneda, Y. and Tsujimoto, Y. (2001) Essential role of voltage-dependent anion channel in various forms of apoptosis in mammalian cells. *J. Cell Biol.* **152**: 237–250.

Shimizu, S., Narita, M. and Tsujimoto, Y. (1999) Bcl-2 family proteins regulate the release of apoptogenic cytochrome c by the mitochondrial channel VDAC. *Nature* **399**: 483–487 [erratum **407**: 76].

Shimizu, S. and Tsujimoto, Y. (2000) Proapoptotic BH3-only Bcl-2 family members induce cytochrome c release, but not mitochondrial membrane potential loss, and do not directly modulate voltage-dependent anion channel activity. *Proc. Natl Acad. Sci. U. S. A.* **97**: 577–582.

Shirogane, T., Fukada, T., Muller, J.M., Shima, D.T., Hibi, M. and Hirano, T. (1999) Synergistic roles for Pim-1 and c-Myc in STAT3-mediated cell cycle progression and antiapoptosis. *Immunity* **11**: 709–719.

Shresta, S., MacIvor, D.M., Heusel, J.W., Russell, J.H. and Ley, T.J. (1995) Natural killer and lymphokine-activated killer cells require granzyme B for the rapid induction of apoptosis in susceptible target cells. *Proc. Natl Acad. Sci. U. S. A.* **92**: 5679–5683.

Shu, H.B., Takeuchi, M. and Goeddel, D.V. (1996) The tumor necrosis factor receptor 2 signal transducers TRAF2 and c-IAP1 are components of the tumor necrosis factor receptor 1 signaling complex. *Proc. Natl Acad. Sci. U. S. A.* **93**: 13973–13978.

Sidrauski, C., Chapman, R. and Walter, P. (1998) The unfolded protein response: an intracellular signalling pathway with many surprising features. *Trends Cell Biol.* **8**: 245–249.

Siegel, R.M., Martin, D.A., Zheng, L., Ng, S.Y., Bertin, J., Cohen, J. and Lenardo, M.J. (1998) Death-effector filaments: novel cytoplasmic structures that recruit caspases and trigger apoptosis. *J. Cell Biol.* **141**: 1243–1253.

Simon, B., Malisan, F., Testi, R., Nicotera, P. and Leist, M. (2002) The disialoganglioside GD3 is released from microglia cells and induces oligodendrocyte apoptosis. *Cell Death Differ.* in press.

Simonian, P.L., Grillot, D.A., Merino, R. and Nunez, G. (1996) Bax can antagonize Bcl-XL during etoposide and cisplatin-induced cell death independently of its heterodimerization with Bcl-XL. *J. Biol. Chem.* **271**: 22764–22772.

Singer, G.G. and Abbas, A.K. (1994) The fas antigen is involved in peripheral but not thymic deletion of T lymphocytes in T cell receptor transgenic mice. *Immunity* **1**: 365–371.

Single, B., Leist, M. and Nicotera, P. (2001) Differential effects of bcl-2 on cell death triggered under ATP-depleting conditions. *Exp. Cell Res.* **262**: 8–16.

Sizemore, N., Leung, S. and Stark, G.R. (1999) Activation of phosphatidylinositol 3-kinase in response to interleukin-1 leads to phosphorylation and activation of the NF-kappaB p65/RelA subunit. *Mol. Cell. Biol.* **19**: 4798–4805.

Slee, E.A., Harte, M.T., Kluck, R.M., *et al.* (1999) Ordering the cytochrome c-initiated caspase cascade: hierarchical activation of caspases-2, -3, -6, -7, -8, and -10 in a caspase-9-dependent manner. *J. Cell Biol.* **144**: 281–292.

Smith, C.A., Farrah, T. and Goodwin, R.G. (1994) The TNF receptor superfamily of cellular and viral proteins: activation, costimulation, and death. *Cell* **76**: 959–962.

Smith, K.G., Strasser, A. and Vaux, D.L. (1996) CrmA expression in T lymphocytes of transgenic mice inhibits CD95 (Fas/APO-1)-transduced apoptosis, but does not cause lymphadenopathy or autoimmune disease. *EMBO J.* **15**: 5167–5176.

Smith, K.M., Gee, L. and Bode, H.R. (2000) HyAlx, an aristaless-related gene, is involved in tentacle formation in hydra. *Development* **127**: 4743–4752.

Smith, T.F. and Waterman, M.S. (1981) Identification of common molecular subsequences. *J. Mol. Biol.* **147**: 195–197.

Smyth, M.J., Thia, K.Y., Cretney, E., Kelly, J.M., Snook, M.B., Forbes, C.A. and Scalzo, A.A. (1999) Perforin is a major contributor to NK cell control of tumor metastasis. *J. Immunol.* **162**: 6658–6662.

Soengas, M.S., Alarcon, R.M., Yoshida, H., Giaccia, A.J., Hakem, R., Mak, T.W. and Lowe, S.W. (1999) Apaf-1 and caspase-9 in p53-dependent apoptosis and tumor inhibition. *Science* **284**: 156–159.

Solan, N.J., Miyoshi, H., Carmona, E.M., Bren, G.D. and Paya, C.V. (2002) RelB cellular regulation and transcriptional activity are regulated by p100. *J. Biol. Chem.* **277**: 1405–1418.

Song, Z., McCall, K. and Steller, H. (1997) DCP-1, a *Drosophila* cell death protease essential for development. *Science* **275**: 536–540.

Song, Z.W., Guan, B., Bergman, A., Nicholson, D.W., Thornberry, N.A., Peterson, E.P. and Steller, H. (2000) Biochemical and genetic interactions between *Drosophila* caspases and the proapoptotic genes rpr, hid, and grim. *Mol. Cell. Biol.* **20**: 2907–2914.

Sorrentino, V. and Rizzuto, R. (2001) Molecular genetics of Ca(2+) stores and intracellular Ca(2+) signalling. *Trends Pharmacol. Sci.* **22**: 459–464.

Spaner, D., Raju, K., Rabinovich, B. and Miller, R.G. (1999) A role for perforin in activation-induced T cell death in vivo: increased expansion of allogeneic perforin-deficient T cells in SCID mice. *J. Immunol.* **162**: 1192–1199.

Speliotes, E.K., Uren, A., Vaux, D. and Horvitz, H.R. (2000) The survivin-like *C. elegans* BIR-1 protein acts with the Aurora-like kinase AIR-2 to affect chromosomes and the spindle midzone. *Mol. Cell* **6**: 211–223.

Sperandio, S., de Belle, I. and Bredesen, D.E. (2000) An alternative, nonapoptotic form of programmed cell death. *Proc. Natl Acad. Sci. U. S. A.* **97**: 14376–14381.

Spiliotis, E.T., Manley, H., Osorio, M., Zuniga, M.C. and Edidin, M. (2000) Selective export of MHC class I molecules from the ER after their dissociation from TAP. *Immunity* **13**: 841–851.

Spooncer, E., Heyworth, C.M., Dunn, A. and Dexter, T.M. (1986) Self-renewal and differentiation of interleukin-3-dependent multipotent stem cells are modulated by stromal cells and serum factors. *Differentiation* **31**: 111–118.

Srinivasula, S.M., Ahmad, M., Fernandes-Alnemri, T. and Alnemri, E.S. (1998) Autoactivation of procaspase-9 by Apaf1-mediated oligomerization. *Mol. Cell* **1**: 949–957.

Srinivasula, S.M., Ahmad, M., Ottilie, S., *et al.* (1997) FLAME-1, a novel FADD-like anti-apoptotic molecule that regulates Fas/TNFR1–induced apoptosis. *J. Biol. Chem.* **272**: 18542–18545.

Srinivasula, S.M., Datta, P., Kobayashi, M., *et al.* (2002) sickle, a novel *Drosophila* death gene in the reaper/hid/grim region, encodes an IAP-inhibitory protein. *Curr. Biol.* **12**: 125–130.

Srinivasula, S.M., Hegde, R., Saleh, A., *et al.* (2001) A conserved XIAP-interaction motif in caspase-9 and Smac/DIABLO regulates caspase activity and apoptosis. *Nature* **410**: 112–116.

Srivastava, R.K., Sollott, S.J., Khan, L., Hansford, R., Lakatta, E.G. and Longo, D.L. (1999) Bcl-2 and Bcl-X(L) block thapsigargin-induced nitric oxide generation, c-Jun NH(2)-terminal kinase activity, and apoptosis. *Mol. Cell. Biol.* **19**: 5659–5674.

Stadelmann, C., Deckwerth, T.L., Srinivasan, A., Bancher, C., Bruck, W., Jellinger, K. and Lassmann, H. (1999) Activation of caspase-3 in single neurons and autophagic granules of granulovacuolar degeneration in Alzheimer's disease. Evidence for apoptotic cell death. *Am. J. Pathol.* **155**: 1459–1466.

Stanfield, G.M. and Horvitz, H.R. (2000) The ced-8 gene controls the timing of programmed cell deaths in *C. elegans. Mol. Cell* **5**: 423–433.

Stanger, B.Z., Leder, P., Lee, T.H., Kim, E. and Seed, B. (1995) RIP: a novel protein containing a death domain that interacts with Fas/APO-1 (CD95) in yeast and causes cell death. *Cell* **81**: 513–523.

Staub, E., Dahl, E. and Rosenthal, A. (2001) The DAPIN family: a novel domain links apoptotic and interferon response proteins. *Trends Biochem. Sci.* **26**: 83–85.

Stehlik, C., de Martin, R., Kumabashiri, I., Schmid, J.A., Binder, B.R. and Lipp, J. (1998) Nuclear factor (NF)-kappaB-regulated X-chromosome-linked iap gene expression protects endothelial cells from tumor necrosis factor alpha-induced apoptosis. *J. Exp. Med.* **188**: 211–216.

Stemerdink, C. and Jacobs, J.R. (1997) Argos and Spitz group genes function to regulate midline glial cell number in *Drosophila* embryos. *Development* **124**: 3787–3796.

Stennicke, H.R., Deveraux, Q.L., Humke, E.W., Reed, J.C., Dixit, V.M. and Salvesen, G.S. (1999) Caspase-9 can be activated without proteolytic processing. *J. Biol. Chem.* **274**: 8359–8362.

Stennicke, H.R. and Salvesen, G.S. (1999) Catalytic properties of the caspases. *Cell Death Differ.* **6**: 1054–1059.

Sternberg, P. W. and Horvitz, H. R. (1981) Gonadal cell lineages of the nematode *Panagrellus redivivus* and implications for evolution by the modification of cell lineage. *Dev. Biol.* **88**: 147–166.

Stoka, V., Turk, B., Schendel, S.L., *et al.* (2001) Lysosomal protease pathways to apoptosis. Cleavage of bid, not pro-caspases, is the most likely route. *J. Biol. Chem.* **276**: 3149–3157.

Stoven, S., Ando, I., Kadalayil, L., Engstrom, Y. and Hultmark, D. (2000) Activation of the *Drosophila* NF-kappaB factor Relish by rapid endoproteolytic cleavage. *EMBO Rep.* **1**: 347–352.

Straarup, E.M., Schousboe, P., Hansen, H.Q., Kristiansen, K., Hoffmann, E.K., Rasmussen, L. and Christensen, S.T. (1997) Effects of protein kinase C activators and staurosporine on protein kinase activity, cell survival, and proliferation in *Tetrahymena thermophila*. *Microbios* **91**: 181–190.

Strasser, A., Harris, A.W., Jacks, T. and Cory, S. (1994) DNA damage can induce apoptosis in proliferating lymphoid cells via p53-independent mechanisms inhibitable by Bcl-2. *Cell* **79**: 329–339.

Strasser, A., O'Connor, L. and Dixit, V.M. (2000) Apoptosis signaling. *Annu. Rev. Biochem.* **69**: 217–245.

Su, Z.Z., Madireddi, M.T., Lin, J.J., Young, C.S., Kitada, S., Reed, J.C., Goldstein, N.I. and Fisher, P.B. (1998) The cancer growth suppressor gene mda-7 selectively induces apoptosis in human breast cancer cells and inhibits tumor growth in nude mice. *Proc. Natl Acad. Sci. U. S. A.* **95**: 14400–14405.

Sugawara, H., Kurosaki, M., Takata, M. and Kurosaki, T. (1997) Genetic evidence for involvement of type 1, type 2 and type 3 inositol 1,4,5-trisphosphate receptors in signal transduction through the B-cell antigen receptor. *EMBO J.* **16**: 3078–3088.

Sulston, J.E. and Horvitz, H.R. (1977) Post-embryonic cell lineages of the nematode, *Caenorhabditis elegans*. *Dev. Biol.* **56**: 110–156.

Sulston, J.E., Schierenberg, E., White, J.G. and Thomson, J.N. (1983) The embryonic cell lineage of the nematode *Caenorhabditis elegans*. *Dev. Biol.* **100**: 64–119.

Sun, C., Cai, M., Gunasekera, A.H., *et al.* (1999) NMR structure and mutagenesis of the inhibitor-of-apoptosis protein XIAP. *Nature* **401**: 818–822.

Sun, X., Majumder, P., Shioya, H., Wu, F., Kumar, S., Weichselbaum, R., Kharbanda, S. and Kufe, D. (2000) Activation of the cytoplasmic c-Abl tyrosine kinase by reactive oxygen species. *J. Biol. Chem.* **275**: 17237–17240.

Sundararajan, R. and White, E. (2001) E1B 19K blocks Bax oligomerization and tumor necrosis factor alpha-mediated apoptosis. *J. Virol.* **75**: 7506–7516.

Susin, S.A., Daugas, E., Ravagnan, L., *et al.* (2000) Two distinct pathways leading to nuclear apoptosis. *J. Exp. Med.* **192**: 571–580.

Susin, S.A., Lorenzo, H.K., Zamzami, N., *et al.* (1999) Molecular characterization of mitochondrial apoptosis-inducing factor. *Nature* **397**: 441–446.

Suter, M., Reme, C., Grimm, C., Wenzel, A., Jäättelä, M., Esser, P., Kociok, N., Leist, M. and Richter, C. (2000) Age-related macular degeneration. The lipofusion component *N*-retinyl-*N*-retinylidene ethanolamine detaches proapoptotic proteins from mitochondria and induces apoptosis in mammalian retinal pigment epithelial cells. *J. Biol. Chem.* **275**: 39625–39630.

Suzuki, M., Youle, R.J., Tjandra, N., Chou, J.J., Li, H., Salvesen, G.S., Yuan, J. and Wagner, G. (2000) Structure of Bax: coregulation of dimer formation and intracellular localization. *Cell* **103**: 645–654.

Suzuki, Y., Imai, Y., Nakayama, H., Takahashi, K., Takio, K. and Takahashi, R. (2001) A serine protease, HtrA2, is released from the mitochondria and interacts with XIAP, inducing cell death. *Mol. Cell* **8**: 613–621.

Szalai, G., Krishnamurthy, R. and Hajnoczky, G. (1999) Apoptosis driven by IP(3)-linked mitochondrial calcium signals. *EMBO J.* **18**: 6349–6361.

Tagami, S., Eguchi, Y., Kinoshita, M., Takeda, M. and Tsujimoto, Y. (2000) A novel protein, RTN-XS, interacts with both Bcl-XL and Bcl-2 on endoplasmic reticulum and reduces their anti-apoptotic activity. *Oncogene* **19**: 5736–5746.

Takahashi, R., Deveraux, Q., Tamm, I., Welsh, K., Assa-Munt, N., Salvesen, G.S. and Reed, J.C. (1998) A single BIR domain of XIAP sufficient for inhibiting caspases. *J. Biol. Chem.* **273**: 7787–7790.

Takanami, T., Sato, S., Ishihara, T., Katsura, I., Takahashi, H. and Higashitani, A. (1998) Characterization of a *Caenorhabditis elegans* recA-like gene Ce-rdh-1 involved in meiotic recombination. *DNA Res.* **5**: 373–377.

Takeda, K., Hayakawa, Y., Smyth, M.J., Kayagaki, N., Yamaguchi, N., Kakuta, S., Iwakura, Y., Yagita, H. and Okumura, K. (2001) Involvement of tumor necrosis factor-related apoptosis-inducing ligand in surveillance of tumor metastasis by liver natural killer cells. *Nat. Med.* **7**: 94–100.

Takei, K., Stukenbrok, H., Metcalf, A., Mignery, G.A., Sudhof, T.C., Volpe, P. and De Camilli, P. (1992) Ca^{2+} stores in Purkinje neurons: endoplasmic reticulum subcompartments demonstrated by the heterogeneous distribution of the InsP3 receptor, Ca(2+)-ATPase, and calsequestrin. *J. Neurosci.* **12**: 489–505.

Talanian, R.V., Quinlan, C., Trautz, S., Hackett, M.C., Mankovich, J.A., Banach, D., Ghayur, T., Brady, K.D. and Wong, W.W. (1997a) Substrate specificities of caspase family proteases. *J. Biol. Chem.* **272**: 9677–9682.

Talanian, R.V., Yang, X., Turbov, J., Seth, P., Ghayur, T., Casiano, C.A., Orth, K. and Froelich, C.J. (1997b) Granule-mediated killing: pathways for granzyme B-initiated apoptosis. *J. Exp. Med.* **186**: 1323–1331.

Talmadge, J.E., Meyers, K.M., Prieur, D.J. and Starkey, J.R. (1980) Role of NK cells in tumour growth and metastasis in beige mice. *Nature* **284**: 622–624.

Tamatani, M., Matsuyama, T., Yamaguchi, A., *et al.* (2001) ORP150 protects against hypoxia/ischemia-induced neuronal death. *Nat. Med.* **7**: 317–323.

Tanaka, M., Itai, T., Adachi, M. and Nagata, S. (1998) Downregulation of Fas ligand by shedding. *Nat. Med.* **4**: 31–36.

Tang, G., Minemoto, Y., Dibling, B., Purcell, N.H., Li, Z., Karin, M. and Lin, A. (2001a) Inhibition of JNK activation through NF-kappaB target genes. *Nature* **414**: 313–317.

Tang, G., Yang, J., Minemoto, Y. and Lin, A. (2001b) Blocking caspase-3-mediated proteolysis of IKKbeta suppresses TNF-alpha-induced apoptosis. *Mol. Cell* **8**: 1005–1016.

Tao, W., Kurschner, C. and Morgan, J.I. (1997) Modulation of cell death in yeast by the Bcl-2 family of proteins. *J. Biol. Chem.* **272**: 15547–15552.

Tartaglia, L.A., Ayres, T.M., Wong, G.H. and Goeddel, D.V. (1993) A novel domain within the 55 kd TNF receptor signals cell death. *Cell* **74**: 845–853.

Technau, U. and Bode, H.R. (1999) HyBra1, a Brachyury homologue, acts during head formation in *Hydra*. *Development* **126**: 999–1010.

Thome, M., Schneider, P., Hofmann, K., *et al.* (1997) Viral FLICE-inhibitory proteins (FLIPs) prevent apoptosis induced by death receptors. *Nature* **386**: 517–521.

Thompson, C.B. (1995) Apoptosis in the pathogenesis and treatment of disease. *Science* **267**: 1456–1462.

Thornberry, N.A., Rano, T.A., Peterson, E.P., *et al.* (1997) A combinatorial approach defines specificities of members of the caspase family and granzyme B. Functional relationships established for key mediators of apoptosis. *J. Biol. Chem.* **272**: 17907–17911.

Thress, K., Evans, E.K. and Kornbluth, S. (1999) Reaper-induced dissociation of a Scythe-sequestered cytochrome c-releasing activity. *EMBO J.* **18**: 5486–5493.

Thress, K., Henzel, W., Shillinglaw, W. and Kornbluth, S. (1998) Scythe: a novel reaper-binding apoptotic regulator. *EMBO J.* **17**: 6135–6143.

Thummel, C.S. (1996) Flies on steroids – *Drosophila* metamorphosis and the mechanisms of steroid hormone action. *Trends Genet.* **12**: 306–310.

Tirasophon, W., Welihinda, A.A., Kaufman, R.J. and Genes, D. (1998) A stress response pathway from the endoplasmic reticulum to the nucleus requires a novel bifunctional protein kinase/endonuclease (Ire1p) in mammalian cells. *Genes Dev.* **12**: 1812–1824.

Tobiume, K., Matsuzawa, A., Takahashi, T., *et al.* (2001) ASK1 is required for sustained activations of JNK/p38 MAP kinases and apoptosis. *EMBO Rep.* **2**: 222–228.

Torgler, C.N., Mariastella, T., Raven, T., Aubry, J.P., Brown, R. and Meldrum, E. (1997) Expression of bak in *S. pombe* results in lethality mediated through interaction with the calnexin homologue cnx1. *Oncogene* **4**: 263–271.

Torii, S., Egan, D.A., Evans, R.A. and Reed, J.C. (1999) Human Daxx regulates Fas-induced apoptosis from nuclear PML oncogenic domains (PODs) *EMBO J.* **18**: 6037–6049.

Torriglia, A., Perani, P., Brossas, J.Y., Chaudun, E., Treton, J., Courtois, Y. and Counis, M.F. (1998) L-DNase II, a molecule that links proteases and endonucleases in apoptosis, derives from the ubiquitous serpin leukocyte elastase inhibitor. *Mol. Cell. Biol.* **18**: 3612–3619 [erratum **18**: 4947].

Tosello-Trampont, A.C., Brugnera, E. and Ravichandran, K.S. (2001) Evidence for a conserved role for CRKII and Rac in engulfment of apoptotic cells. *J. Biol. Chem.* **276**: 13797–13802.

Tournier, C., Hess, P., Yang, D.D., *et al.* (2000) Requirement of JNK for stress-induced activation of the cytochrome c-mediated death pathway. *Science* **288**: 870–874.

Trapani, J.A., Davis, J., Sutton, V.R. and Smyth, M.J. (2000) Proapoptotic functions of cytotoxic lymphocyte granule constituents in vitro and in vivo. *Curr. Opin. Immunol.* **12**: 323–329.

Travers, K.J., Patil, C.K., Wodicka, L., Lockhart, D.J., Weissman, J.S. and Walter, P. (2000) Functional and genomic analyses reveal an essential coordination between the unfolded protein response and ER-associated degradation. *Cell* **101**: 249–258.

Troy, C.M., Rabacchi, S.A., Hohl, J.B., Angelastro, J.M., Greene, L.A. and Shelanski, M.L. (2001) Death in the balance: alternative participation of the caspase-2 and -9 pathways in neuronal death induced by nerve growth factor deprivation. *J. Neurosci.* **21**: 5007 5016.

Tsujimoto, Y., Cossman, J., Jaffe, E. and Croce, C.M. (1985) Involvement of the bcl-2 gene in human follicular lymphoma. *Science* **228**: 1440–1443.

Tsujimoto, Y., Finger, L.R., Yunis, J., Nowell, P.C. and Croce, C.M. (1984)

Cloning of the chromosome breakpoint of neoplastic B cells with the t(14;18) chromosome translocation. *Science* **226**: 1097–1099.

Tsujimoto, Y. and Shimizu, S. (2000) VDAC regulation by the Bcl-2 family of proteins. *Cell Death Differ.* **7**: 1174–1181.

Tuazon, P.T. and Traugh, J.A. (1991) Casein kinase I and II – multipotential serine protein kinases: structure, function, and regulation. *Adv. Second Messenger Phosphoprotein Res.* **23**: 123–164.

Turk, V., Turk, B. and Turk, D. (2001) Lysosomal cysteine proteases: facts and opportunities. *EMBO J.* **20**: 4629–4633.

Turmaine, M., Raza, A., Mahal, A., Mangiarini, L., Bates, G.P. and Davies, S.W. (2000) Nonapoptotic neurodegeneration in a transgenic mouse model of Huntington's disease. *Proc. Natl Acad. Sci. U. S. A.* **97**: 8093–8097.

Tzagoloff, A. (1970) Assembly of the mitochondrial membrane system. 3. Function and synthesis of the oligomycin sensitivity-conferring protein of yeast mitochondria. *J. Biol. Chem.* **245**: 1545–1551.

Tzung, S.P., Kim, K.M., Basañez, G., Giedt, C.D., Simon, J., Zimmerberg, J., Zhang, K.Y. and Hockenbery, D.M. (2001) Antimycin A mimics a cell-death-inducing Bcl-2 homology domain 3. *Nat. Cell Biol.* **3**: 183–191.

Uhlmann, E.J., D'Sa-Eipper, C., Subramanian, T., *et al.* (1996) Deletion of a non-conserved region of Bcl-2 confers a novel gain of function: suppression of apoptosis with concomitant cell proliferation. *Cancer Res.* **56**: 2506–2509.

Urano, F., Bertolotti, A. and Ron, D. (2000a) IRE1 and efferent signaling from the endoplasmic reticulum. *J. Cell Sci.* **113** (Pt 21): 3697–3702.

Urano, F., Wang, X., Bertolotti, A., Zhang, Y., Chung, P., Harding, H.P. and Ron, D. (2000b) Coupling of stress in the ER to activation of JNK protein kinases by transmembrane protein kinase IRE1. *Science* **287**: 664–666.

Uren, A.G., O'Rourke, K., Aravind, L.A., Pisabarro, M.T., Seshagiri, S., Koonin, E.V. and Dixit, V.M. (2000) Identification of paracaspases and metacaspases: two ancient families of caspase-like proteins, one of which plays a key role in MALT lymphoma. *Mol. Cell* **6**: 961–967.

Vairo, G., Innes, K.M., Adams, J.M., Chou, J.J., Li, H., Salvesen, G.S., Yuan, J. and Wagner, G. (1996) Bcl-2 has a cell cycle inhibitory function separable from its enhancement of cell survival. *Oncogene* **13**: 1511–1519.

Vairo, G., Soos, T.J., Upton, T.M., *et al.* (2000) Bcl-2 retards cell cycle entry through p27(Kip1), pRB relative p130, and altered E2F regulation. *Mol. Cell. Biol.* **20**: 4745–4753.

Van Antwerp, D.J., Martin, S.J., Kafri, T., Green, D.R. and Verma, I.M. (1996) Suppression of TNF-alpha-induced apoptosis by NF-kappaB. *Science* **274**: 787–789.

Van de Craen, M., Vandenabeele, P., Declercq, W., *et al.* (1997) Characterization of seven murine caspase family members. *FEBS Lett.* **403**: 61–69.

van der Biezen, E.A. and Jones, J.D. (1998) The NB-ARC domain: a novel signalling motif shared by plant resistance gene products and regulators of cell death in animals. *Curr. Biol.* **8**: R226–227.

van Eijk, M. and de Groot, C. (1999) Germinal center B cell apoptosis requires both caspase and cathepsin activity. *J. Immunol.* **163**: 2478–2482.

van Loo, G., van Gurp, M., Depuydt, B., *et al.* (2002) The serine protease Omi/HtrA2 is released from mitochondria during apoptosis. Omi interacts with caspase inhibitor XIAP and induces enhanced caspase activity. *Cell Death Differ.* **9**: 20–26.

van Loo, G.S. (2001) Endonuclease G: a mitochondrial protein released in apoptosis and involved in caspase-independent DNA degradation. *Cell Death Differ.* **8**: 1136–1142.

Van Parijs, L., Ibraghimov, A. and Abbas, A.K. (1996) The roles of costimulation and Fas in T cell apoptosis and peripheral tolerance. *Immunity* **4**: 321–328.

Vander Heiden, M.G., Chandel, N.S., Li, X.X., Schumacker, P.T., Colombini, M. and Thompson, C.B. (2000) Outer mitochondrial membrane permeability can regulate coupled respiration and cell survival. *Proc. Natl Acad. Sci. U. S. A.* **97**: 4666–4671.

Vander Heiden, M.G., Chandel, N.S., Schumacker, P.T. and Thompson, C.B. (1999) Bcl-xL prevents cell death following growth factor withdrawal by facilitating mitochondrial ATP/ADP exchange. *Mol. Cell* **3**: 159–167.

Vander Heiden, M.G., Chandel, N.S., Williamson, E.K., Schumacker, P.T. and Thompson, C.B. (1997) Bcl-xL regulates the membrane potential and volume homeostasis of mitochondria. *Cell* **91**: 627–637.

Vander Heiden, M.G., Li, X.X., Gottleib, E., Hill, R.B., Thompson, C.B. and Colombini, M. (2001) Bcl-xL promotes the open configuration of the voltage-dependent anion channel and metabolite passage through the outer mitochondrial membrane. *J. Biol. Chem.* **276**: 19414–19419.

Vander Heiden, M.G. and Thompson, C.B. (1999) Bcl-2 proteins: regulators of apoptosis or of mitochondrial homeostasis? *Nat. Cell. Biol.* **1**: 209–216.

Varfolomeev, E.E., Boldin, M.P., Goncharov, T.M. and Wallach, D. (1996) A potential mechanism of 'cross-talk' between the p55 tumor necrosis factor receptor and Fas/APO1: proteins binding to the death domains of the two receptors also bind to each other. *J. Exp. Med.* **183**: 1271–1275.

Varfolomeev, E.E., Schuchmann, M., Luria, V., *et al.* (1998) Targeted disruption of the mouse Caspase 8 gene ablates cell death induction by the TNF receptors, Fas/Apo1, and DR3 and is lethal prenatally. *Immunity* **9**: 267–276.

Varghese, J., Radhika, G. and Sarin, A. (2001) The role of calpain in caspase activation during etoposide-induced apoptosis in T cells. *Eur. J. Immunol.* **31**: 2035–2041.

Varkey, J., Chen, P., Jemmerson, R. and Abrams, J.M. (1999) Altered cytochrome C display precedes apoptotic cell death in *Drosophila. J. Cell Biol.* **144**: 701–710.

Vaux, D.L. (1997) CED-4 – the third horseman of apoptosis. *Cell* **90**: 389–390.

Vaux, D.L., Cory, S. and Adams, J.M. (1988) Bcl-2 gene promotes haemopoietic cell survival and cooperates with c-myc to immortalize pre-B cells. *Nature* **335**: 440–442.

Vaux, D.L., Weissman, I.L. and Kim, S.K. (1992) Prevention of programmed cell death in *Caenorhabditis elegans* by human bcl-2. *Science* **258**: 1955–1957.

Veis, D.J., Sorenson, C.M., Shutter, J.R. and Korsmeyer, S.J. (1993) Bcl-2-deficient mice demonstrate fulminant lymphoid apoptosis, polycystic kidneys, and hypopigmented hair. *Cell* **75**: 229–240.

Vercammen, D., Beyaert, R., Denecker, G., Goossens, V., Van Loo, G., Declercq, W., Grooten, J., Fiers, W. and Vandenabeele, P. (1998a) Inhibition of caspases increases the sensitivity of L929 cells to necrosis mediated by tumor necrosis factor. *J. Exp. Med.* **187**: 1477–1485.

Vercammen, D., Brouckaert, G., Denecker, G., Van de Craen, M., Declercq, W., Fiers, W. and Vandenabeele, P. (1998b) Dual signaling of the Fas receptor: initiation of both apoptotic and necrotic cell death pathways. *J. Exp. Med.* **188**: 919–930.

Verhagen, A.M., Coulson, E.J. and Vaux, D.L. (2001) Inhibitor of apoptosis proteins and their relatives: IAPs and other BIRPs. *Genome Biol.* **2**: REVIEWS3009.

Verhagen, A.M., Ekert, P.G., Pakusch, M., Silke, J., Connolly, L.M., Reid, G.E., Moritz, R.L., Simpson, R.J. and Vaux, D.L. (2000) Identification of DIABLO, a mammalian protein that promotes apoptosis by binding to and antagonizing IAP proteins. *Cell* **102**: 43–53.

Verhagen, A.M., Silke, J., Ekert, P.G., *et al.* (2002) HtrA2 promotes cell death through its serine protease activity and its ability to antagonize inhibitor of apoptosis proteins. *J. Biol. Chem.* **277**: 445–454.

Verheij, M., Bose, R., Lin, X.H., *et al.* (1996) Requirement for ceramide-initiated SAPK/JNK signalling in stress-induced apoptosis. *Nature* **380**: 75–79.

Verma, S., Zhao, L.J. and Chinnadurai, G. (2001) Phosphorylation of the proapoptotic protein BIK: mapping of phosphorylation sites and effect on apoptosis. *J. Biol. Chem.* **276**: 4671–4676.

Vernooy, S.Y., Copeland, J., Ghaboosi, N., Griffin, E.E., Yoo, S.J. and Hay, B.A. (2000) Cell death regulation in *Drosophila*: conservation of mechanism and unique insights. *J. Cell Biol.* **150**: 69–76.

Virdee, K., Parone, P.A., Tolkovsky, A.M., Chou, J.J., Li, H., Salvesen, G.S., Yuan, J. and Wagner, G. (2000) Phosphorylation of the pro-apoptotic protein BAD on serine 155, a novel site, contributes to cell survival. *Curr. Biol.* **10**: 1151–1154.

Vito, P., Lacana, E. and D'Adamio, L. (1996) Interfering with apoptosis: Ca(2+)-binding protein ALG-2 and Alzheimer's disease gene ALG-3. *Science* **271**: 521–525.

Voehringer, D.W., Hirschberg, D.L., Xiao, J., Lu, Q., Roederer, M., Lock, C.B., Herzenberg, L.A. and Steinman, L. (2000) Gene microarray identification of redox and mitochondrial elements that control resistance or sensitivity to apoptosis. *Proc. Natl Acad. Sci. U. S. A.* **97**: 2680–2685.

Vogelstein, B., Lane, D. and Levine, A.J. (2000) Surfing the p53 network. *Nature* **408**: 307–310.

Volbracht, C., Fava, E., Leist, M. and Nicotera, P. (2001a) Calpain inhibitors prevent nitric oxide-triggered excitotoxic apoptosis. *Neuroreport* **12**: 3645–3648.

Volbracht, C., Leist, M., Kolb, S.A. and Nicotera, P. (2001b) Apoptosis in caspase-inhibited neurons. *Mol. Med.* **7**: 36–48.

von Ahsen, O., Renken, C., Perkins, G., Kluck, R.M., Bossy-Wetzel, E. and Newmeyer, D.D. (2000a) Preservation of mitochondrial structure and function after Bid- or Bax-mediated cytochrome c release. *J. Cell Biol.* **150**: 1027–1036.

Von Ahsen, O., Waterhouse, N.J., Kuwana, T., Newmeyer, D.D. and Green, D.R. (2000b) The 'harmless' release of cytochrome c. *Cell Death Differ.* **7**: 1192–1199.

von Mering, M., Wellmer, A., Michel, U., Bunkowski, S., Tlustochowska, A., Bruck, W., Kuhnt, U. and Nau, R. (2001) Transcriptional regulation of caspases in experimental pneumococcal meningitis. *Brain Pathol.* **11**: 282–295.

Vucic, D., Kaiser, W.J., Harvey, A.J. and Miller, L.K. (1997) Inhibition of Reaper-induced apoptosis by interaction with inhibitor of apoptosis proteins (IAPs) *Proc. Natl Acad. Sci. U. S. A.* **94**: 10183–10188.

Vucic, D., Kaiser, W.J. and Miller, L.K. (1998) Inhibitor of apoptosis proteins physically interact with and block apoptosis induced by *Drosophila* proteins HID and GRIM. *Mol. Cell. Biol.* **18**: 3300–3309.

Wadgaonkar, R., Phelps, K.M., Haque, Z., Williams, A.J., Silverman, E.S. and Collins, T. (1999) CREB-binding protein is a nuclear integrator of nuclear factor-kappaB and p53 signaling. *J. Biol. Chem.* **274**: 1879–1882.

Wahl, G.M. and Carr, A.M. (2001) The evolution of diverse biological responses to DNA damage: insights from yeast and p53. *Nat. Cell. Biol.* **3**: E277–E286.

Walczak, H., Degli-Esposti, M.A., Johnson, R.S., *et al.* (1997) TRAIL-R2: a novel apoptosis-mediating receptor for TRAIL. *EMBO J.* **16**: 5386–5397.

Walczak, H., Miller, R.E., Ariail, K., *et al.* (1999) Tumoricidal activity of tumor necrosis factor-related apoptosis-inducing ligand in vivo. *Nat. Med.* **5**: 157–163.

Walker, N.P.C., Talanian, R.V., Brady, K.D., *et al.* (1994) Crystal structure of the cysteine protease interleukin-1? converting enzyme: a (p20/p10)2 homodimer. *Cell* **78**: 343–352.

Walworth, N.C. (2000) Cell-cycle checkpoint kinases: checking in on the cell cycle. *Curr. Opin. Cell Biol.* **12**: 697–704.

Wang, C.Y., Mayo, M.W. and Baldwin, A.S., Jr. (1996) TNF- and cancer therapy-induced apoptosis: potentiation by inhibition of NF-kappaB. *Science* **274**: 784–787.

Wang, C.Y., Mayo, M.W., Korneluk, R.G., Goeddel, D.V. and Baldwin, A.S., Jr. (1998) NF-kappaB antiapoptosis: induction of TRAF1 and TRAF2 and c-IAP1 and c-IAP2 to suppress caspase-8 activation. *Science* **281**: 1680–1683.

Wang, D., Westerheide, S.D., Hanson, J.L. and Baldwin, A.S., Jr. (2000) Tumor necrosis factor alpha-induced phosphorylation of RelA/p65 on Ser529 is controlled by casein kinase II. *J. Biol. Chem.* **275**: 32592–32597.

Wang, G.Q., Wieckowski, E., Goldstein, L.A., *et al.* (2001) Resistance to granzyme B-mediated cytochrome c release in Bak-deficient cells. *J. Exp. Med.* **194**: 1325–1337.

Wang, H.G., Pathan, N., Ethell, I.M., *et al.* (1999) Ca^{2+}-induced apoptosis through calcineurin dephosphorylation of BAD. *Science* **284**: 339–343.

Wang, J., Zheng, L., Lobito, A., *et al.* (1999) Inherited human caspase 10 mutations underlie defective lymphocyte and dendritic cell apoptosis in autoimmune lymphoproliferative syndrome type II. *Cell* **98**: 47–58.

Wang, J.L., Liu, D., Zhang, Z.J., Shan, S., Han, X., Srinivasula, S.M., Croce, C.M., Alnemri, E.S. and Huang, Z. (2000) Structure-based discovery of an organic compound that binds Bcl-2 protein and induces apoptosis of tumor cells. *Proc. Natl Acad. Sci. U. S. A.* **97**: 7124–7129.

Wang, K., Gross, A., Waksman, G. and Korsmeyer, S.J. (1998) Mutagenesis of the BH3 domain of BAX identifies residues critical for dimerization and killing. *Mol. Cell. Biol.* **18**: 6083–6089.

Wang, K., Yin, X.M., Chao, D.T., Milliman, C.L. and Korsmeyer, S.J. (1996) BID: a novel BH3 domain-only death agonist. *Genes Dev.* **10**: 2859–2869.

Wang, K.K. (2000) Calpain and caspase: can you tell the difference? *Trends Neurosci.* **23**: 20–26.

Wang, L., Miura, M., Bergeron, L., Zhu, H. and Yuan, J. (1994) Ich-1, an Ice/ced-3-related gene, encodes both positive and negative regulators of programmed cell death. *Cell* **78**: 739–750.

Wang, N.S., Unkila, M.T., Reineks, E.Z. and Distelhorst, C.W. (2001) Transient expression of wild-type or mitochondrially targeted Bcl-2 induces apoptosis, whereas transient expression of endoplasmic reticulum-targeted Bcl-2 is protective against Bax-induced cell death. *J. Biol. Chem.* **276**: 44117–44128.

Wang, S.L., Hawkins, C.J., Yoo, S.J., Müller, H.A. and Hay, B.A. (1999) The *Drosophila* caspase inhibitor DIAP1 is essential for cell survival and is negatively regulated by HID. *Cell* **98**: 453–463.

Wang, X.Z., Harding, H.P., Zhang, Y., Jolicoeur, E.M., Kuroda, M. and Ron, D. (1998a) Cloning of mammalian Ire1 reveals diversity in the ER stress responses. *EMBO J.* **17**: 5708–5717.

Wang, X.Z., Kuroda, M., Sok, J., Batchvarova, N., Kimmel, R., Chung, P., Zinszner, H. and Ron, D. (1998b) Identification of novel stress-induced genes downstream of chop. *EMBO J.* **17**: 3619–3630.

Wang, Z.G., Ruggero, D., Ronchetti, S., Zhong, S., Gaboli, M., Rivi, R. and Pandolfi, P.P. (1998) PML is essential for multiple apoptotic pathways. *Nat. Genet.* **20**: 266–272.

Watanabe-Fukunaga, R., Brannan, C.I., Copeland, N.G., Jenkins, N.A. and Nagata, S. (1992) Lymphoproliferation disorder in mice explained by defects in Fas antigen that mediates apoptosis. *Nature* **356**: 314–317.

Waterhouse, N.J., Finucane, D.M., Green, D.R., *et al.* (1998) Calpain activation is upstream of caspases in radiation-induced apoptosis. *Cell Death Differ.* **5**: 1051–1061.

Waterhouse, N.J., Goldstein, J.C., von Ahsen, O., Schuler, M., Newmeyer, D.D. and Green, D.R. (2001) Cytochrome c maintains mitochondrial transmembrane potential and ATP generation after outer mitochondrial membrane permeabilization during the apoptotic process. *J. Cell Biol.* **153**: 319–328.

Watt, W., Koeplinger, K.A., Mildner, A.M., Heinrikson, R.L., Tomasselli, A.G. and Watenpaugh, K.D. (1999) The atomic-resolution structure of human caspase-8, a key activator of apoptosis. *Struct. Fold Des.* **7**: 1135–1143.

Weber, C.H. and Vincenz, C. (2001a) The death domain superfamily: a tale of two interfaces? *Trends Biochem. Sci.* **26**: 475–481.

Weber, C.H. and Vincenz, C. (2001b) A docking model of key components of the DISC complex: death domain superfamily interactions redefined. *FEBS Lett.* **492**: 171–176.

Webster, G.A. and Perkins, N.D. (1999) Transcriptional cross talk between NF-kappaB and p53. *Mol. Cell. Biol.* **19**: 3485–3495.

Wei, M.C., Lindsten, T., Mootha, V.K., Weiler, S., Gross, A., Ashiya, M., Thompson, C.B. and Korsmeyer, S.J. (2000) tBID, a membrane-targeted death ligand, oligomerizes BAK to release cytochrome c. *Genes Dev.* **14**: 2060–2071.

Wei, M.C., Zong, W. X., Cheng, E.H., *et al.* (2001) Proapoptotic BAX and BAK: a requisite gateway to mitochondrial dysfunction and death. *Science* **292**: 727–730.

Wei, Y., Fox, T., Chambers, S.P., Sintchak, J., Coll, J.T., Golec, J.M., Swenson, L., Wilson, K.P. and Charifson, P.S. (2000) The structures of caspases-1, -3, -7 and -8 reveal the basis for substrate and inhibitor selectivity. *Chem. Biol.* **7**: 423–432.

Weinstein, E.J., Grimm, S. and Leder, P. (1998) The oncogene heregulin induces apoptosis in breast epithelial cells and tumors. *Oncogene* **17**: 2107–2113.

Welburn, S.C., Lillico, S. and Murphy, N.B. (1999) Programmed cell death in procyclic form: *Trypanosoma brucei rhodesiense* – identification of differentially expressed genes during con A-induced death. *Mem. Inst. Oswaldo Cruz.* **94**: 229–234.

Welburn, S.C. and Murphy, N.B. (1998) Prohibitin and RACK homologues are up-regulated in trypanosomes induced to undergo apoptosis and in naturally occurring terminally differentiated forms. *Cell Death Differ.* **5**: 615–622.

Weller, M., Malipiero, U., Groscurth, P. and Fontana, A. (1995) T cell apoptosis induced by interleukin-2 deprivation or transforming growth factor-beta. *Exp. Cell Res.* 2: modulation by the phosphatase inhibitors okadaic acid and calyculin A. *Exp. Cell Res.* **221**: 395–403.

Wellington, C.L. and Hayden, M.R. (2000) Caspases and neurodegeneration: on the cutting edge of new therapeutic approaches. *Clin. Genet.* **57**: 1–10.

Westendorp, M.O., Frank, R., Ochsenbauer, C., Stricker, K., Dhein, J., Walczak, H., Debatin, K.M. and Krammer, P.H. (1995) Sensitization of T cells to CD95-mediated apoptosis by HIV-1 Tat and gp120. *Nature* **375**: 497–500.

White, K., Grether, M.E., Abrams, J.M., Young, L., Farrell, K. and Steller, H. (1994) Genetic control of programmed cell death in *Drosophila*. *Science* **264**: 677–683.

White, K., Tahaoglu, E. and Steller, H. (1996) Cell killing by the *Drosophila* gene reaper. *Science* **271**: 805–807.

Wiens, M., Koziol, C., Hassanein, H.M., Muller, I.M. and Muller, W.E. (1999) A homolog of the putative tumor suppressor QM in the sponge *Suberites domuncula*: downregulation during the transition from immortal to mortal (apoptotic) cells. *Tissue Cell* **31**: 163–169.

Wiens, M., Krasko, A., Muller, C.I. and Muller, W.E. (2000) Molecular evolution of apoptotic pathways: cloning of key domains from sponges (Bcl-2 homology domains and death domains) and their phylogenetic relationships. *J. Mol. Evol.* **50**: 520–531.

Wiley, S.R., Schooley, K., Smolak, P.J., *et al.* (1995) Identification and characterization of a new member of the TNF family that induces apoptosis. *Immunity* **3**: 673–682.

Williams, S.S., French, J.N., Gilbert, M., Rangaswami, A.A., Walleczek, J. and Knox, S.J. (2000) Bcl-2 overexpression results in enhanced capacitative calcium entry and resistance to SKF-96365-induced apoptosis. *Cancer Res.* **60**: 4358–4361.

Wilson, K.P., Black, J.A., Thomson, J.A., *et al.* (1994) Structure and mechanism of interleukin-1 beta converting enzyme. *Nature* **370**: 270–275.

Wing, J.P., Karres, J.S., Ogdahl, J.L., Zhou, L., Schwartz, L.M. and Nambu, J.R. (2002) *Drosophila* Sickle is a novel Grim-Reaper cell death activator. *Curr. Biol.* **12**: 131–135.

Wolf, B.B., Goldstein, J.C., Stennicke, H.R., Beere, H., Amarante-Mendes, G.P., Salvesen, G.S. and Green, D.R. (1999) Calpain functions in a caspase-independent manner to promote apoptosis-like events during platelet activation. *Blood* **94**: 1683–1692.

Wolozin, B., Iwasaki, K., Vito, P., *et al.* (1996) Participation of presenilin 2 in apoptosis: enhanced basal activity conferred by an Alzheimer mutation. *Science* **274**: 1710–1713.

Wolter, K.G., Hsu, Y.T., Smith, C.L., Nechushtan, A., Xi, X.G. and Youle, R.J. (1997) Movement of Bax from the cytosol to mitochondria during apoptosis. *J. Cell Biol.* **139**: 1281–1292.

Woo, M., Hakem, R., Soengas, M.S., *et al.* (1998) Essential contribution of caspase 3/CPP32 to apoptosis and its associated nuclear changes. *Genes Dev.* **12**: 806–819.

Wood, D.E. and Newcomb, E.W. (1999) Caspase-dependent activation of calpain during drug-induced apoptosis. *J. Biol. Chem.* **274**: 8309–8315.

Wood, D.E., Thomas, A., Devi, L.A., Berman, Y., Beavis, R.C., Reed, J.C. and Newcomb, E.W. (1998) Bax cleavage is mediated by calpain during drug-induced apoptosis. *Oncogene* **17**: 1069–1078.

Wright, S.C., Schellenberger, U., Wang, H., Kinder, D.H., Talhouk, J.W. and Larrick, J.W. (1997) Activation of CPP32-like proteases is not sufficient to trigger apoptosis: inhibition of apoptosis by agents that suppress activation of AP24, but not CPP32-like activity. *J. Exp. Med.* **186**: 1107–1117.

Wright, S.C., Schellenberger, U., Wang, H., Wang, Y. and Kinder, D.H. (1998) Chemotherapeutic drug activation of the AP24 protease in apoptosis: requirement for caspase 3-like-proteases. *Biochem. Biophys. Res. Commun.* **245**: 797–803.

Wu, D., Chen, P.J., Chen, S., Hu, Y., Nunez, G. and Ellis, R.E. (1999) *C. elegans* MAC-1, an essential member of the AAA family of ATPases, can bind CED-4 and prevent cell death. *Development* **126**: 2021–2031.

Wu, D., Wallen, H.D. and Nunez, G. (1997) Interaction and regulation of subcellular localization of CED-4 by CED-9. *Science* **275**: 1126–1129.

Wu, F., Lukinius, A., Bergstrom, M., Eriksson, B., Watanabe, Y. and Langstrom, B. (1999) A mechanism behind the antitumour effect of 6-diazo-5-oxo-L-norleucine (DON): disruption of mitochondria. *Eur. J. Cancer* **35**: 1155–1161.

Wu, G., Chai, J., Suber, T.L., Wu, J.W., Du, C., Wang, X. and Shi, Y. (2000) Structural basis of IAP recognition by Smac/DIABLO. *Nature* **408**: 1008–1012.

Wu, G.S., Burns, T.F., McDonald, E.R., III, *et al.* (1997) KILLER/DR5 is a DNA damage-inducible p53-regulated death receptor gene. *Nat. Genet.* **17**: 141–143.

Wu, J.W., Cocina, A.E., Chai, J., Hay, B.A. and Shi, Y. (2001) Structural analysis of a functional DIAP1 fragment bound to grim and hid peptides. *Mol. Cell* **8**: 95–104.

Wu, X. and Deng, Y. (2002) Bax and bh3-domain-only proteins in p53-mediated apoptosis. *Front. Biosci.* 7: D151–D156.

Wu, Y.C. and Horvitz, H.R. (1998) The *C. elegans* cell corpse engulfment gene ced-7 encodes a protein similar to ABC transporters. *Cell* 93: 951–960.

Wu, Y.C., Stanfield, G.M. and Horvitz, H.R. (2000) NUC-1, a *Caenorhabditis elegans* DNase II homolog, functions in an intermediate step of DNA degradation during apoptosis. *Genes Dev.* 14: 536–548.

Wu, Y.C., Tsai, M.C., Cheng, L.C., Chou, C.J. and Weng, N.Y. (2001) *C. elegans* CED-12 acts in the conserved crkII/DOCK180/Rac pathway to control cell migration and cell corpse engulfment. *Dev. Cell* 1: 491–502.

Wyllie, A.H. (1980) Glucocorticoid-induced thymocyte apoptosis is associated with endogenous endonuclease activation. *Nature* 284: 555–556.

Wyllie, A.H. and Golstein, P. (2001) More than one way to go. *Proc. Natl Acad. Sci. U. S. A.* 98: 11–13.

Xiang, J., Chao, D.T. and Korsmeyer, S.J. (1996) BAX-induced cell death may not require interleukin 1 beta-converting enzyme-like proteases. *Proc. Natl Acad. Sci. U. S. A.* 93: 14559–14563.

Xiao, T., Towb, P., Wasserman, S.A. and Sprang, S.R. (1999) Three-dimensional structure of a complex between the death domains of Pelle and Tube. *Cell* 99: 545–555.

Xu, J., Foy, T.M., Laman, J.D., Elliott, E.A., Dunn, J.J., Waldschmidt, T.J., Elsemore, J., Noelle, R.J. and Flavell, R.A. (1994) Mice deficient for the CD40 ligand. *Immunity* 1: 423–431.

Xu, Q. and Reed, J.C. (1998) Bax inhibitor-1, a mammalian apoptosis suppressor identified by functional screening in yeast. *Mol. Cell* 1: 337–346.

Xu, X., Leo, C., Jang, Y., *et al.* (2001) Dominant effector genetics in mammalian cells. *Nat. Genet.* 27: 23–29.

Xu, Y., Tao, X., Shen, B., Horng, T., Medzhitov, R., Manley, J.L. and Tong, L. (2000) Structural basis for signal transduction by the Toll/interleukin-1 receptor domains. *Nature* 408: 111–115.

Xue, L., Fletcher, G.C. and Tolkovsky, A.M. (2001) Mitochondria are selectively eliminated from eukaryotic cells after blockade of caspases during apoptosis. *Curr. Biol.* 11: 361–365.

Xue, L., Fletcher, G.C. and Tolkovsky, A.M. (1999) Autophagy is activated by apoptotic signalling in sympathetic neurons: an alternative mechanism of death execution. *Mol. Cell. Neurosci.* 14: 180–198.

Yamaki, M., Umehara, T., Chimura, T. and Horikoshi, M. (2001) Cell death with predominant apoptotic features in *Saccharomyces cerevisiae* mediated by deletion of the histone chaperone ASF1/CIA1. *Genes Cells* 6: 1043–1054.

Yamamoto, A., Lucas, J.J. and Hen, R. (2000) Reversal of neuropathology and motor dysfunction in a conditional model of Huntington's disease. *Cell* 101: 57–66.

Yamamoto, K., Ichijo, H., Korsmeyer, S.J., Chou, J.J., Li, H., Salvesen, G.S., Yuan, J. and Wagner, G. (1999) BCL-2 is phosphorylated and inactivated by an ASK1/Jun N-terminal protein kinase pathway normally activated at G(2)/M. *Mol. Cell. Biol.* 19: 8469–8478.

Yamashima, T. (2000) Implication of cysteine proteases calpain, cathepsin and caspase in ischemic neuronal death of primates. *Prog. Neurobiol.* **62**: 273–295.

Yan, M., Marsters, S.A., Grewal, I.S., Wang, H., Ashkenazi, A. and Dixit, V.M. (2000) Identification of a receptor for BLyS demonstrates a crucial role in humoral immunity. *Nat. Immunol.* **1**: 37–41.

Yang, E., Zha, J., Jockel, J., Boise, L.H., Thompson, C.B. and Korsmeyer, S.J. (1995) Bad, a heterodimeric partner for Bcl-XL and Bcl-2, displaces Bax and promotes cell death. *Cell* **80**: 285–291.

Yang, X., Chang, H.Y. and Baltimore, D. (1998a) Autoproteolytic activation of pro-caspases by oligomerization. *Mol. Cell* **1**: 319–325.

Yang, X., Chang, H.Y. and Baltimore, D. (1998b) Essential role of CED-4 oligomerization in CED-3 activation and apoptosis. *Science* **281**: 1355–1357.

Yang, X., Khosravi-Far, R., Chang, H.Y. and Baltimore, D. (1997) Daxx, a novel Fas-binding protein that activates JNK and apoptosis. *Cell* **89**: 1067–1076.

Yaoita, H., Ogawa, K., Maehara, K. and Maruyama, Y. (2000) Apoptosis in relevant clinical situations: contribution of apoptosis in myocardial infarction. *Cardiovasc. Res.* **45**: 630–641.

Ye, J., Rawson, R.B., Komuro, R., Chen, X., Dave, U.P., Prywes, R., Brown, M.S. and Goldstein, J.L. (2000) ER stress induces cleavage of membrane-bound ATF6 by the same proteases that process SREBPs. *Mol. Cell* **6**: 1355–1364.

Ye, Y., Meyer, H.H. and Rapoport, T.A. (2001) The AAA ATPase Cdc48/p97 and its partners transport proteins from the ER into the cytosol. *Nature* **414**: 652–656.

Yeh, W.C., Hakem, R., Woo, M. and Mak, T.W. (1999) Gene targeting in the analysis of mammalian apoptosis and TNF receptor superfamily signaling. *Immunol. Rev.* **169**: 283–302.

Yeh, W.C., Itie, A., Elia, A.J., *et al.* (2000) Requirement for Casper (c-FLIP) in regulation of death receptor-induced apoptosis and embryonic development. *Immunity* **12**: 633–642.

Yeh, W.C., Pompa, J.L., McCurrach, M.E., *et al.* (1998) FADD: essential for embryo development and signaling from some, but not all, inducers of apoptosis. *Science* **279**: 1954–1958.

Yeh, W.C., Shahinian, A., Speiser, D., *et al.* (1997) Early lethality, functional NF-kappaB activation, and increased sensitivity to TNF-induced cell death in TRAF2-deficient mice. *Immunity* **7**: 715–725.

Yin, X.M., Oltvai, Z.N. and Korsmeyer, S.J. (1994) BH1 and BH2 domains of Bcl-2 are required for inhibition of apoptosis and heterodimerization with Bax. *Nature* **369**: 321–323.

Yin, X.M., Wang, K., Gross, A., Zhao, Y., Zinkel, S., Klocke, B., Roth, K.A. and Korsmeyer, S.J. (1999) Bid-deficient mice are resistant to Fas-induced hepatocellular apoptosis. *Nature* **400**: 886–891.

Yoneda, T., Imaizumi, K., Oono, K., Yui, D., Gomi, F., Katayama, T. and Tohyama, M. (2001) Activation of caspase-12, an endoplastic reticulum (ER) resident caspase, through tumor necrosis factor receptor-associated factor 2-dependent mechanism in response to the ER stress. *J. Biol. Chem.* **276**: 13935–13940.

Yoshida, H., Haze, K., Yanagi, H., Yura, T. and Mori, K. (1998a) Identification of the cis-acting endoplasmic reticulum stress response element responsible for transcriptional induction of mammalian glucose-regulated proteins. Involvement of basic leucine zipper transcription factors. *J. Biol. Chem.* **273**: 33741–33749.

Yoshida, H., Kong, Y.Y., Yoshida, R., Elia, A.J., Hakem, A., Hakem, R., Penninger, J.M. and Mak, T.W. (1998a) Apaf1 is required for mitochondrial pathways of apoptosis and brain development. *Cell* **94**: 739–750.

Yoshida, H., Matsui, T., Yamamoto, A., Okada, T. and Mori, K. (2001) XBP1 mRNA is induced by ATF6 and spliced by IRE1 in response to ER stress to produce a highly active transcription factor. *Cell* **107**: 881–891.

Youn, H.D., Sun, L., Prywes, R. and Liu, J.O. (1999) Apoptosis of T cells mediated by Ca^{2+}-induced release of the transcription factor MEF2. *Science* **286**: 790–793.

Yu, J., Zhang, L., Hwang, P.M., Kinzler, K.W. and Vogelstein, B. (2001) PUMA induces the rapid apoptosis of colorectal cancer cells. *Mol. Cell* **7**: 673–682.

Yuan, J. and Horvitz, H.R. (1992) The *Caenorhabditis elegans* cell death gene ced-4 encodes a novel protein and is expressed during the period of extensive programmed cell death. *Development* **116**: 309–320.

Yuan, J., Shaham, S., Ledoux, S., Ellis, H.M. and Horvitz, H.R. (1993) The *C. elegans* cell death gene ced-3 encodes a protein similar to mammalian interleukin-1 beta-converting enzyme. *Cell* **75**: 641–652.

Yuan, J. and Yankner, B.A. (2000) Apoptosis in the nervous system. *Nature* **407**: 802–809.

Yuan, J.Y. and Horvitz, H.R. (1990) The *Caenorhabditis elegans* genes ced-3 and ced-4 act cell autonomously to cause programmed cell death. *Dev. Biol.* **138**: 33–41.

Yuan, Z.M., Huang, Y., Whang, Y., Sawyers, C., Weichselbaum, R., Kharbanda, S. and Kufe, D. (1996) Role for c-Abl tyrosine kinase in growth arrest response to DNA damage. *Nature* **382**: 272–274.

Yuan, Z.M., Shioya, H., Ishiko, T., *et al.* (1999) p73 is regulated by tyrosine kinase c-Abl in the apoptotic response to DNA damage. *Nature* **399**: 814–817.

Yuste, V.J., Sanchez-Lopez, I., Sole, C., Encinas, M., Bayascas, J.R., Boix, J. and Comella, J.X. (2002) The prevention of the staurosporine-induced apoptosis by Bcl-X(L), but not by Bcl-2 or caspase inhibitors, allows the extensive differentiation of human neuroblastoma cells. *J. Neurochem.* **80**: 126–139.

Zamzami, N. and Kroemer, G. (2001) The mitochondrion in apoptosis: how Pandora's box opens. *Nat. Rev. Mol. Cell. Biol.* **2**: 67–71.

Zang, Y., Beard, R.L., Chandraratna, R.A. and Kang, J.X. (2001) Evidence of a lysosomal pathway for apoptosis induced by the synthetic retinoid CD437 in human leukemia HL-60 cells. *Cell Death Differ.* **8**: 177–185.

Zanke, B.W., Lee, C., Arab, S. and Tannock, I.F. (1998) Death of tumor cells after intracellular acidification is dependent on stress-activated protein kinases (SAPK/JNK) pathway activation and cannot be inhibited by Bcl-2 expression or interleukin 1beta-converting enzyme inhibition. *Cancer Res.* **58**: 2801–2808.

Zeuner, A., Eramo, A., Peschle, C. and De Maria, R. (1999) Caspase activation without death. *Cell Death Differ.* **6**: 1075–1080.

Zha, H., Fisk, H.A., Yaffe, M.P., Mahajan, N., Herman, B. and Reed, J.C. (1996) Structure-function comparisons of the proapoptotic protein Bax in yeast and mammalian cells. *Mol. Cell. Biol.* **16**: 6494–6508.

Zha, J., Harada, H., Osipov, K., Jockel, J., Waksman, G. and Korsmeyer, S.J. (1997) BH3 domain of BAD is required for heterodimerization with BCL-XL and proapoptotic activity. *J. Biol. Chem.* **272**: 24101–24104.

Zha, J., Harada, H., Yang, E., Jockel, J. and Korsmeyer, S.J. (1996) Serine phosphorylation of death agonist BAD in response to survival factor results in binding to 14-3-3, not BCL-X(L) *Cell* **87**: 619–628.

Zha, J., Weiler, S., Oh, K.J., Wei, M.C. and Korsmeyer, S.J. (2000) Posttranslational *N*-myristoylation of BID as a molecular switch for targeting mitochondria and apoptosis. *Science* **290**: 1761–1765.

Zhai, D., Miao, Q., Xin, X. and Yang, F. (2001) Leakage and aggregation of phospholipid vesicles induced by the BH3-only Bcl-2 family member, BID. *Eur. J. Biochem.* **268**: 48–55.

Zhang, H., Heim, J. and Meyhack, B. (1998) Redistribution of Bax from cytosol to membranes is induced by apoptotic stimuli and is an early step in the apoptotic pathway. *Biochem. Biophys. Res. Commun.* **251**: 454–459.

Zhang, J., Cado, D., Chen, A., Kabra, N.H. and Winoto, A. (1998) Fas-mediated apoptosis and activation-induced T-cell proliferation are defective in mice lacking FADD/Mort1. *Nature* **392**: 296–300.

Zhang, K.Z., Westberg, J.A., Holtta, E., Andersson, L.C., Chou, J.J., Li, H., Salvesen, G.S., Yuan, J. and Wagner, G. (1996) BCL2 regulates neural differentiation. *Proc. Natl Acad. Sci. U. S. A.* **93**: 4504–4508.

Zhang, L., Yu, J., Park, B.H., Kinzler, K.W. and Vogelstein, B. (2000) Role of BAX in the apoptotic response to anticancer agents. *Science* **290**: 989–992.

Zhang, Y., Goodyer, C. and LeBlanc, A. (2000) Selective and protracted apoptosis in human primary neurons microinjected with active caspase-3, -6, -7, and -8. *J. Neurosci.* **20**: 8384–8389.

Zhao, R., Gish, K., Murphy, M., Yin, Y., Notterman, D., Hoffman, W.H., Tom, E., Mack, D.H. and Levine, A.J. (2000) Analysis of p53-regulated gene expression patterns using oligonucleotide arrays. *Genes Dev.* **14**: 981–993.

Zhao, Y. and Elder, R.T. (2000) Yeast perspectives on HIV-1 VPR. *Front. Biosci.* **5**: D905–D916.

Zhao, Y., Yu, M., Chen, M., Elder, R.T., Yamamoto, A. and Cao, J. (1998) Pleiotropic effects of HIV-1 protein R (Vpr) on morphogenesis and cell survival in fission yeast and antagonism by pentoxifylline. *Virology* **246**: 266–267.

Zheng, L., Fisher, G., Miller, R.E., Peschon, J., Lynch, D.H. and Lenardo, M.J. (1995) Induction of apoptosis in mature T cells by tumour necrosis factor. *Nature* **377**: 348–351.

Zheng, T.S., Hunot, S., Kuida, K. and Flavell, R.A. (1999) Caspase knockouts: matters of life and death. *Cell Death Differ.* **6**: 1043–1053.

Zheng, Y., Ouaaz, F., Bruzzo, P., Singh, V., Gerondakis, S. and Beg, A.A. (2001) NF-kappa B RelA (p65) is essential for TNF-alpha-induced fas expression but dispensable for both TCR-induced expression and activation-induced cell death. *J. Immunol.* **166**: 4949–4957.

Zhivotovsky, B., Orrenius, S., Brustugun, O.T. and Doskeland, S.O. (1998) Injected cytochrome c induces apoptosis. *Nature* **391**: 449–450.

Zhong, H., SuYang, H., Erdjument-Bromage, H., Tempst, P. and Ghosh, S. (1997) The transcriptional activity of NF-kappaB is regulated by the IkappaB-associated PKAc subunit through a cyclic AMP-independent mechanism. *Cell* **89**: 413–424.

Zhou, L., Hashimi, H., Schwartz, L.M. and Nambu, J.R. (1995) Programmed cell death in the *Drosophila* central nervous system midline. *Curr. Biol.* **5**: 784–790.

Zhou, L., Song, Z., Tittel, J. and Steller, H. (1999) HAC-1, a *Drosophila* homolog of APAF-1 and CED-4 functions in developmental and radiation-induced apoptosis. *Mol. Cell* **4**: 745–755.

Zhou, X.M., Liu, Y., Payne, G., *et al.* (2000) Growth factors inactivate the cell death promoter BAD by phosphorylation of its BH3 domain on Ser155. *J. Biol. Chem.* **275**: 25046–25051.

Zhou, Z., Caron, E., Hartwieg, E., Hall, A. and Horvitz, H.R. (2001a) The *C. elegans* PH domain protein CED-12 regulates cytoskeletal reorganization via a Rho/Rac GTPase signaling pathway. *Dev. Cell* **1**: 477–489.

Zhou, Z., Hartwieg, E. and Horvitz, H.R. (2001b) CED-1 is a transmembrane receptor that mediates cell corpse engulfment in *C. elegans*. *Cell* **104**: 43–56.

Zhu, D.M. and Uckun, F.M. (2000) Cathepsin inhibition induces apoptotic death in human leukemia and lymphoma cells. *Leuk. Lymphoma* **39**: 343–354.

Zhu, L., Ling, S., Yu, X.D., Venkatesh, L.K., Subramanian, T., Chinnadurai, G. and Kuo, T.H. (1999) Modulation of mitochondrial Ca(2+) homeostasis by Bcl-2. *J. Biol. Chem.* **274**: 33267–33273.

Ziauddin, J. and Sabatini, D.M. (2001) Microarrays of cells expressing defined cDNAs. *Nature* **411**: 107–110.

Zihler (1972) Some aspects of gametogenesis and fertilisation in *Hydra*. *Wilhelm Roux Archiv* **169**: 239–267.

Zimmermann, K.C., Ricci, J.E., Droin, N.M. and Green, D.R. (2002) The role of ARK in stress-induced apoptosis in *Drosophila* cells. *J. Cell Biol.* **156**: 1077–1087.

Zong, W.X., Edelstein, L.C., Chen, C., Bash, J. and Gelinas, C. (1999) The prosurvival Bcl-2 homolog Bfl-1/A1 is a direct transcriptional target of NF-kappaB that blocks TNFalpha-induced apoptosis. *Genes Dev.* **13**: 382–387.

Zornig, M., Hueber, A.O. and Evan, G. (1998) p53-dependent impairment of T-cell proliferation in FADD dominant-negative transgenic mice. *Curr. Biol.* **8**: 467–470.

Zou, H., Henzel, W.J., Liu, X., Lutschg, A. and Wang, X. (1997) Apaf1, a human protein homologous to *C. elegans* CED-4, participates in cytochrome c-dependent activation of caspase-3. *Cell* **90**: 405–413.

Zou, H., Li, Y., Liu, X. and Wang, X. (1999) An APAF-1.cytochrome c multimeric complex is a functional apoptosome that activates procaspase-9. *J. Biol. Chem.* **274**: 11549–11556.

Zuppini, A., Groenendyk, J., Cormack, L.A., Shore, G., Opas, M., Bleackley, R.C. and Michalak, M. (2002) Calnexin deficiency and endoplasmic reticulum stress-induced apoptosis. *Biochemistry (Mosc.)* **41**: 2850–2858.

Index

Note: Page references in **bold** refer to Tables; those in *italics* refer to Figures